IEE ELECTRICAL MEASUREMENT SERIES 7

Series Editors: A. E. Bailey
Dr. O. C. Jones
Dr. A. C. Lynch

Radio frequency & microwave power measurement

Other volumes in this series:

Volume 1 **Lock-in amplifiers: principles and applications** M. L. Meade
Volume 2 **Microwave impedance measurement** P. I. Somlo and J. D. Hunter
Volume 3 **Microwave measurement** A. E. Bailey (Editor)
Volume 4 **The current comparator** W. J. M. Moore and P. N. Miljanic
Volume 5 **Principles of microwave measurements** G. H. Bryant
Volume 6 **Instrumentation and Analytical Science** Department of Instrumentation and Analytical Science: UMIST
Volume 7 **Radio frequency and microwave power measurement** A. Fantom

Radio frequency & microwave power measurement

A. Fantom

Peter Peregrinus Ltd. on behalf of the Institution of Electrical Engineers

Published by: Peter Peregrinus Ltd., London, United Kingdom

Peter Peregrinus Ltd.,
Michael Faraday House,
Six Hills Way, Stevenage,
Herts. SG1 2AY, United Kingdom

British Library Cataloguing in Publication Data

Fantom, A.
Radio frequency and microwave power measurement
1. Microwaves. Measurement
I. Title II. Series
621.381'37

ISBN 0 86341 120 7

Printed in England by BPCC Wheatons Ltd., Exeter

Contents

Preface

Power meters are among the most basic of the measurement tools available to the radio-frequency and microwave engineer. They enable one to determine not only power but also a number of other quantities describing signal amplitudes, such as voltage and current. This book presents a wide-ranging account of the many types of power meter which have been developed and of the methods used for calibrating them. It is assumed that the measurements are carried out in guided wave structures such as coaxial lines or waveguides. The material falls into four parts, excluding the first chapter which is concerned with background information and fundamental aspects. Chapters 2 to 5 deal with calorimeters, whose uses include the important role of primary standards. Chapters 6 to 10 describe the various non-calorimetric types of power meter, which comprise bolometers, thermoelectric and diode instruments, force-operated instruments, and many other less well known devices. Calibration techniques and related topics are treated in Chapters 11 to 13, while Chapters 14 to 16 cover the remaining areas of pulsed power measurement, voltage and current measurements, and international intercomparison of standards. Within the individual subject headings an outline is given of the historical evolution of the devices discussed and of the basis of relevant physical effects. Aspects relating to automation and computer-corrected measurements are included where necessary but these topics are too wide to be treated adequately in a book devoted to a single measurement quantity. Continuous wave sinusoidal signals are assumed in the main. Each chapter is provided with a fairly generous list of references, although a comprehensive bibliography would run into several thousands rather than the hundreds cited.

Terminology can be a cause of confusion. The names commonly given to measuring instruments and other devices often reflect the intended use rather than the way in which the instruments function. This can act as a barrier to understanding when one wishes to discuss operating principles, since the same device may appear under different names. For example, an instrument referred to as a power meter may in fact be a voltmeter which is simply calibrated in terms of power and thus reads correctly only for a particular load impedance. A reflectometer may also be known as a transfer instrument when used for calibrating one power meter against another rather than for measuring reflection coefficients. On the other hand, names which attempt to describe the nature of devices rather than their use tend to have a less immediate impact – for example lossless, reciprocal, matched four-port instead of directional coupler. In general the aim throughout the book has been to adopt names which will be most familiar and meaningful to the majority of readers, even though these may not always be pedantically accurate descriptions.

The level of knowledge assumed of the reader corresponds roughly to that of a graduate electrical engineer or physicist. Many of the basic concepts, however, are not particularly difficult to grasp, although their application may on occasion lead to a bewildering degree of complexity. The problem in such cases is to avoid becoming confused by details and thereby losing sight of the general picture.

A number of people have helped in the preparation of the book. Mr J. Hansen-Addy of the National Physical Laboratory (NPL) provided the benefit of his experience in constructing and calibrating power meters, and in addition read through several drafts of the manuscript. Discussions with Mr R.N. Clarke of NPL, who also read through one complete draft and assisted in the proof-reading, greatly helped in clarifying my thoughts on a variety of topics including impedance measurement. The following kindly gave up their time to review individual sections:

Mr R.W. Yell, Mr J.McA. Steele, Dr R.B.D. Knight, Mr L.T.J. Hinton, Dr B.W. Petley and Dr J.C. Gallop of NPL

Dr L.C. Oldfield, Mr J.P. Ide, Mr E.J. Griffin, Mr F.L. Warner and Mr P.J. Skilton of the Royal Signals and Radar Establishment (RSRE), Malvern

Mr G.R. Orford of the Services Electrical Standards Centre (SESC), Bromley

Dr I. Yokoshima and Dr T. Inoue of the Electrotechnical Laboratory (ETL), Japan

Dr A. Jurkus and Mr R.F. Clark of the National Research Council (NRC), Canada

Professor A.L. Cullen of University College, London

To all these and many others whose influence has helped to shape the book go my sincere thanks. Finally, my thanks are due to Helen Fantom for assistance in typing and to my parents, who provided continuous encouragement over a period of several years.

Alan E. Fantom

Hampton Hill

Acknowledgments

Reproduction of Figs. 3.5, 3.6, 4.4, 5.5, 5.6, 5.7, 6.4, 6.5, 6.12, 7.4, 10.3, 14.8 is by permission of the Controller, HMSO.

Reproduction of Figs. 4.3, 5.3, 5.4, 14.5, 15.4 is by permission of the National Bureau of Standards.

Reproduction of Figs. 9.6, 9.7, 9.8, 9.9, 9.11, 11.10, 15.12 is by permission of the IEE.

Reproduction of Figs. 3.11, 4.5, 6.19, 6.20, 7.3, 7.9, 9.10, 10.2, 10.5, 12.12, 12.15, 14.7, 15.2, 15.3, 15.6, 15.7, 15.8, 16.3, 16.4 is by permission of the IEEE.

Reproduction of Figs. 6.9, 6.18, 7.6, 7.7, 8.2, 8.3, 8.8, 8.9 is by permission of Hewlett-Packard Inc.

Reproduction of Figs. 16.1, 16.2 is by permission of 'Metrologia'.

Reproduction of Fig. 16.5 is by permission of the BIPM.

Reproduction of Fig. 11.3 is by permission of the Narda Corporation.

Reproduction of Figs. 7.8, 14.2 is by permission of 'Microwaves and RF'.

Reproduction of Fig. 2.6 is by permission of Polytechnic Press, Brooklyn.

Chapter 1

Background and fundamentals

1.1 Background

This book deals with the measurement of power from 1 MHz to 200 GHz, a frequency range which covers roughly the radio-frequency and microwave regions of the electromagnetic spectrum. Radio-frequency and microwave techniques are sometimes thought of as two separate subjects occupying two distinct frequency ranges. Microwaves, according to their currently recommended definition [1], lie between 300 MHz and 300 GHz, with wavelengths from 1 m to 1 mm. They are often termed decimetre, centimetre and millimetre waves. Their exploitation dates from around 1940 and at that time the techniques adopted stood in sharp contrast to conventional lower frequency radio engineering methods. The use of waveguides, the need to take into account propagation delays and to think nearly always in terms of travelling waves, all served to make microwaves a distinct subject. However, since then the distinction has become less significant. For example, coaxial lines are used widely at microwave frequencies [2,3], whilst for precise measurements propagation delays must be taken into account at frequencies well below 300 MHz – typically from 30 MHz upwards. Moreover, the frequency range covered by many power meters and other instruments lies partly in the rf region and partly in the microwave region. Thus in reality one is dealing with a continuous spectrum. Nevertheless, for some purposes it is convenient to divide this spectrum into sub-ranges. Often a break is made at 1 GHz, which coincides with the highest frequency at which one normally attempts to measure voltage and current directly. Another significant frequency is 40 GHz, which for many years was very roughly the highest frequency at which coaxial lines were in regular use.

Generally speaking, accuracy diminishes with increasing frequency. This is due to such factors as repeatability of connectors, skin effect, and mechanical tolerances in relation to the wavelength. At microwave frequencies a 'precision' measurement often means an uncertainty of half a percent, whereas a precision dc measurement is more likely to mean a part per million.

Electrical networks may be represented by various sets of parameters. Impedance and admittance parameters are commonly used for lumped element circuits where the wavelength is long compared with the size of typical components. When travelling waves are involved, scattering parameters are more appropriate and it is this representation which will be mainly employed in this book. For a summary of network parameters see ref. [4] and Appendix 1.

One can distinguish two broad situations in which power measurements are carried out. In the first, power itself is the electrical quantity of interest. For example, the signal-to-noise performance of a radio communication system or radar installation depends on the transmitted power; similarly the heating effect of a microwave oven depends on the dissipated power. In the second situation power is measured in order to deduce some other electrical quantity which would be more difficult to measure directly. Thus voltage and current are often obtained by indirect methods based on power and impedance measurements. The number of possible applications of this second kind is relatively large and includes not only measurements in coaxial lines and waveguides but also some measurements in free space. Field strength meters, for instance, are often calibrated by means of a standard electromagnetic field which is set up by feeding a measured power into an antenna of known gain.

Three aspects of measurement which demand equal attention are standards, everyday commercial instruments, and methods of calibrating or comparing one instrument against another. Each has its own history of development. Basic standards are required in order to realise measurement quantities from their definition. Their evolution tends to follow the general level of commercial activity and, since it was the lower frequencies which in the main were exploited first, so also standards were first set up for these lower frequencies. The first international comparison of microwave power standards was at a frequency of approximately 10 GHz [5]. Around 1970 a growing need was felt for standards above 40 GHz [6,7], and by the early 1980s frequencies of 100 GHz and above were reached, although in only a small number of laboratories [8]. An important period was from 1960 to 1970, as it was in this period that much of the earlier pioneering work of the 1940s and 1950s was consolidated to form the basis of the subject of precision microwave measurements [9]. For example, precision coaxial connectors and many other precision components were developed at this time [10].

The evolution of commercial power meters was greatly influenced by developments in thin film techniques and semiconductor technology. Whereas from the 1940s to the 1960s the thermistor was the most popular commercial power meter [11], the 1970s saw the introduction into the microwave region of thin film thermoelectric power meters and of diode power meters employing low barrier height Schottky barrier diodes [12]. Previously, both thermoelectric and diode power meters had been limited to below 1 GHz. By 1980 the incorporation of microprocessors into these instruments enabled some of their drawbacks to be overcome – for example, the non-linear relation between the indicated and actual power could be compensated for by corrections, although the inherent need for such corrections made the instruments less attractive for precise measurements.

The influence of the microprocessor on calibration and comparison methods was even greater than its influence on commercial power meter design. The introduction of relatively inexpensive desk-top computers in the 1970s caused a reappraisal of established comparison techniques and a consequent shift in the relative importance of different methods. Previously, the incentive had been to strive to make the hardware as nearly perfect as possible. For example, it was highly desirable to use components with low reflection coefficients, as this avoided the need for tuners or other means of reducing mismatch errors. In contrast, the new approach was to compensate for the imperfections of the apparatus by the application of previously determined correction factors, a task which could be delegated to the computer. Automated network analysers and six-port reflectometers use such techniques. Although these instruments have the potential for greater convenience and accuracy, they also brought with them a growing problem of the assessment and evaluation of their measurement uncertainties. The trend continued into the 1980s, with greater emphasis on computer-centred measurement systems rather than individual instruments and components. As a result standards laboratories became increasingly concerned with the assessment of the accuracy of complex systems, with rather less time devoted to the basic standards on which their calibration depends.

In discussing uncertainties of measuring instruments one must bear in mind that agreement has not yet been reached on a standard way of calculating and expressing these. One should therefore exercise caution when comparing the uncertainties of different instruments or techniques, since the figures may not have been arrived at on the same basis. For rf and microwave measurements the trend is to treat all uncertainties statistically at a confidence level of at least 95 percent. Expressed on this basis the most accurate power measurements, which are those carried out using calorimeters, have uncertainties of roughly 0.1 percent at 1 GHz, 0.3 percent at 40 GHz, and 0.5 percent at 100 GHz. In other areas of measurement, such as the determination of fundamental physical constants, uncertainties are more usually quoted in terms of one standard deviation, whilst for some dc and audio-frequency electrical measurements the worst case method of simple arithmetic addition for combining the systematic contributions to uncertainty is used rather than the statistical approach. Further details are given in refs. [13–15] and in Appendix 2.

1.2 Guide to the literature

References on specific aspects of rf and microwave measurements are included in the appropriate chapters of this book, but some general sources of information will be mentioned here. A number of useful books have been published [16–30]. Of these, Montgomery [17] describes the measurement techniques developed in the pioneering days of microwaves during the Second World War (1939–1945), whilst the three-volume Handbook of Microwave Measurements [22] edited by Sucher and Fox and published in 1963 summarises the state of microwave measurements at that time. Laverghetta's book [27] on microwave measurement practice provides a very readable summary of the subject, although written at a rather elementary level. More specialised topics are treated in Warner's book on Microwave Attenuation Measurement [28], which also contains a wealth of background information, and in Somlo and Hunter's book on Microwave Impedance Measurement [30]. Reviews describing measurement techniques at various stages of evolution are those by Nergaard [31,32], Clayton et al. [33], Moreno and Lundstrom [34], Schrack [35], Rumfelt and Elwell [36,37], Beatty [38], Steele et al. [39], Engen [40], Jurkus and Stumper [41], Bayer et al. [42], and Fantom [43]. Special issues of the Proceedings of the IEEE devoted to rf and microwave measurements are those of June 1967 [9], April 1978 [44] and January 1986 [45]. The International Scientific Radio Union (URSI) publishes at three-yearly intervals a review of Progress in Radio Science containing a section on rf and microwave metrology [46]. The national standards laboratories of several countries produce bulletins, journals, or occasional publications, examples of which are given in refs. [47–52]. A standard reference work and bibliographical guide to the more traditional aspects of microwaves is Harvey [53], published in 1963. Other relevant publications are [12,54–69].

1.3 The basis of electrical measurements

1.3.1 Definitions of units

The watt – The unit of electrical power, the watt, is defined to be equal to the mechanical watt. In principle, therefore, one could measure electrical power by converting it to mechanical power, which could then be determined from force and velocity measurements. Standards for force and velocity are derived from mass, length and time standards. In practice, less direct methods are used, but these all have the same fundamental aim of relating electrical power to mechanical quantities.

The ampere – The ampere is a purely electrical quantity derived, like the watt, from mechanical measurements [70,71]. It is defined as that constant current which, if maintained in two straight parallel conductors of infinite length, of negligible circular cross section and placed 1 metre apart in vacuum, would produce between these conductors a force equal to 2×10^{-7} newton per metre of length.

The volt – The volt is dependent on the definitions of the watt and the ampere. It is defined as the difference in potential between two points of a conducting wire carrying a constant current of 1 ampere when the power dissipated between these points is equal to 1 watt.

The ohm – The unit of resistance, the ohm, is the electrical resistance between two points of a conductor when a constant potential difference of 1 V, applied to these points, produces in the conductor a current of 1 A, the conductor not being the seat of any electromotive force.

1.3.2 Realisation of dc units

The traditional method of realising the ampere is by measuring the force between two current-carrying coils, which is carried out using a current balance [72,73]. The ohm was realised for many years by means of a calculable mutual inductor [74], but this method was superseded in the 1970s by the calculable capacitor method based on the Thompson–Lampard theorem [75,76]. Realisations of both the ampere and the ohm by these methods entail a great deal of very laborious and painstaking work and are carried out only infrequently. For everyday use more convenient standards of voltage and resistance are maintained. Until the 1970s these were normally groups of Weston mercury–cadmium standard cells and wire-wound resistors respectively.

The discovery of the ac Josephson effect in 1962 made it possible to begin to move towards maintained electrical standards which are based on quantum effects and depend only on the values of fundamental constants [77]. In a Josephson voltage standard, which operates at cryogenic temperatures, a superconducting weak link is irradiated by a microwave signal. The voltage–current characteristic displays steps and the increment in voltage corresponding to one step can be calculated from the frequency of the microwave signal and the Josephson constant $2e/h$, where e is the electronic charge and h is Planck's constant. Once the value of this constant is known, or an agreed value is assigned to it, standard voltages can be reproduced accurately by any laboratory with the necessary equipment [78–85]. In 1973 the best estimate of $2e/h$ was 483 594 GHz/V [86] and the Consultative Committee for Electricity (CCE) suggested this value to maintain the volt. The uncertainty in this figure (2.6 parts per million standard deviation) was about two orders of magnitude worse than the precision with which standard voltages could be intercompared. To remedy this situation several approaches were attempted which aimed at a more accurate determination of $2e/h$. One such approach was the absolute determination of the volt using a liquid electrometer [87,88]. A second scheme was the moving coil method [89–91], which is a more direct way of relating electrical and mechanical power (Fig.1.1). In this method the measurement is carried out in two stages. First, a direct current is passed through a coil in a strong magnetic field and the force exerted on the coil is measured. The current is then switched

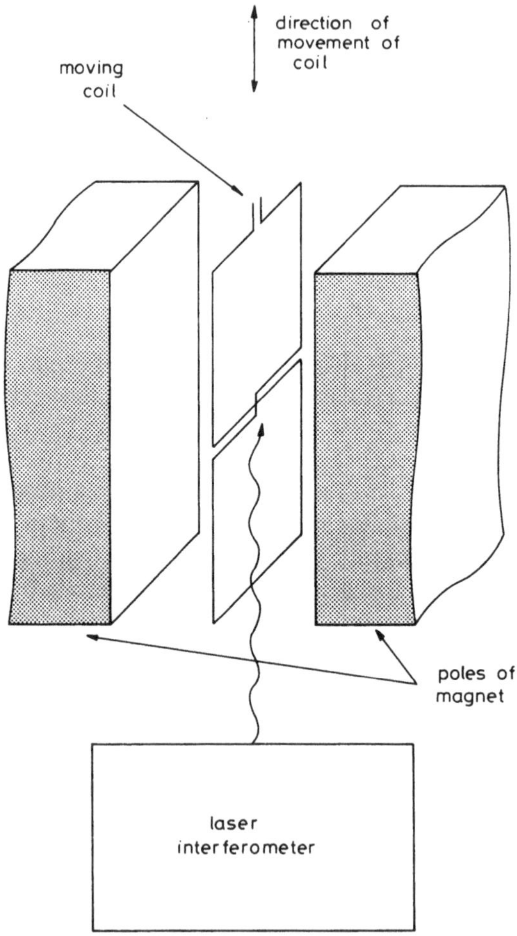

Fig. 1.1 *Principle of the moving coil method of relating dc electrical power to mechanical power. The two halves of the coil are connected in opposition in order to obtain an approximately linear relation between magnetic flux linkage and displacement*

off and the coil is moved in the magnetic field, giving rise to an induced emf. By equating the product of current and emf to the product of force and velocity one can derive the electrical watt in terms of the mechanical watt. The technique is a virtual work method, because no actual work is done. A third scheme, classed as a real work method, is based on the levitation of a superconductor [92]. Measurements using the newer methods suggested that the 1973 best estimate of 483 594 GHz/V for $2e/h$ was approximately 8 parts per million too low [71,93] and a more recent value is 483 597.67 GHz/V with an uncertainty of 0.3 ppm standard deviation [94]. This value has since been adopted by the CCE.

A further advance towards quantum standards, announced in 1980, was the discovery [95] that the Hall resistance $\mathcal{R}_H$ measured in metal oxide semiconductor field effect transistors (MOSFETS) at cryogenic temperatures around 1 K and for strong magnetic flux densities of about 10 T is quantised in submultiple of h/e^2. Thus

$$\mathcal{R}_H = \frac{h}{ie^2} \tag{1.1}$$

where i is an integer. As the magnetic field is increased, there is a plateau over which $\mathcal{R}_H$ remains constant. This effect, known as the quantum Hall effect, can be used to maintain the ohm if the fundamental constant h/e^2 is known [96–101]. The value of this constant is 25 812.8056 Ω with an uncertainty of 0.045 ppm standard deviation [94].

If the volt and the ohm are maintained by the Josephson and quantum Hall effects respectively, then power measurements derived from them will depend only on Planck's constant h and frequency. The value of h recommended at the 1973 least squares adjustment [86] was 6.626176×10^{-34} J Hz^{-1}, with an uncertainty of 5.4 ppm (one standard deviation), but it became apparent that this was approximately 15 ppm too high [71] and a more recent value is $6.6260755 \times 10^{-34}$ J Hz^{-1} with an uncertainty of 0.6 ppm standard deviation [94]. This more recent value has also been adopted by the CCE.

1.3.3. Realisation of rf and microwave units

Fig. 1.2 summarises the main ways in which power, voltage and current can be derived at rf and microwave frequencies [62,102]. In setting up power standards at these frequencies one has two main practical choices. The first and most common method is to derive power measurements from dc power standards. The instruments which enable this to be done are known as rf/dc transfer instruments or rf/dc substitution instruments. The usual procedure is to convert the high frequency power into heat in a calorimeter (see Chapter 2). Then the amount of dc power needed to produce the same temperature rise as the high frequency power (the dc substituted power) is determined. This dc power, after the application of some corrections, is equal to the rf or microwave power to be measured.

The alternative approach is to derive the high frequency power measurement more or less directly from mechanical measurements. Methods based on this approach are often called absolute methods and they make use of some of the force-operated instruments described in Chapter 9. The easiest of these instruments to describe, but not the easiest to use in practice, is the radiation pressure wattmeter [103]. If a beam of electromagnetic radiation is reflected by a perfectly conducting sheet, the power in the beam can be calculated from the force on the sheet and the velocity of propagation of the radiation. This effect was first observed for light beams in free space, but was later demonstrated for microwaves travelling inside a waveguide (see Chapter 9). The small forces

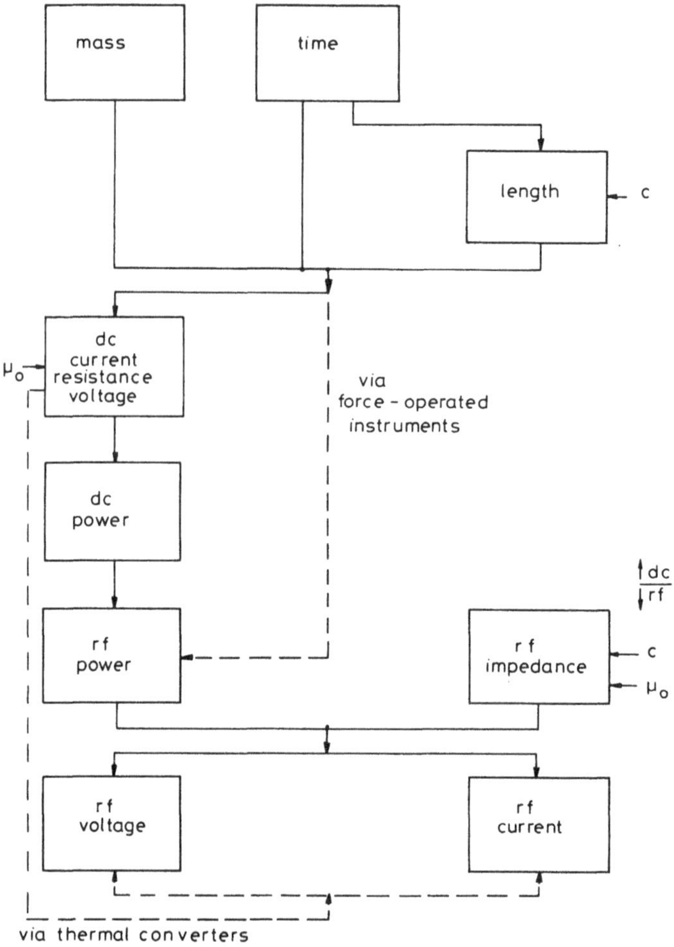

Fig. 1.2 *Simplified chart showing routes by which rf power, voltage, and current are derived. Solid lines indicate main routes and broken lines alternative routes. The defined constants c and μ_0 are respectively the speed of light and the permeability of free space*

which must be measured are easily swamped by thermal convection currents in the surrounding air, unless the instrument is evacuated, and this makes the radiation pressure wattmeter rather impractical. However, an alternative force-operated instrument known as the torque vane wattmeter [104] suffers less from this problem, although it is easily affected by mechanical vibrations. Even so, force-operated instruments are neither the most accurate nor the most convenient of devices, and the extent of their use has been very limited. RF/dc transfer standards based on force-operated instruments are also possible, but are even less frequent (see Chapter 15).

Impedance standards [30] for microwave frequencies are usually transmission lines of calculable characteristic impedance and of length which is a known fraction of a wavelength, for example a quarter wave line. At lower frequencies below about 300 MHz coaxial capacitors whose capacitance can be calculated from the dimensions are used. For coaxial lines consisting of perfect conductors the characteristic impedance Z_0 is given by:

$$Z_0 = \frac{1}{2\pi}\sqrt{\frac{\mu}{\varepsilon}} \ln\left(\frac{b}{a}\right) \tag{1.2}$$

and the capacitance C per unit length by:

$$C = \frac{2\pi\,\varepsilon}{\ln\left(\frac{b}{a}\right)} \tag{1.3}$$

where a and b are the radii of the inner and outer conductors, μ is the permeability of the medium and ε the permittivity of the medium. In vacuum the permeability μ_0 is a defined constant, equal to $4\pi \times 10^{-7}$ H/m, and the permittivity ε_0 is related to this value and the speed of light c by the equation:

$$c = \frac{1}{\sqrt{\mu_0 \varepsilon_0}} \tag{1.4}$$

Prior to 1983, the speed of light was an experimentally determined constant which depended on the definition of the metre in terms of the standard orange line of krypton 86 [105]. Since 1983, c has been a defined constant by means of which the metre is derived from frequency standards [106–108]. Its value is 299 792 458 m/s.

The characteristic impedance for practical conductors is modified by the skin effect, which is equivalent to a series resistance with an additional inductance. The resistive component is calculated from the skin depth δ, while the additional inductanceis calculated from the recession depth Δr. The way in which δ and Δr vary with frequency for both single material and plated conductors is treated in ref. [30] and in Chapter 15. From power and impedance standards one can derive voltage and current, provided that attention is paid to the definition of these quantities at high frequencies. The definition of the direct voltage between two points is that it is the work done in moving a unit charge from one point to the other. This definition is not unique at high frequencies, because the work done may depend on the path chosen, and one must therefore specify the allowable paths. For waveguides the voltage, current and impedance may be defined in more than one way (see ref. [109] and Chapter 15). One can avoid this problem by working in terms of quantities which are dimensionless ratios, such as voltage reflection coefficient or voltage standing wave radio.

Referring to Fig. 1.2, for voltage and current at frequencies above 100 MHz the normal derivation route is via rf power and impedance. Below 100 MHz however these can also be derived directly from dc voltage and current respectively by means of carefully evaluated thermal converters employing thermoelements or, in the case of voltage, from specially constructed thermistor voltage standards (see Chapter 15). Attenuation has been omitted from Fig. 1.2, since this depends only on the ability to measure ratios of power, voltage or current. RF impedance can be derived absolutely from a knowledge of the defined constants c and μ_0 without need for any other standards since, although equation 1.2 contains two lengths, only the ratio of these lengths is required.

1.4 References

1 Gardiol, F.E., 'Introduction to microwaves', Artech House, 1984.

2 Bryant, J.H., 'Coaxial lines; US historical perspective', *IEEE Trans.*, **MTT-32** 9, pp. 981–982, Sept. 1984.

3 Sladek, N.J., and Jewsch, R.L., 'Standardization of coaxial connectors in the IEC', *Proc. IEEE,* **74**, 1, pp. 14–18, Jan. 1986.

4 Beatty, R.W., and Kerns, D.M., 'Relationships between different kinds of network parameters, not assuming reciprocity or equality of the waveguide or transmission line characteristic impendances', *Proc. IEEE,* **52**, 1, p. 84 and 4, p. 426, 1964.

5 Engen, G.F., 'International intercomparison of standards for microwave power measurement', *IEEE Trans.,* **MTT-13**, 5, pp. 713–715, Sept. 1965.

6 Fellers, R.G., 'Radio measurements and techniques in the frequency range above 30 GHz', Proc. URSI General Assembly (Commission 1: Radio measurements and standards), Ottawa, Vol. 2, pp. 221–231, 1969.

7 Davis, R.T., 'NBS establishes millimeter measurement standards', *Microwaves*, **11**, 4, pp. 9–10, April 1972.

8 'NBS Service for power standards 94–95 GHz', *Microwave J.*, p. 64, Nov. 1980.

9 Special issue on rf measurements, *Proc. IEEE*, **55**, 6, June 1967.

10 'IEEE Standard for precision coaxial connectors', *IEEE Trans.,* **IM-17**, 3, pp. 206–218, Sept. 1968.

11 'Microwave power measurement', Application Note 64, Hewlett-Packard, 1965.

12 'Fundamentals of rf and microwave power measurements', Application Note 64-1, Hewlett-Packard, Aug. 1977.

13 Dietrich, C.F., 'Uncertainty, calibration and probability', Adam Hilger, 1973.

14 Harris, I.A., and Hinton, L.J.T., 'The expression of uncertainty in electrical measurements', Guidance Publication 3003 – Issue 5, National Measurement Accreditation Service, 1987.

15 Hinton, L.J.T., 'Uncertainty and confidence in measurements', in 'Microwave measurement', Bailey, A.E. (ed.), Peter Peregrinus, 1985.

16 Radio Research Laboratory Staff, 'Very high frequency techniques', Vol. II, Chap. 24, McGraw-Hill, 1947.

17 Montgomery, C.G., 'Technique of microwave measurements', McGraw-Hill 1947.

18 Barlow, H.M., and Cullen, A.L., 'Microwave measurements', Constable, 1950.

19 Terman, F.E., and Pettit, J.M., 'Electronic measurements', Chap. 2, McGraw-Hill, 1952.

20 King, D.D., 'Measurements at centimeter wavelengths', Boston Tech., 1953.

21 Ginzton, E.L., 'Microwave measurements', McGraw-Hill, 1957.

22 Sucher, M., and Fox, J., 'Handbook of microwave measurements', (3 vols, 3rd edition), Polytechnic Press, 1963.
23 Lance, A.L., 'Introduction to microwave theory and measurements', McGraw-Hill, 1964.
24 Adam, S.F., 'Microwave theory and applications', Chap. 5, Prentice Hall, 1969.
25 Oliver, B.M., and Cage, J.M., 'Electronic measurements and instrumentation', McGraw-Hill, 1971.
26 Lane, J.A., 'Microwave power measurement', Peter Peregrinus, 1972.
27 Laverghetta, T., 'Microwave measurements and techniques', Artech House, 1976.
28 Warner, F.L. 'Microwave attenuation measurement,' Peter Peregrinus, 1977.
29 Bailey, A.E. (ed.), 'Microwave measurement', Peter Peregrinus, 1985.
30 Somlo, P.I., and Hunter, J.D., 'Microwave impedance measurement', Peter Peregrinus, 1985.
31 Nergaard, L.S., 'Electrical measurements at wavelengths less than two meters', *Proc. IRE,* **24**, 9, p. 1207, Sept. 1936.
32 Nergaard, L.S., 'A survey of ultra-high-frequency measurements', *RCA Rev.*, **3**, p. 156, 1938.
33 Clayton, R.J., Houldin, J.E., Lamont, H.R.L., and Willshaw, W.E., 'Radio measurements in the decimetre and centimetre wavebands', *J. IEE*, **93**, Pt. III, 22, pp. 97–125, Mar. 1946.
34 Moreno, T., and Lundstrom, O.C., 'Microwave power measurement', *Proc. IRE*, **35**, 5, p. 514, May 1947.
35 Schrack, R.A., 'Radio-frequency power measurements', NBS Circ. 536, Mar. 1953.
36 Rumfelt, A.Y., and Elwell, L.B., 'Radio frequency power measurements', *Proc. IEEE,* **55**, 6, pp. 837–850, June 1967.
37 Rumfelt, A.Y., and Elwell, L.B., 'Correction to "Radio Frequency Power Measurements", *Proc. IEEE,* **55**, pp. 837–850, June 1967', *Proc. IEEE,* **56**, 12, p. 2159, Dec. 1967.
38 Beatty, R.W., 'Coaxial power, impedance and attenuation calibration at the National Bureau of Standards', *Microwave J.*, **11**, 5, pp. 65–75, May 1968.
39 Steele, J.McA., Ditchfield, C.R., and Bailey, A.E., 'Electrical standards of measurement, Part 2: R.F. and microwave standards', *Proc. IEE*, **122**, 10R, pp. 1037–1053, Oct. 1975.
40 Engen, G.F., 'Advances in microwave measurement science', *Proc. IEEE*, **66**, 4, pp. 374–384, April 1978.
41 Jurkus, A.P., and Stumper, U., 'National standards and standard measurement systems for impedance and reflection coefficient', *Proc. IEEE*, **74**, 1, pp. 39–45, Jan. 1986.
42 Bayer, H., Warner, F.L., and Yell, R.W., 'Attenuation and Ratio – National Standards', *Proc. IEEE*, **74**, 1, pp. 46–59, Jan. 1986.
43 Fantom, A.E., 'Power and energy: national standards', *Proc. IEEE,* **74**, 1, pp. 94–101, Jan. 1986.
44 Special Issue on Electromagnetic Metrology, *Proc. IEEE*, **66**, 4, April 1978.
45 Special issue on radio measurement methods and standards, *Proc. IEEE,* **74**, 1, Jan. 1986.
46 'Review of radio science', International Union of Radio Science (URSI), Brussels (published every 3 years).
47 NBS Handbook 77, 'Precision measurement and calibration: electricity and electronics', National Bureau of Standards, 1959.
48 'Precision measurement and calibration, electricity–radio frequency', NBS Special Publication SP300, Vol. 4, June 1970.
49 Kamper, R.A. (ed.), 'Metrology for electromagnetic technology: a bibliography of NBS publications', NBS IR 79–1625.
50 'Calibration and related measurement services of the National Bureau of Standards', Rep. No. NBS-SP-250, Nat. Bur. Stand., 1982.
51 Bayer, H. (ed.) 'Special developments and methods for high frequency metrology', PTB-Berichte PTB-E-24, June 1983.

52 Bayer, H., 'Actual position and development tendencies in the subject of high frequency measurement technology', Proc. 49th PTB Seminar, PTB-E-24, pp. 23–26, June 1983 (in German).

53 Harvey, A.F., 'Microwave Engineering', Academic Press, 1963.

54 Crowley-Milling, M.C., Gordon, D.S., Miller, C.W., and Saxon, G., 'The measurement of power at centimetric and decimetric wavelengths', *J. IEE*, **93**, Pt. IIIA, 9, pp. 1452–1456, Mar.–May 1946.

55 Gaffney, F.J., 'Microwave measurements and test equipments', *Proc. IRE*, **34**, 10, p. 775, Oct. 1946.

56 Jones, T.I., and Colebrook, F.M., 'An appraisal of laboratory radio measuring techniques', *J. IEE* (Radio Sec.), **95**, pp. 315–316, Pt. III, Sept. 1948.

57 Larson, R.E., 'Microwave measurements in the electronic calibration centre', *Proc. IEE*, **109**, Pt. B, Suppl. 23, pp. 644–650, May 1962.

58 Engen, G.F., 'A survey of the microwave measurement techniques employed at the National Bureau of Standards', *Proc. IEE*, **109**, Pt. B. Suppl. 23, pp. 734–739, May 1962.

59 Cullen, A.L., 'Microwave power measurement', IEE Paper No. 3799E, May 1962.

60 Desch, R.F., and Larson, R.E., 'Bolometric microwave power calibration techniques at the National Bureau of Standards', *IEEE Trans.*, **IM-12**, 1, pp. 29–33, June 1963.

61 Clark, R.F., Jurkus, A., and Pattenson, C.F., 'Calibration facility for RF quantities offers central tie-in for testing labs', *Canad. Electron. Engin.*, **10**, pp. 40–45, 1966.

62 Selby, M.C., 'The system of electromagnetic quantities at 30 kHz to 1 GHz', *Metrologia,* **2**, 1, pp. 37–45, Jan. 1966.

63 Selby, M.C., 'Progress in the United States on electromagnetic standards and measurements at 30 kHz to 1 GHz, 1963 through 1965', *IEEE Trans.*, **IM-16**, 2, pp. 172–178, June 1967.

64 Giachino, G., and Rietto, G., 'Microwave calibration at the National Electrotechnical Institute', *Alta Frequenza*, **37**, pp. 699–711, 1968 (in Italian).

65 Blouet, J., 'High-frequency metrology', *Revue Générale de l'Electricité,* **78**, pp. 392–397, 1969 (in French).

66 Erard, L., 'Primary standards at high frequency', *Revue Géneralé de l'Electricité,* **85**, pp. 899–904, Nov. 1976 (in French).

67 Adam, S.F., 'Microwave instrumentation: and historical perspective', *IEEE Trans.*, **MTT-32**, 9, pp. 1157–1161. Sept. 1984.

68 Kamper, R.A., 'Uncertainty charts for RF and microwave measurements', *Proc. IEEE*, **74**, 1, pp. 27–32, Jan. 1986.

69 Hoer, C.A., 'Summary of NBS calibration services and systems', *Proc. IEEE*, **74**, 1, pp. 32–35, Jan. 1986.

70 'Definitions of Basic (=Base) SI Units', *Metrologia,* **4**, 3, p. 147, Oct. 1968 (see corrigendum *Metrologia* 1969, p. 33).

71 Taylor, B.N., 'Electrical units, fundamental constants and the 1983 least squares adjustment', *IEEE Trans.*, **IM-34**, 2, pp. 155–163, June 1985.

72 Vigoureux, P., 'A determination of the ampere', *Metrologia,* **1**, pp. 3–7, Jan. 1965.

73 Dix, C.H., and Bailey, A.E., 'Electrical standards of measurement, part 1, DC and low-frequency standards', *Proc. IEE,* **122,** 10R, pp. 1018–1036, Oct. 1975.

74 Harrison, P.W., and Rayner, G.H., 'A primary standard of mutual inductance', *Metrologia,* **3**, 1, pp. 1–12, Jan. 1967.

75 Thompson, A.M., 'An absolute determination of resistance based on a calculable standard of capacitance', *Metrologia,* **4**, 1, pp. 1–7, Jan. 1968.

76 Jones, R.G., and Kibble, B.P., 'Realization of the SI ohm at NPL', *IEEE Trans.*, **IM-34**, 2, pp. 181–184, June 1985.

77 Josephson, B.D., 'Possible new effects in superconducting tunnelling', *Phys. Lett.*, **1**, pp. 251–253, 1962.

78 Taylor, B.N., Parker, W.H., Langenberg, D.N., and Denenstein, A., 'On the use of the AC Josephson effect to maintain standards of electromotive force', *Metrologia,* **3**, 4, pp. 89–98, Oct. 1967.

79 Petley, B.W., and Morris, K., 'A measurement of 2e/h by the ac Josephson effect', *Metrologia,* **6**, 2, pp. 46–51, April 1970.

80 Finnegan, T.F., and Denenstein, A., 'A new transfer and maintenance voltage standard using superconducting tunnel junctions', *Metrologia*, **7**, 4, pp. 167, Oct. 1971.

81 Gallop. J.C., and Petley, B.W., 'A new NPL determination of 2e/h': *Metrologia,* **8**, 4, pp. 129–132, 1972.

82 Field, B.F., Finnegan, T.F., and Toots, J., 'Volt maintenance at NBS via 2e/h: a new definition of the NBS volt', *Metrologia*, **9**, pp. 155–166, 1973.

83 Harvey, I.K., 'Cryogenic ac Josephson effect emf standard using a superconducting current comparator', *Metrologia*, **12**, 2, pp. 47–54, 1976.

84 Niemayer, J., Hinken, J.H., and Kautz, R.L., 'Near-zero bias arrays of Josephson tunnel junctions providing standard voltages up to 1 V', *IEEE Trans.*, **IM-34**, 2, pp. 185–187, June 1985.

85 Petley, B.W., 'The fundamental physical constants and the frontier of measurement', Adam Hilger, 1985.

86 Cohen, E.R., and Taylor, B.N., 'The 1973 least-squares adjustment of the fundamental constants', *J. Phys. Chem. Ref. Data*, **2**, 4, pp. 663–734, 1973.

87 Clothier, W.K., 'A proposal for an absolute liquid electrometer', *Metrologia*, **1**, 4, pp. 181–184, Oct. 1965.

88 Sloggett, G.J., Clothier, W.K., Currey, M.F., Benjamin, D.J., and Bairnsfather, H., 'Absolute determination of the volt using a liquid electrometer', *IEEE Trans.*, **IM-34**, 2, pp. 187–191, June 1985.

89 Kibble, B.P., Smith, R.C., and Robinson, I.A., 'The NPL moving-coil ampere determination', *IEEE Trans.*, **IM-32**, 1, pp. 141–143, Mar. 1983.

90 Olsen, P.T., Bower, V.E., Phillips, W.D., Williams, E.R., and Jones, G.R. Jr., 'The NBS absolute ampere experiment', *IEEE Trans.*, **IM-34**, 2, pp. 175–181, June 1985.

91 Kibble, B.P., Robinson, I.A., and Bellis, J.H., 'A realisation of the SI ampere by the NPL moving coil balance', Conference on Precision Electromagnetic Measurements, Conference Digest Addendum, Gaithersburg, June 1986.

92 Kibble, B.P., 'Realising the ampere by levitating a superconducting mass – a suggested procedure', *IEEE Trans.*, **IM-32**, 1, p. 144, Mar. 1983.

93 Taylor, B.N., 'Is the present realisation of the absolute ampere in error?', *Metrologia*, **12**, 2, pp. 81–83, 1976.

94 Cohen, E.R., and Taylor, B.N., 'The 1986 adjustment of the fundamental physical constants', Codata Bulletin No. 63, Nov. 1986.

95 von Klitzing, K., Dorda, G., and Pepper, M., 'New method for high accuracy determination of the fine-structure constant based on the quantised Hall resistance', *Phys. Rev. Lett.*, **45**, pp. 494–496, 1980.

96 Bliek, L., Braun, E., Melchert, F., Schlapp, W., Warnecke, P., and Weinmann, G., 'Measurement of the fine structure constant by means of the quantized Hall resistance', *Metrologia*, **19**, pp. 83–84, 1983.

97 Cage, M.E., Dziuba, R.F., and Field, B.F., 'A test of the quantum Hall effect as a resistance standard', *IEEE Trans.*, **IM-34**, 2, pp. 301–303, June 1985.

98 Bliek, L., Braun, E., Melchert, F., Warnecke, P., Schlapp, W., Weimann, G., Ploog, K., Ebert, G., and Dorda, G.E., 'High precision measurements of the quantized Hall resistance at the PTB', *IEEE Trans.*, **IM-34**, 2, pp. 304–305, June 1985.

99 Wada, T., Shida, K., Nishinaka, H., and Igarashi, T., 'Study of the quantized Hall effect as a resistance standard at ETL', *IEEE Trans.*, **IM-34**, 2, pp. 306–309, June 1985.

100 Hartland, A., Davies, G.J., and Wood, D.R., 'A measurement system for the determination of h/e^2 in terms of the SI ohm and the maintained ohm at the NPL', *IEEE Trans.*, **IM-34**, 2, pp. 309–314, June 1985.
101 van der Wel, W., Harmans, K.J.P.M., Kaarls, R., and Mooij, J.E., 'Application of the quantized Hall effect to a new resistance standard at VSL', *IEEE Trans.*, **IM-34**, 2, pp. 314–316, June 1985.
102 Beatty, R.W., 'The system of electromagnetic quantities at frequencies above 1 GHz', *Metrologia*, **2**, 1, pp. 46–54, Jan. 1966.
103 Cullen, A.L., 'Absolute power measurement at microwave frequencies', *Proc. IEE*, **99**, Pt. IV, 4, pp. 100–110, April 1952.
104 Cullen, A.L., and Stephenson, I.M., 'A torque-operated wattmeter for 3-cm microwaves', *Proc. IEE*, **99**, 12, pp. 294–301, Dec. 1952.
105 Giacomo, P., 'News from the BIPM: some changes in the SI', *IEEE Trans.*, **IM-34**, 2, pp. 116–117, June 1985.
106 Terrien, J., 'International agreement on the velocity of light', *Metrologia*, **10**, 1, p. 9, 1974.
107 Giacomo, P., 'Laser frequency measurements and the redefinition of the meter', *IEEE Trans.*, **IM-32**, 1, pp. 244–246, Mar. 1983.
108 'Documents concerning the new definition of the metre', *Metrologia*, **19**, 4, pp. 163–177, 1984.
109 Kerns, D.M., 'Definitions of v, i, Z, Y, a, b, Γ, and S', *Proc. IEEE*, **55**, 6, pp. 892–900, June 1967.

Part A

Calorimeters

Chapter 2

Introduction to calorimeters

Calorimeters are heat-measuring instruments. They form the basis of the vast majority of primary standards for rf and microwave power and, in addition, some types also find use as secondary standards. One problem which all have in common is the conflict between accuracy on the one hand and convenience factors such as adequate sensitivity and rapid response on the other hand. The main reason for this conflict is the limited range of thermal resistances of available materials, which forces compromises on the designer. The result is that calorimeters intended for use as primary standards tend to have long time constants, typically in the range 1 to 10 minutes, and are often very bulky, whilst commercially made secondary instruments have unknown errors which are difficult to evaluate without calibration against a more accurate instrument. Some calorimeter designs have succeeded in narrowing the gap between the two types [1,2], but this gap is still considerable. In the past it was common to find calorimetric power meters used as everyday instruments on the bench [3], but for most routine purposes the preference today is for the speed and sensitivity obtainable from thermistors, thermoelectric and diode power meters. Consequently for non-standards work the calorimeter has tended to be displaced by these instruments, in spite of its potentially high accuracy and stability of calibration.

A calorimeter musts ideally be insensitive to the spatial distribution of the sources of the heat within it. If this aim is achieved, then power can in principle be measured at any frequency simply by dissipating it in the calorimeter and determining thecorresponding dc substituted power: that is, the dc power giving the same reading as the unknown power. Unfortunately the task of designing a calorimeter which is completely insensitive to heat distribution is one which is never completely realised and the best that can be achieved is to build an instrument which has known correction factors. These correction factors are evaluated from a combination of subsidiary measurements and calculation. They can be collected together into a single parameter, the effective efficiency, which is defined as:

$$\text{effective efficiency} = \frac{\text{dc substituted power}}{\text{rf power absorbed at input of calorimeter}} \qquad (2.1)$$

Generally speaking, effective efficiency is a more convenient quantity with which to deal than rf power: it is a relatively stable parameter for most instruments and, being dimensionless, is independent of the system of units used and the method of realisation of these units. For the most accurate primary standard calorimeters, effective efficiency can be evaluated with an uncertainty of roughly 0.1 percent at 1 GHz, 0.2 percent at 40 GHz and 0.5 percent at 100 GHz [4]. The corresponding uncertainties in the measured value of the absorbed rf or microwave power will naturally be somewhat higher than these figures, since they depend also on the uncertainty of the dc standards used and on the repeatability of the connectors (see Chapter 13). One can, if necessary, measure dc power to tens of parts per million, but for the applications of interest here an uncertainty in the range 0.01–0.1 percent normally suffices. Consequently secondary instruments may be used and direct access to the primary dc standards is not required.

2.1 Commonly used calorimeter types

Although the principle of measuring power by means of its heating effect is one of the oldest methods, present-day rf and microwave calorimeters have their origins in the developments of the 1940s and 1950s, when many different types were investigated to meet the newly arisen need for reliable measurements of signal levels in the microwave range [5–8]. Of the various designs considered during this period, three main versions have remained in use, although with many improvements; these are the dry load calorimeter, the microcalorimeter, and the flow calorimeter. The first two are referred to as static types. Substantial refinements of all three types were made during the 1960s, a period of general rapid progress in the field of precision rf and microwave measurements.

The essential components of a calorimeter, as illustrated in Fig. 2.1 for the dry load type, are a load in which the power is dissipated, a thermally isolating transmission line which connects the input to the load, and a temperature sensor. Many calorimeters use the dual load principle, in which one of the loads functions as the absorber while a second load acts as a temperature reference (Fig. 2.2). The temperature sensor registers the difference in the temperatures of the two loads. In theory the effects of external temperature fluctuations cancel because of symmetry, although if the surroundings are not at a uniform temperature the temperature gradients may still cause errors. Sometimes the reference load is replaced by a body having the same thermal properties as the rf load but not necessarily identical with it.

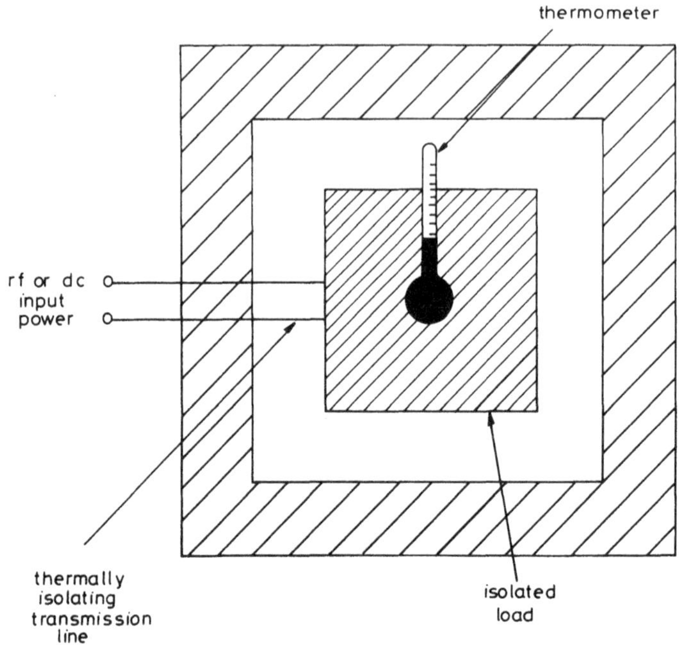

Fig. 2.1 *Principle of the dry load calorimeter*

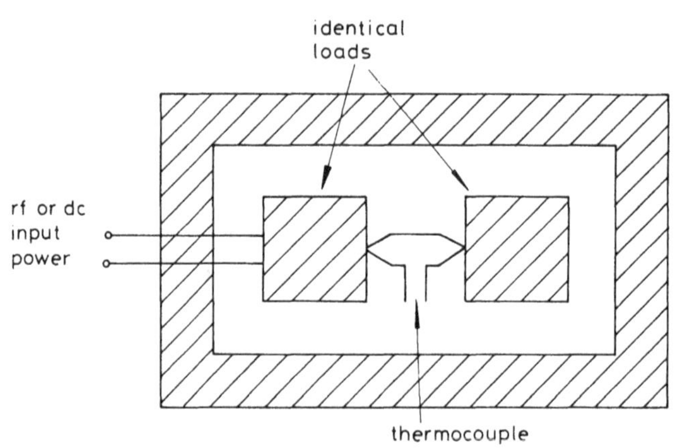

Fig. 2.2 *Dual load calorimeter using thermocouple as differential temperature sensor*

The absorbing element of the load of a dry load calorimeter is usually a thin film resistor, although lossy dielectrics are also used for waveguide versions. The temperature sensor is mounted on the outside of the load in a position where it cannot be influenced directly by the electromagnetic fields. Such an arrangement is one of the distinguishing features of a calorimeter and is essential for high accuracy.

The microcalorimeter (Fig. 2.3) is the most widely used calorimeter type for primary standards applications. Strictly speaking, it is not a power meter but an instrument for directly determining the effective efficiency of a bolometer mount. The principlewas originally devised for the calibration of metal wire bolometers, but thermistors and film bolometers can also be calibrated by this method. Before the start of the measurement, the bolometer mount is inserted into the calorimeter, where it acts as the load. When the measurement is complete the bolometer is removed and it can then be used as a calibrated reference standard.

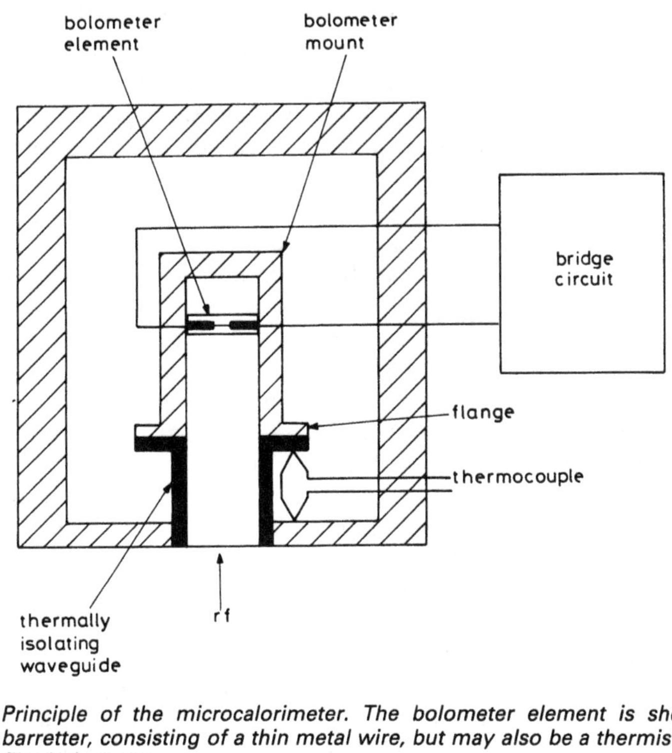

Fig. 2.3 *Principle of the microcalorimeter. The bolometer element is shown as a barretter, consisting of a thin metal wire, but may also be a thermistor or thin film bolometer*

The third of the three main calorimeter types, the flow calorimeter (Fig. 2.4), has been declining in popularity for many years. The power to be measured heats a fluid flowing through the load. An indication of the power is given by the rise in temperature of the fluid as it passes from the inlet to the outlet. Waveguide versions are usually directly heated and employ water as the working fluid. Coaxial flow calorimeters which have been built for the lower

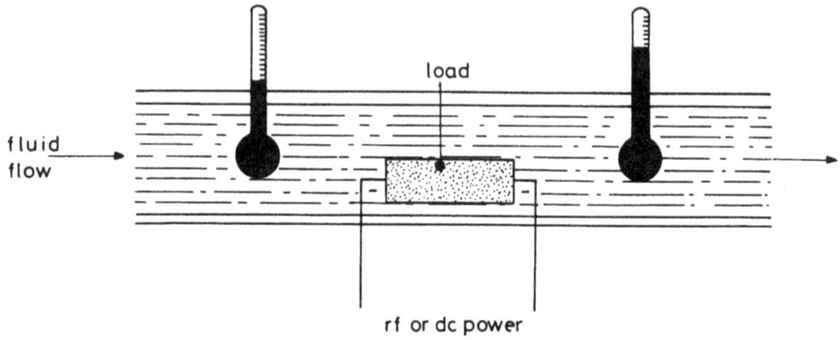

Fig. 2.4 *Principle of the flow calorimeter*

frequencies have in the main been indirectly heated ones, with oil as the fluid. Air can also be used, but the use of gases creates an additional problem because of heating due to the compressibility. Flow calorimeters can handle much higher powers than most static types and their power rating can be increased by raising the flow rate. Their main application is for powers of several watts or more. By way of comparison, the power range for static types is 1 mW to 10 W. For standards applications the modern tendency is to derive all power measurements from static calorimeters operating in the region of 10mW rather than building separate flow calorimeters for the higher levels. It is therefore essential to have access to good attenuation standards.

The temperature rise which must be measured in a calorimeter is usually less than 1 deg C. For example, at 10 mW the change in temperature registered by the sensor will typically be about a tenth of a degree C, although the temperature rise of the interior of the load will be much greater than this. In order to measure power with a resolution of 0.01 percent it is necessary to read the temperature to ten millionths of a degree and to do this a stable environment and a sensitive temperature sensor are required. The difficulties of measuring such small rises in temperature are not trivial and calorimeters are consequently amongst the least sensitive of power-measuring devices. Usually thermopiles, resistance thermometers, and sometimes thermistors are employed. A common type of thermopile is the multi-junction plated wire type (Fig. 2.5), which has also been used in multi-junction ac/dc transfer instruments for audio frequencies [9]. It is made by winding a constantan wire around a glass or plastic former and partially plating the wire with copper. Although the shunting effect of the constantan underneath the copper reduces the sensitivity slightly, this effect is not serious. Up to four hundred junction pairs can be made on one former in this way.

2.2 Some other calorimeter types

Among the types which were investigated in the early microwave years was the adiabatic calorimeter. If power P is applied to a body of thermal capacity, C,

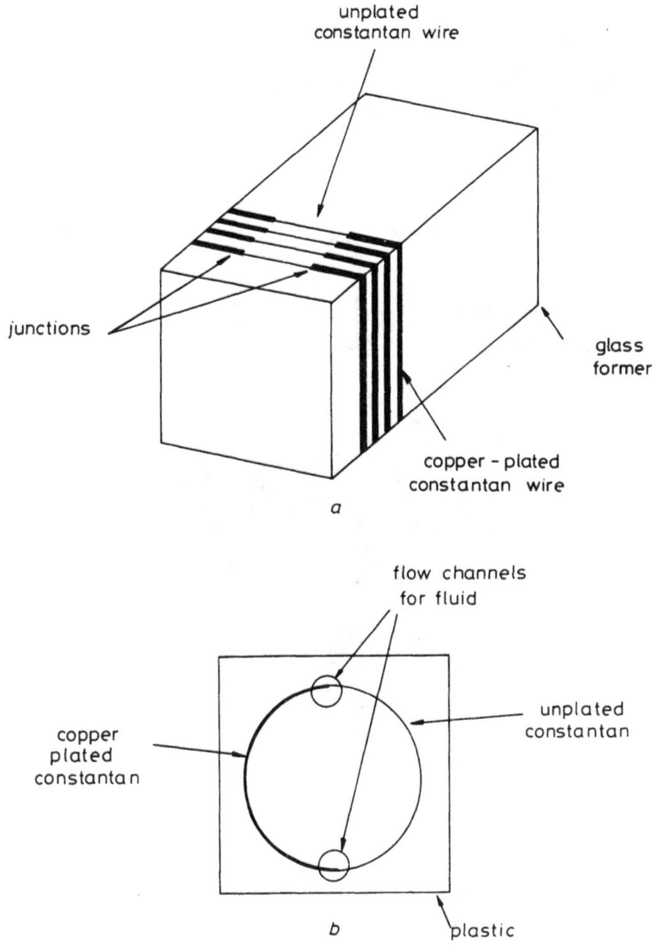

Fig. 2.5 *Plated wire type of thermopile used in many dry load calorimeters (a) and version for flow calorimeters (b)*

the temperature T follows approximately an exponential curve:

$$T = RP\,(1 - \exp\,(-t/RC)) \tag{2.2}$$

where R is the thermal resistance and t the time. Whereas in non-adiabatic calorimeters the final steady state temperature is taken as an indication of the power, for the adiabatic mode of operation it is the initial rate of rise of temperature which is used. The applied power is equal to the product of the rate of rise of temperature and the heat capacity of the load. Heat losses are small because the temperature of the load does not rise appreciably above its surroundings during the measurement. This represents an advantage for absolute calorimetry but it is less important in rf/dc substitution calorimetry, for which the effects of such losses cancel. The effective efficiency of a calorimeter when used adiabatically will not necessarily be the same as when it is used under steady state conditions.

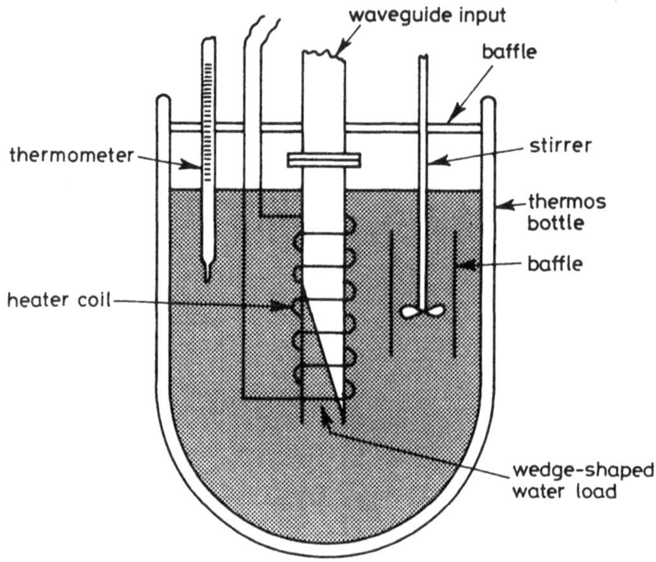

Fig. 2.6 *Adiabatic calorimeter*

Two adiabatic calorimeters are referred to by Carlin [10]. In one of these (Fig. 2.6) the calorimetric fluid (water) is contained in a covered Dewar flask which isolates it thermally from the surroundings. A wedge-shaped water load is formed by an open-ended waveguide with tapered walls containing a sealed inclined glass partition. Low-frequency power is applied for calibration purposes to a heating coil wound around the waveguide inside the calorimeter. A stirrer is used to ensure thorough mixing of the water. In the second adiabatic calorimeter, a dry waveguide termination is housed inside an evacuated metal box which itself is contained in an agitated bath. Heat losses by radiation and conduction are prevented by maintaining a zero temperature differential between the metal box and the termination. This is accomplished by means of a servo-loop which controls a circulating supply of hot water to the bath. When rf power is applied, the temperatures of both the load and surrounding metal box rise linearly with time in a parallel manner.

Another calorimeter type which is no longer in use is the change-of-state calorimeter. The heat to be measured melts a quantity of a solid, or boils off a quantity of a liquid. The medium is everywhere in thermal equilibrium and therefore all of the incoming heat is used to produce the change of state. The power P is given by:

$$P = \frac{MH}{t} \tag{2.3}$$

where M is the mass undergoing a change of state in time t and H the coefficient of latent heat. As with other calorimeters, the change-of-state calorimeter is normally used as an rf/dc substitution instrument. Ref. [11] describes an ice calorimeter using this principle.

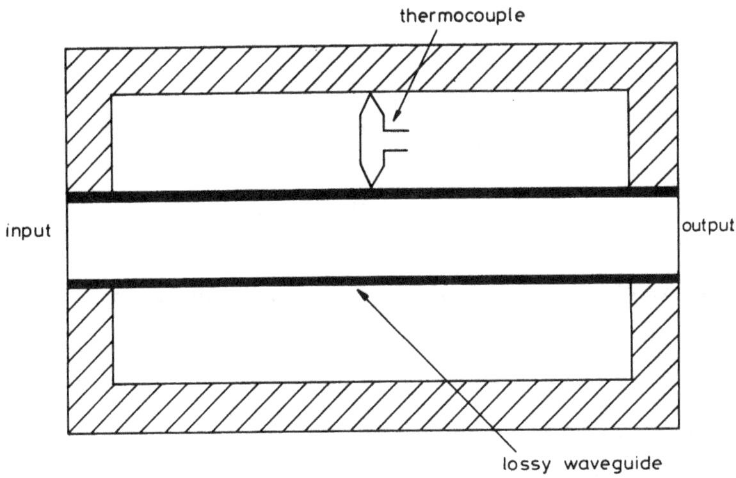

Fig. 2.7 *A simple feed-through calorimeter*

It is of course possible to construct feed-through calorimeters which absorb a part of the power and transmit the rest rather than completely absorbing the power to be measured [12,13]. Fig 2.7 shows a simplified waveguide instrument in which a proportion of the power is absorbed in the walls of a waveguide transmission line. However, little use has been made of such instruments in standards laboratories. They present a problem which does not arise with the terminating type of calorimeter: namely, the power reading is usually correct only for one particular load impedance. A much more common way of making a feed-through power meter is to connect a terminating power meter to the side arm of a directional coupler. The assembly of coupler and power meter is calibrated as one unit by connecting a calorimeter to its output, as described in Chapter 11. By adopting this approach the designer of the directional coupler can concentrate on obtaining the required directional properties and the designer of the calorimeter which is used to calibrate the assembly can concentrate on obtaining a good rf/dc transfer performance.

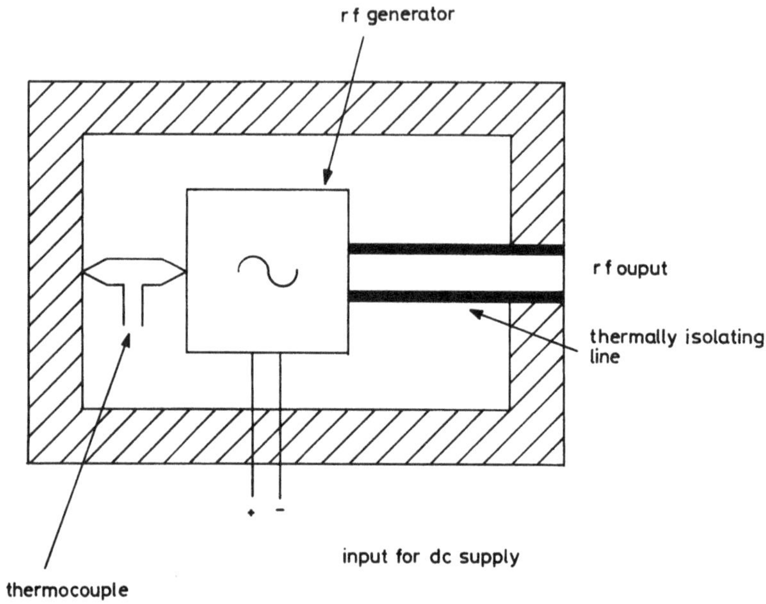

Fig. 2.8 *A possible method of measuring source power calorimetrically*

Beside the terminating and feed-through types of calorimeter there is in principle a third way of obtaining a calorimetric power standard. This is to measure the output power of a generator by determining the difference between the dc power drawn from the power supply and the rate of production of heat in the generator itself (Fig. 2.8). Theoretically this method gives a direct measure of the power delivered to the load, irrespective of the generator or load impedances, although in practice the effects of power absorbed in the thermally isolating line will not be load-independent. The technique has been used to measure the output power of radio transmitters [14] by making use of the rise in temperature of the cooling water as a convenient indication of the heat produced, but it is not used for accurate measurements.

Most methods of evaluating the correction factors of calorimeters entail disassembly of the instrument, a process which is time-consuming and which also makes it necessary to check that the calorimeter is functioning correctly after reassembly. Consequently the procedure is normally carried out no more than once a year and at a limited number of frequencies. Methods which do not require disassembly would be both more convenient and more suitable for automation, making it practicable to carry out the evaluation each time the calorimeter is used. They could also enable a wider range of constructional techniques to be used and might make miniaturisation possible, which could shorten the response time. However, in spite of the apparent advantages, a satisfactory method of achieving this remains to be found.

2.3 References

1 Crawford, M.L., 'A new RF-DC substitution calorimeter controlled reference power', *IEEE Trans.*, **IM-17**, 4, pp. 378–384, 1968.
2 Fantom, A.E., 'An improved coaxial calorimetric rf power meter for use as a primary standard', *Proc. IEE*, **126, 9**, pp. 849–854, Sept. 1979.
3 Serchuk, A., 'Microwave calorimeters, a product survey', *Microwaves*, **2**, pp. 48–51, Sept. 1963.
4 Fantom, A.E., 'Power and energy: national standards', *Proc. IEEE*, **74**, 1, pp. 94–101, Jan. 1986.
5 Griesheimer, R.N., 'Microwave power measurements', Chapter 3 in 'Technique of microwave measurements', ed. C.G. Montgomery, McGraw-Hill, 1947.
6 Barlow, H.M., and Cullen, A.L., 'Microwave measurements', Constable, 1950.
7 King, D.D., 'Measurements at centimeter wavelengths', Boston Tech., 1953.
8 Ginzton, E.L., 'Microwave measurements', McGraw-Hill, 1957.
9 Wilkins, F.J., Deacon, T.A., and Becker, R.S., 'Multijunction thermal converter; an accurate ac-dc transfer instrument', *Proc. IEE*, **112**, 4, p. 794, April 1965.
10 Carlin, J.J., 'Measurement of power', Vol. 1, Chapter 3 in 'Handbook of microwave measurements', ed. M. Sucher and J. Fox, Polytechnic Press, 1963.
11 Ginnings, D.C., and Corruccini, R.J., 'An improved ice calorimeter', *J. Res. NBS*, **38**, 583, RP 1796, 1947.
12 Brady, M.M., 'In-line waveguide calorimeter for high-power measurement', *IRE Trans.*, **MTT-10**, 5, pp. 359–366, Sept. 1962.
13 Brady, M.M., 'Correction to "In-line waveguide calorimeter for high-power measurements" – Accounting for transverse waveguide wall currents', *IEEE Trans.*, **MTT-11**, 2, p. 152, Mar. 1963.
14 Pession, G., and Gorio, T., 'Measurement of power and efficiency of radio transmitting apparatus', *Proc. IRE*, **19**, 3, pp. 377–400, 1931.

Chapter 3

Dry load calorimeters

3.1 Coaxial dry load calorimeters

Some early work on coaxial dry load calorimeters is described in two papers by Sucher and Carlin [1] and by James and Sweet [2]. These papers describe a set of rugged broad-band dual load calorimetric power meters, both coaxial and waveguide, which were intended for operation at frequencies from zero to 75 GHz. In comparison with later instruments the accuracy was very modest, with an uncertainty of 2 percent for the coaxial version and 1–2.5 percent for the waveguide versions. Nevertheless, these designs established the general direction in which subsequent instruments followed. New precision loads and connectors developed in the 1960s led to a new generation of coaxial calorimeters with better performance and with uncertainties below 0.5 percent for frequencies up to 8 GHz. Two particular designs dating from this period are described by Jurkus [3] and by Jurkus and Clark [4] respectively. These instruments, which were fitted with 14 mm precision connectors and were developments of an earlier version fitted with an N type connector [5], were amongst the first to make use of the tractorial load (see [6] and Appendix 3). The use of this type of load, consisting of a cylindrical resistor and a tapered outer conductor which follows a tractrix curve, largely avoided the problem of fringing fields from which transverse film loads suffered [7]. The calorimeters operated at power levels up to 100 mW and 10 W respectively. Fig. 3.1 shows the general arrangement of the 100 mW version, which had a maximum useful frequency of 8.5 GHz, although it was evaluated initially only up to 6 GHz. This calorimeter, of which a 7 mm version was later produced [8], has frequently been quoted in the literature and it will be used to illustrate the design principles and methods of evaluating the effective efficiency. The calorimeter was of the dual load type, with identical rf and reference loads. It was housed in a massive aluminium double thermal enclosure. The absorbing element of the load was a thin film resistor consisting of a carbon film deposited on a high thermal conductivity dielectric substrate. The wall of the load was of leaded copper, and the connector, which for analysis purposes was

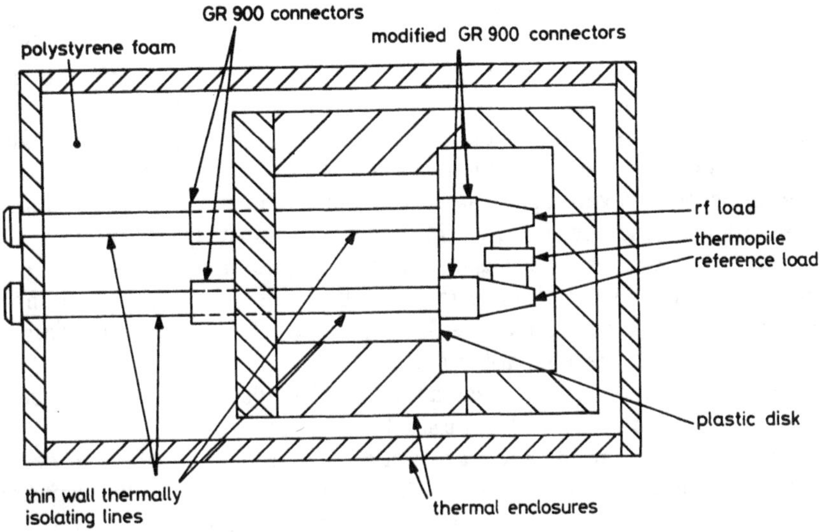

Fig. 3.1 *Coaxial dual load dry calorimeter of the Jurkus type*

regarded as part of the load, had an outer conductor of coin silver (an alloy consisting of ninety percent silver and ten percent copper). The temperature sensor was of the multi-junction plated wire thermopile type, which allows a large number of junctions to be obtained (see Chapter 2). The loads were isolated from the surroundings by German silver lines having a wall thickness of 0.25 mm. German silver, an alloy of copper, zinc and nickel, was chosen because of its relatively high thermal resistivity compared with other metals, although stainless steel is also suitable. Each line consisted of two sections, both fitted with connectors and attached by a flange to the inner and outer thermal enclosures respectively. Removal of the 50 Ω termination allowed the attenuation of the line to be measured. The outer surfaces of the walls of the loads were tapered so as to minimise the mass and hence the time constant. A further reduction in time constant was obtained by means of a thermal shunt connected between the two loads, the thermal resistance of this shunt being chosen so as to achieve a compromise between adequate sensitivity and reasonably short time constant. Even so, the reading time (defined in this case as ten times the time constant) was 25 minutes. The construction of the ten watt version was roughly similar to the 100 mW calorimeter except that an annular reference body replaced the reference load. The thermally isolating lines were also thinner (0.1 mm) and the thermopile construction was different, since fewer junctions were required.

The load of a calorimeter is a critical element. It is desirable that it should be well matched electrically and have a very small equivalence error: that is, equal dissipated rf and dc powers must produce the same temperature reading.

One of the larger contributions to the equivalence error is Peltier heating or cooling of the load during the dc part of the measurement, this error being typically 0.2 percent at 10 mW [9]. Usually, however, the Peltier effect is excluded from the quoted figures of equivalence error or effective efficiency, because it is assumed that it will be cancelled out to a first approximation when the calorimeter is used in practice. This is done by repeating the dc measurement with revered polarity and taking the mean of the two readings. Second order thermoelectric errors, in which Thomson heating plays a role, do not cancel on reversal, but it is generally thought that these are small in comparison with the total uncertainty. They are taken into account in audio-frequency thermal converters, for which the other uncertainty contributions are smaller [10] (see Chapter 15).

The achievement of a low equivalence error depends on a number of factors. The distribution of the power dissipated in the load should ideally be frequency-independent and the tractorial load gives a very good approximation to this condition (Appendix 3). The tractorial load is also especially attractive because the condition for frequency-independent distribution of power coincides with the condition for low reflection coefficient. The temperature distribution is smoothed out by the use of thick walls of high thermal conductivity, but these metals are relatively soft and their alloys have better machining properties. Attention must be paid to the positioning of the temperature sensor in relation to the various paths by which the heat may leave the load. Ideally all the heat would be forced to flow along a single path, so that a measurement of the temperature drop across the thermal resistance of this path would give an accurate indication of the total power [7,11]. This is not always possible but an improvement can be obtained by using several temperature sensors and taking a mean value, preferably a weighted mean [12].

3.1.1 Evaluation of equivalence error of load

In practice the equivalence error of a calorimeter load can never be exactly zero and it must be carefully evaluated in order to ascribe uncertainty limits. The following method is described in ref. [4] and illustrated in Fig. 3.2. If the power dissipation distribution in the resistor varies with frequency, then, for a given total power, the temperature T_{max} at the input end of the resistor will also vary with frequency. As a result the heat lost down the inner conductor, which is not detected by the thermopile, will be frequency-dependent. The temperature T_{max} cannot be measured directly, but it is possible to measure the average temperature T_{mean} of the resistor by monitoring its dc resistance and to deduce T_{max} from this, as shown in Fig. 3.2. The procedure is to supply power to the calorimeter at several frequencies up to 8 GHz and to measure the dc resistance simultaneously, keeping the thermopile output constant. It is assumed that at rf frequencies the power distribution, if not uniform, can be represented by a linear slope. In ref. [4] it is claimed that by the use of this

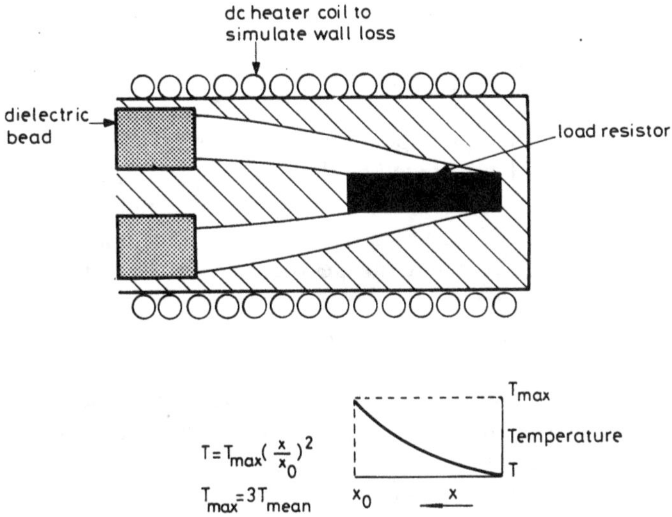

Fig. 3.2 *Method of assessment of load equivalence error used by Jurkus and Clark. The curve shows the situation where the power absorption in the load resistor is uniform. Different power distributions in the load resistor result in different mean temperatures for the same total power. Thus a linearly tapering power distribution can be distinguished from a uniform one by measuring the dc resistance. The dc heater wire enables the effect of wall losses to be evaluated*

method an upper limit of 0.01 percent can be assigned to the error at 8 GHz. A second contribution to the equivalence error is the fact that some power is dissipated in the walls of the load. To assess this effect it is necessary to know the sensitivity of the thermopile to heat generated in this region, which can be measured by attaching a separate dc heater to the outside of the load. Measurements described in ref. [4] showed that the sensitivity of the thermopile to heat generated in this region was 0.9 times the sensitivity to heat generated in the resistor. Since at 8 GHz the power absorbed in the walls was estimated to be a maximum of 0.1 percent of the total power, the resultant error was $(1 - 0.9) \times 0.1$ percent or 0.01 percent.

An alternative method of investigating temperature distributions in the load would be to modify the thermal properties, for example by putting in a PTFE ring, and observe the effect on the effective efficiency. The difficulty here lies in not disturbing the calorimeter. A further method, which can be used with some calorimeters, is to study the shape of the response curve with time [13]. This can be a useful technique if plated plastic thermally isolating lines are used, which have appreciable time constants. Any difference in the rf and dc power distributions results in a difference in the curve of response against time for rf and dc inputs. Consequently, by carrying out calibrations using a range of different reading times and checking that the results agree, limits can be put on the equivalence error.

3.1.2 Power loss in thermally isolating line

The power absorbed by the thermally isolating line represents an appreciable proportion of the total input power to a coaxial calorimeter, typically about 0.1 percent at dc rising to between 0.5 percent and 5 percent at 8 GHz. The proportion of power absorbed depends on the efficiency of the line: that is, the power delivered to the load divided by the power absorbed at the input to the line. The reading of the calorimeter must be corrected for this effect by multiplying by the ratio of the line efficiency at dc to the line efficiency at the frequency of interest. In principle the electrical properties of the line could be calculated for any frequency from the dc resistivity of the material and the dimensions; Post gives a theory for a line composed of a single material [14] and multilayer lines are treated by Somlo and Hunter [15]. But in practice calculation is not sufficiently accurate for the rf behaviour and measurements must be made. One way of determining the efficiency of a two-port network is from a simple insertion loss measurement, using a power meter as a load (Fig. 3.3). Assume for the sake of explanation that the dc loss may be neglected. If P_1 and P_2 are the power meter readings with and without the line inserted, then for the case when the components are matched the ratio P_1/P_2 gives the required efficiency. The assumption of matched components may be adequate for many cases but a complete measurement requires that the reflection coefficients be taken into account. This is done by determining a factor M, where

$$\text{efficiency} = M\frac{P_1}{P_2} \tag{3.1}$$

and M is equal to

$$M = \frac{1-|\Gamma_l|^2}{1-|\Gamma_{l'}|^2}\,\frac{|1-\Gamma_s\Gamma_{l'}|^2}{|1-\Gamma_s\Gamma_l|^2} \tag{3.2}$$

The reflection coefficients Γ_s,Γ_l, Γ'_l are those of the source (or equivalent source for a levelled system as explained in Chapter 11), the load power meter, and the thermally isolating line when terminated by the load power meter. A further correction factor, designated by K, is necessary to allow for the fact that the efficiency of a two-port network depends on the load reflection coefficient. This arises because, although equation 3.1 gives the efficiency when the two-port is terminated by the power meter used to measure it, when this power meter is replaced by the actual load of the calorimeter the efficiency changes. Its new value is given by:

$$\text{efficiency} = M\,K\frac{P_1}{P_2} \tag{3.3}$$

where

$$K = \frac{1-|\Gamma_{l'}|^2}{1-|\Gamma_l|^2}\,\frac{1-|\Gamma_{cal\ load}|^2}{1-|\Gamma_{cal\ input}|^2}\,\frac{|1-s_{22}\Gamma_l|^2}{|1-s_{22}\Gamma_{cal\ load}|^2} \tag{3.4}$$

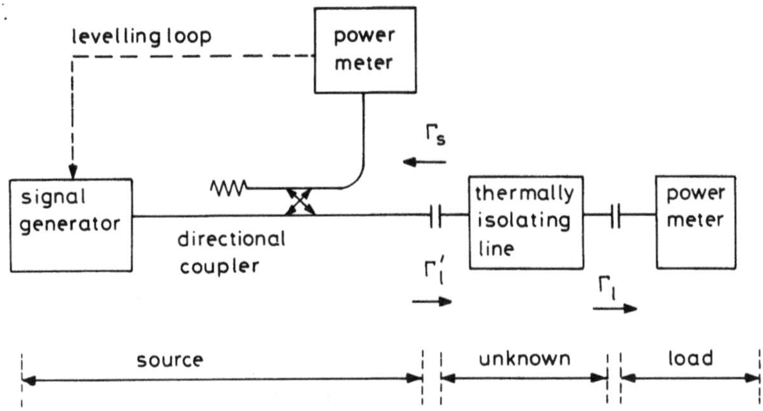

Fig. 3.3 *Determination of power absorption in thermally isolating line by insertion loss measurement. For accurate measurements the reflection coefficients must be taken into account. Γ_s is the equivalent source reflection coefficient, which depends only on the properties of the directional coupler*

where the scattering parameter s_{22} refers to the output end of the line. The factors M and K could be combined into a single constant, with some simplification through cancellation of the first terms in equations 3.2 and 3.4. However, it is often convenient to keep them separate, because there are occasions when K, but not necessarily M, is assumed equal to unity. The maximum error caused by this assumption can be evaluated with the help of the realisability conditions for a passive two-port, that is the conditions that express the fact that the power absorbed by the two-port must be positive for all combinations of terminations. Chapter 13 gives details of these conditions. If it is possible to keep all reflection coefficients well below 0.01, then mismatch effects can often be ignored altogether. This was the usual practice until the 1970s, since it avoided the tedious reflection coefficient measurement and the complex arithmetic which would otherwise be required. Further information on efficiency measurement is given in Section 13.2 of Chapter 13, which deals with adapter measurements.

Not all of the power absorbed in the thermally isolating line is lost. The portion of this power which is absorbed in the part of the line nearest to the output end will make a contribution to the heating of the load and the calorimeter will consequently be more efficient than would be expected from measurements of the line efficiency alone. The method used by Jurkus [3] to take this effect into account is as follows. By separating the power absorbed by the calorimeter into two parts one can write:

$$V = K_1P_L + K_2P_A \tag{3.5}$$

where V is the thermopile output voltage, P_L the power dissipated in the load, P_A the power dissipated in the coaxial line, K_1 the sensitivity to load power and K_2 the sensitivity to power dissipated in the coaxial line. Also

$$P_L + P_A = AP_L \tag{3.6}$$

where A is the attenuation of the line expressed as a power ratio. To determine K_1 and K_2 it is necessary to take two sets of readings and to substitute these into equation 3.5. Two simultaneous equations are thereby obtained, which can be solved for K_1 and K_2. Both sets of readings are obtained from measurements made either with dc or preferably with low-frequency ac at 50 or 60 Hz. The use of ac avoids errors due to Peltier heating, which for these measurements may be quite large because of the low resistance of the line at low frequencies. For the first measurement the low frequency power is applied to the calorimeter in the normal way and the thermopile reading noted. In this case most of the power is dissipated in the load. Then the resistive element of the load is replaced by a short circuit and the measurement is repeated. Virtually the whole of the power is now absorbed in the line. The short circuit must not disturb the thermopile or change its sensitivity. It is essential in carrying out the second measurement that the distribution of the absorbed low-frequency power throughout the line should be the same as for rf. The conditions for this are (*a*) straight tubular inner and outer conductors with thickness constant along their length, (*b*) inner and outer conductors of the same thickness, and (*c*) a low-frequency current block to prevent current from flowing in the heat sink. A typical value for the ratio K_2/K_1 is 0.2 to 0.3. For the higher frequencies a method also exists for carrying out the measurement in one operation without the need to measure K_1 and K_2 separately. This is discussed in Section 3.2.2 under waveguide dry load calorimeters.

3.1.3 Alternative coaxial designs

Since the time of refs. [3] and [4] various refinements and alternative designs of coaxial dry load calorimeter have been produced. That described in ref. [9] follows very closely the Jurkus design [3], except that the method of mounting the thermopiles was changed to avoid problems of poor thermal contact between the load and thermopile junctions. Feedback was applied to this calorimeter, as also later to Jurkus's calorimeter, in order to reduce the response time. The output of the thermopile was amplified and fed back to the reference load so that the two loads were brought to the same temperature, as illustrated in Fig. 3.4. In theory, if symmetry exists between the two loads then the dc feedback power is equal to the rf power being measured, apart from the correction for the effective efficiency of the calorimeter and the thermoelectric reversal error. However, in practice there will be some asymmetry and it is necessary to calibrate out its effects by applying a known dc power to the rf load.

An area where there has been a great variety of approaches is in the

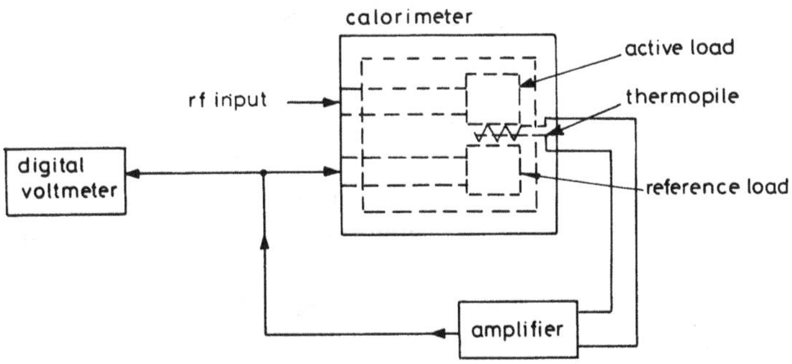

Fig. 3.4 *A common method of applying feedback to a twin load calorimeter. The power fed to the reference load is proportional to the square of the thermopile output voltage*

construction of the thermally isolating lines. The reason for this is that many compromises are called for, with the result that there is no single obvious best design and different workers have accordingly emphasised different aspects. From an electrical point of view the attenuation should clearly be a small as possible, since corrections are then minimised and interpolation between frequencies is easier. This aim conflicts with the need for good thermal isolation. The minimum acceptable isolation depends both on the expected ambient temperature fluctuations and on the thermal mass of the housing of the calorimeter, which acts as a heat sink. In theory one could compensate for poor thermal isolation by increasing the heat capacity of the housing, but in practice this requires that the lines should be in very good thermal contact with the heat sink. Experience has shown that, even for the outer conductor, adequate thermal contact is not easy to achieve and for the inner conductor it is extremely difficult, as there is no direct thermal connection.

The commonly used thin-walled German silver lines are far from optimal. The thickness is usually 0.1 to 0.3 mm, which at the higher frequencies is many times the skin depth, with the result that that part of the metal which is not close to the conducting surfaces worsens the thermal performance without contributing to the electrical conduction. Silver plating of the conducting surfaces improves the ratio of thermal to electrical resistance, although the theoretical behaviour of the attenuation as a function of frequency is then more complicated [15]. Thin-walled tubes of other materials such as stainless steel, brass or electroformed copper may be used. The latter two are even further from the optimum. To obtain higher ratios of thermal to electrical resistance many workers have used plated electrical insulators, usually glass or, if poorer rigidity can be tolerated, a plastic such as acrylic resin. Boström [16] describes a calorimeter using gold plated boro-silicate glass for the inner conductor and gold plated polystyrene for the outer. The thickness of the gold was 17 to 29 μm. The method of carrying out the plating is described in detail in ref. [17] Crawford [18] also used gold plated glass for the inner conductor. His calorimeter, constructed using 7 mm line, was intended as a convenient portable instrument. By the application of feedback it achieved a response time of 2 minutes, which was considerably shorter than many calorimeters.

Plastic lines cannot be made as long as glass ones by reason of their poor rigidity, but this disadvantage is compensated for by the fact that their higher thermal resistivity allows the same isolation to be obtained with a shorter length. To make full use of this fact, however, the plating must be reduced to around 1 μm for operation up to 8 GHz; otherwise heat conduction via the metal dominates. Such a line was used for the inner conductor of the calorimeter shown in Fig. 3.5 [13,19,20], which will be described here briefly. Not only the input line but also the inner conductor of the taper leading into the tractorial load was of gold plated plastic. Using this technique, an effective efficiency of at least 99.5 percent was achieved at all frequencies up to 8 GHz. The inner conductor of the line, together with the input connector, could be removed for measurement of the loss. The loss measurement was carried out with the aid of a specially made test outer conductor, the small loss of which was assumed equal to that of the outer conductor of the line of the calorimeter itself. The uncertainty in effective efficiency was 0.2 percent or better for frequencies up to 8 GHz. An unconventional linearised feedback circuit [21] was employed (Fig. 3.6), in which dc power was fed to two auxiliary heaters, one on each load. The power to one heater was increased as the power to the other was reduced. This caused the squared terms in the equation describing the feedback loop to cancel. The advantage of such an arrangement is that the loop gain is independent of power level, avoiding a problem which occurs for the more conventional type of feedback shown in Fig. 3.4, where the loop gain drops to zero at zero input power, making zeroing very slow.

A 7 mm calorimeter which did not use the tractorial load was described by Hollway and Muller [22]. An elaborate thermal circuit was developed for this calorimeter and used to examine the effect of taking the reading before the temperature had stabilised. The connectors were enlarged for better thermal contact with the heat sink. The theoretical attenuation of a multi-layer line was derived from a specially written computer program [23].

A commercial coaxial calorimeter which has been found useful as a secondary standard is that of ref. [24]. This also uses a resistance thermometer as the temperature sensor, with nickel windings, and operates to 15 GHz. Its response time is less than one minute and its power range approximately 3 mW to 300 mW. The effective efficiency at 8 GHz is approximately 98.5 percent and at 100 MHz approximately 99.9 percent.

3.2 Waveguide dry load calorimeters

Waveguide calorimeters possess a number of advantages over their coaxial counterparts. The absence of an inner conductor simplifies both the

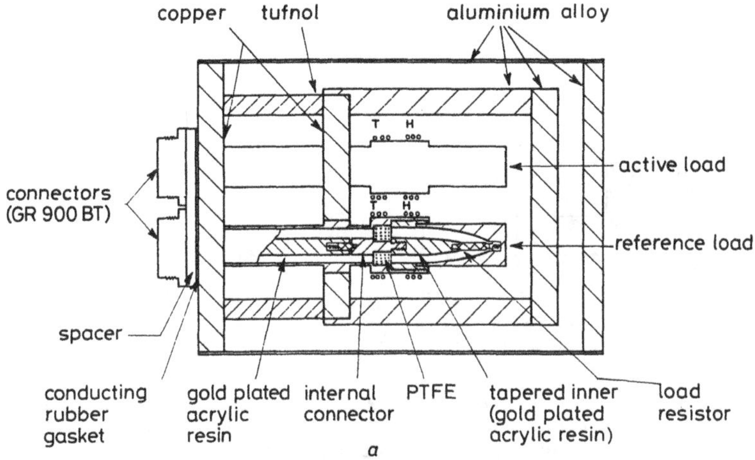

Fig. 3.5 *Coaxial dry load calorimeter due to Fantom. Plated plastic is used for the inner conductors of the thermally isolating lines. T denotes a resistance thermometer and H an auxiliary dc feedback heater*

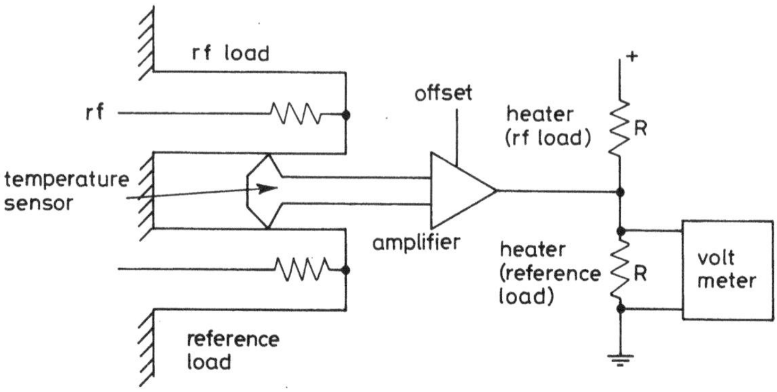

Fig. 3.6 *Linearised feedback circuit. The difference between the powers dissipated in the heater resistors R, which are attached to the outside of the two loads, is linearly related to the temperature sensor output voltage*

mechanical construction and the thermal equivalent circuit. For a given frequency the line losses are lower and the repeatability of flange couplings is better than that of coaxial connectors. Well matched loads can be made using a simple tapered absorbing element of appropriate length in relation to the wavelength.

The main difficulties are associated with the equivalence error of the load. For waveguide loads, in contrast to coaxial ones, low reflection coefficient and frequency-independent distribution of dissipated power do not necessarily go hand in hand. Consequently the temperature distribution in the absorber is likely to vary over the operating frequency band, making it impossible to achieve an exact match between rf and dc temperature distributions except at a single frequency. It is therefore the more important that the performance of the load should be thoroughly investigated. In some waveguide calorimeters the dc power is dissipated directly in the rf absorber element, whilst in others a separate dc heater is incorporated. Both thin film [1,2] and volume-absorbing loads [25–27] have been used. A complication is the necessity to provide a dc input connection to the load without interfering with the rf properties.

Fig. 3.7 shows the general design of the early waveguide instruments described in refs. [1] and [2]. These instruments operated at frequencies up to 75 GHz with claimed uncertainties in the range 1 to 2.5 percent. The absorbing element was a longitudinally mounted tapered resistive strip mounted in a silver waveguide having approximately 0.4 mm thick walls. Silver possesses not only a higher thermal conductivity than other suitable materials but also a smaller heat capacity per unit volume. The temperature sensor was a thermopile which was positioned to respond to the average temperature along the length of the waveguide. DC power was dissipated directly in the resistive film and its distribution controlled by a longitudinal slit (made by scratching the film) extending part way from the back towards the tapered end (see Fig.

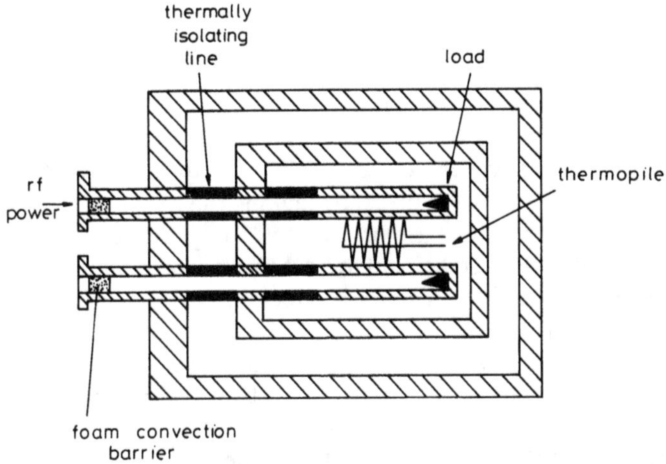

Fig. 3.7 *Waveguide dual load dry calorimeter of the type described by Carlin and Sucher and by James and Sweet. The thermally isolating lines may be of the plated insulator type or electroformed. The loads are thin film resistors*

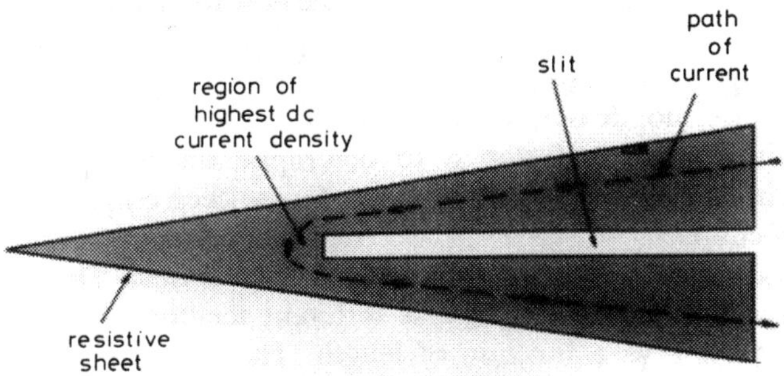

Fig. 3.8 *A method of controlling the dc power distribution in a waveguide thin film load element. The current flows around the slit. Maximum current density occurs near the end of the slit*

3.8). The dc current circulated around this barrier and the temperature maximum for dc occurred near the end of the slit. The thermally isolating lines were of the plated insulator type for some of the instruments and of the electroformed variety for others, although earlier calorimeters had used an air gap for isolation. Protection against ambient temperature fluctuations was obtained by the use of a double thermal housing. For the higher-frequency versions a modified form of construction was adopted, namely a symmetrical Y arrangement of the type used for some microcalorimeters (see Chapter 4, Fig. 4.3).

3.2.1 Equivalence error of load

The equivalence error of the load in Fig. 3.7 depends on both the thermal characteristics of the calorimeter and the disposition of the temperature sensors. Heat will escape from the load partly via the walls and partly via the two ends. The thermopile gives an accurate indication of the heat lost via the walls alone. Thus for the chosen arrangement of temperature sensor the equivalence error is caused by the fraction of the heat which escapes from the ends of the load, this being different for rf and dc inputs. The end losses can be reduced by making the waveguide containing the absorbing element long in relation to the absorber itself and also by making the thermal resistance of the thermally isolating line as high as possible, but compromises must be made in practice. Attention must also be paid to heat lost by convection currents along the inside of the waveguide. A foam convection barrier is normally used to suppress these.

In view of the difficulties of predicting thermal behaviour, it is important to check experimentally whether the design objectives regarding the equivalence error of the load have been met. This is sometimes done on a worst-case basis, by assuming that the dc power is dissipated at the mid-point of the load while the rf power is dissipated entirelyat one or other of the two ends. This, however, leads to an unduly pessimistic figure and if the relevant temperature distributions can be ascertained it is better to estimate the actual equivalence error. The first step in doing this is to determine how the calorimeter responds to different heat distributions in the absorber. For this part of the measurement the absorbing element is replaced by a movable point-like dc heater and the thermopile output is noted as a function of the axial position of the dc heater. The second step is to determine the actual temperature distributions in the load. Two methods for this have been employed. One is to attach thermocouples to the load element, taking care to ensure that the leads of the thermocouples are at right angles to the electric field. The second is to fabricate a set of similar loads, but of different lengths, and measure the attenuation of these as a function of length. The second method is more suitable for the higher frequencies, where the absorber is too small for the attachment of thermocouples. In order to make use of the measured temperature distributions the assumption must be made that they can be taken as representing the impressed power distribution for the silver waveguide. A weighted mean calorimeter reading is calculated for the rf and dc distributions by taking the power dissipated at each point in the load and multiplying this by the thermopile sensitivity for heat generated at that point. In this way the ratio of the thermopile sensitivity for heat generated at that point. In this way the ratio of the thermopile outputs for equal rf and dc powers absorbed in the load can be calculated. On the basis of such experiments it was concluded that the equivalence error of the loads of the waveguide calorimeters described in ref. [2] was 0.3 percent or less.

3.2.2 Yokoshima's calorimeter

A dry load calorimeter for the frequency range 60 to 90 GHz, with an uncertainty of 0.7 percent, is described by Yokoshima [26]. In this instrument the main temperature sensor is a resistance thermometer consisting of 0.1 mm diameter copper wire. A thermistor is attached to the input end of the load and functions as a second temperature sensor, enabling a correction to be made for the heat lost via the thermally isolating line. This two-thermometer arrangement [12] allows more freedom in the calorimeter design than the conventional arrangement, which requires that the end losses from the load must be small for low equivalence error. The principle is illustrated in Fig. 3.9, using thermocouples to represent the temperature sensors for clarity. The lines and load are manufactured from a single length of 0.5 mm thick German silver waveguide, silver-plated internally and with a gold flash applied. At the load end a 50 mm long 7 mm diameter copper sleeve surrounds the guide to provide temperature smoothing. The absorbing element of the load is a 25 mm long

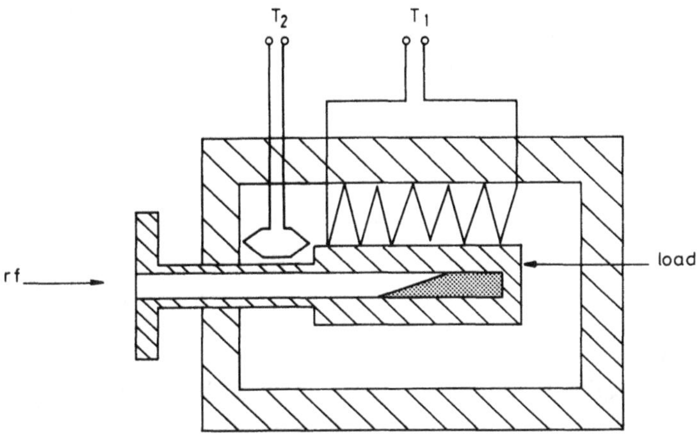

Fig. 3.9 *Two-thermometer method of reducing load equivalence error in a waveguide dry load calorimeter, as described by Fantom and Yokoshima. The improvement in accuracy is obtained by taking a weighted mean of temperatures T_1 and T_2. Although the temperature sensors shown are thermocouples, in practice T_1 could be a metal wire resistance thermometer and T_2 a thermistor*

tapered polyiron wedge. A dc heater, consisting of a small thermistor bead, is embedded in this absorber by drilling a 0.5 mm diameter hole and fixing with adhesive. Feedback of the linear type shown in Fig. 3.6 is used with the calorimeter, reducing the response time from 25 minutes to 2 minutes.

The outputs T_1 and T_2 of the resistance thermometer and the thermistor respectively are combined to give a single power reading P:

$$P = K_1T_1 + K_2T_2 \tag{3.7}$$

where K_1 and K_2 are constants whose ratio must be chosen to give the smallest equivalence error for the load. The correct value for the ratio K_2/K_1 is the value which minimises the dependence of P on the temperature distribution and this can be found by using the movable dc heater technique described above. The actual values of K_1 and K_2, as opposed to their ratio, are determined when a known dc power is applied during normal use of the calorimeter. In practice the ratio T_2/T_1 as a function of frequency is pre-measured as part of the evaluation process, so that in normal use only the resistance thermometer T_1 need be read and a correction based on the known ratio T_2/T_1 applied.

Yokoshima's method of correcting for power absorbed in the thermally isolating line differs from that described in Section 3.1 for the coaxial dry load calorimeter. In the coaxial case two separate measurements are necessary. First, the power absorbed in the thermally isolating line was determined by measuring the rf attenuation. Then the effect of this power dissipation on the thermopile reading was measured, using low-frequency power. In Yokoshima's method the two parts of the measurement are carried out simultaneously using rf power. The procedure yields directly the combined correction factor for the thermally isolating line and hence the effective efficiency of the calorimeter. A simplified explanation is as follows. The calorimeter is first connected to a matched source of power P_0 and the reading P noted, as given by equation 3.7. In this case

$$P = P_0\,[(1 - |\Gamma_i|^2) + \varepsilon] \tag{3.8}$$

where Γ_i is the input reflection coefficient of the calorimeter and ε is negative. The absorbing element of the load is then replaced by a sliding short circuit (Fig. 3.10). The power absorbed in the line is now slightly less than twice the previous value, since both the forward and reflected waves dissipate power. Assuming that the thermally isolating line can be treated as a matched attenuator with transmission coefficient s_{21} and that the incident power remains unchanged, ε is now replaced by $(1 + |s_{21}|^2)\varepsilon$ in equation 3.8. Since for a short-circuited matched attenuator $s_{21}{}^2$ is equal to minus the input reflection coefficient Γ_{is}, we have:

$$P_s = P_0\,[(1 - |\Gamma_{is}|^2) + (1 + |\Gamma_{is}|)\varepsilon] \tag{3.9}$$

where P_s and Γ_{is} are the values of P and Γ_i corresponding to the short circuit.

Eliminating P_0 from equations 3.8 and 3.9 enables the unknown ε to be found. One way of doing this which also provides an insight into the basis of the method is to rewrite equation 3.9 as:

$$\varepsilon = \frac{\dfrac{P_s}{P_0} - (1 - |\Gamma_{is}|^2)}{1 + |\Gamma_{is}|} \tag{3.10}$$

and solve by iteration. That is, ε is first put equal to zero in equation 3.8 as a first estimate and the corresponding value of P_0 found, which is then substituted in equation 3.10 to give a second estimate for ε. This new value of ε

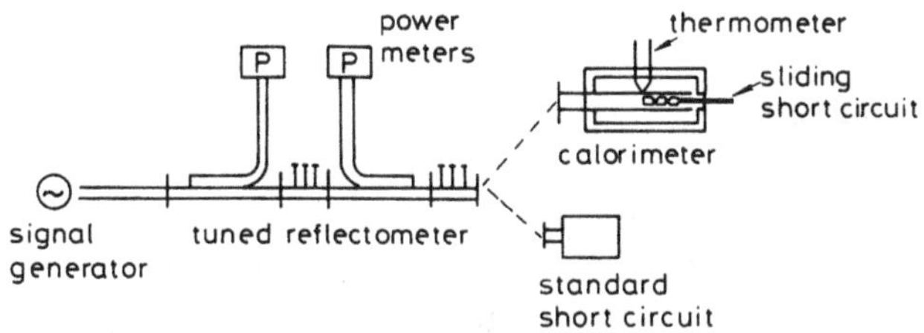

Fig. 3.10 *Yokoshima's method of measuring the correction for loss in the thermally isolating line of a waveguide dry load calorimeter. The power indicated by the calorimeter and its reflection coefficient are recorded simultaneously. Cyclic variations with the position of the short circuit must be averaged out*

is in turn substituted back in equation 3.8 and so on. In practice the line does not behave exactly as a matched attenuator because, in addition to any reflection at the flange, its loss takes the form of a distributed series resistance which changes the characteristic impedance. The result is an error which varies cyclically with the phase of the short circuit, but this error can be averaged out by displacing the short circuit through a quarterwavelength and repeating the whole measurement. The method is not suitable for the lower frequencies, where a displacement of a quarter wavelength would take the short circuit outside the load. An advantage of Yokoshima's method is that the effect of losses in the sliding short circuit is very much reduced, because such losses cancel approximately. This can be seen from equation 3.10, where the loss in the short circuit produces approximately equal changes in the two terms of the numerator, so that their difference remains almost constant. It will be appreciated that the underlying principle in this method is the measurement of the same incident power P_0 with the load in two different states (low reflection and high reflection) and subsequent equation of the two results.

3.2.3 Other designs

A calorimeter which differs from the Yokoshima approach was developed by Vowinkel [27]. The object of the design was to combine relative simplicity with good accuracy, convenience, and broad bandwidth. The claimed overall uncertainty was less than 1 percent, including mismatch and flange repeatability, and the response time with feedback 40 seconds, making the instrument suitable for general purpose use. The temperature sensors were thermistors, which were chosen because of their high sensitivity in relation to thermocouples and metal resistance thermometers. At one time the thermistor was regarded as unsuitable for such applications because of poor long-term stability and non-linear temperature dependence, but there have been improvements in stability over the years [28] and the problem of non-linear temperature dependence can be reduced by feedback.

The inner and outer jackets of the instrument are made of copper, whilst the thermally isolating lines and load are formed from a single length of coin silver waveguide. The two thermally isolating sections are made by thinning the guide down to a thickness of 0.3 mm. In the region of the load the guide is left to its normal thickness in order to achieve smoothing of the temperature distribution. The absorbing elements are made of carbon-loaded plastic having a resistivity of 90 Ω cm. The shape is that of a pyramid with its apex in the centre of the waveguide (Fig. 3.11) and this special shape has the advantage that any difference in the rf and dc temperature distributions is not directly transferred to the waveguide walls, but rather such differences are smoothed out to a certain extent before the heat is conducted to the end of the absorbers and then to the waveguide walls. An insulated wire is inserted into the absorber along its axis to supply the dc current. The length of this wire is chosen to bring the rf and dc temperature distributions as closely into coincidence as possible. Although the dc power distribution can be calculated, the rf distribution must be deduced from experiment. This was done by measuring the attenuation of sections of absorber of different lengths. The method has some inaccuracies because of the reflection at the broad end of the shortened absorber. To estimate the remaining equivalence error, the temperature distribution along the outside of the waveguide walls in the region

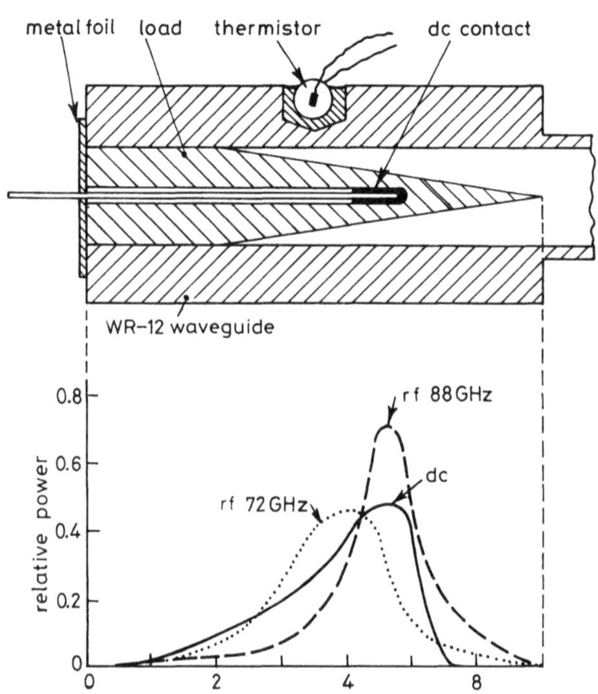

Fig. 3.11 *Pyramid-shaped load used by Vowinkel and corresponding heat distribution for dc, 72 GHz, and 88 GHz*

of the loads was checked for dc and rf power. Within the measurement uncertainty of 0.2 percent no difference was found and this value was taken as the upper limit for the equivalence error. The correction for the power absorbed in the lines was not measured directly, but was deduced from the previously determined attenuation constant of the waveguide and the heat equation. The calorimeter was also shown to be usable at higher frequencies, where the waveguide is overmoded, and even at optical frequencies, where the agreement with other methods was within 5 percent.

Keen [29] describes a calorimeter for the band 90 to 140 GHz with a claimed uncertainty of 4 percent. The heating effect of the power dissipated in the waveguide walls was found by removing the absorbing element of the load and allowing power to be radiated, after tuning with a fine wire. The uncertainties of this unusual method have not been thoroughly evaluated.

At even higher frequencies a calorimeter for the band 110 to 170 GHz was developed by Nemoto, Inoue, Sugiura, and Toda [30] (Fig. 3.12). This used Bi–Te thermocouples as temperature sensors. The load was operated at constant temperature by means of a feedback loop. A Bi–Te thermoelectric cooler was incorporated, which was set to a fixed value of cooling power so as to remove most of the heat generated in the load. The load and the input parts of the waveguide were made of copper, whilst the thermally isolating part was made of nickel. To eliminate reflection and loss at the junctions, all parts were made as one body by a multi-process electro-forming technique. The inner

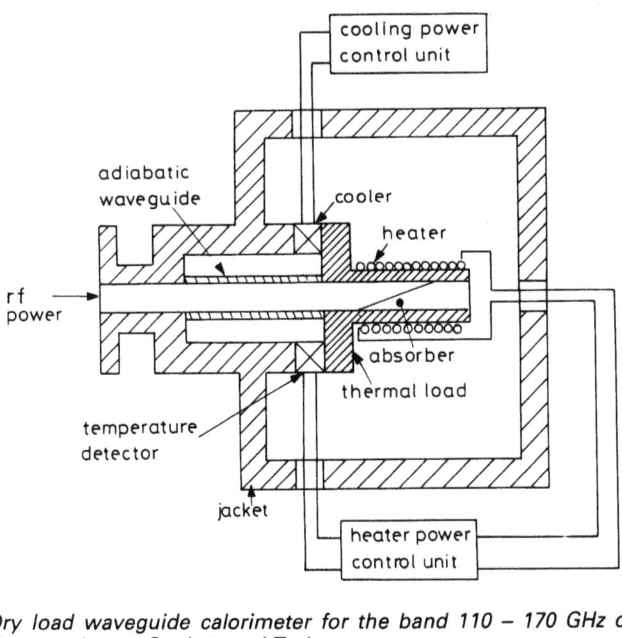

Fig. 3.12 *Dry load waveguide calorimeter for the band 110 – 170 GHz described by Nemoto, Inoue, Sugiura and Toda*

surface was gold-plated throughout (2–3 μm). The absorber was made of ferrite composite whose forward edge was tapered. A manganin heater wire of 0.06 mm diameter was wound on the load and moulded by a high thermal conductive bond. The correction factor was determined experimentally using an additional heater in the thermal load. The overall uncertainty of this instrument was 2 percent.

A commercial millimetre wave calorimeter design, now no longer manufactured, is described in ref. [31]. In this instrument the absorber is a tapered Ni–Cr evaporated thin film on a beryllia substrate. The substrate is brought out through a slit in the centre of the broad wall of the waveguide and a thermopile attached to it. This construction makes the thermopile reading much more sensitive to distribution of heat in the load than for the calorimeters described previously, so that the instrument is not suitable as a primary standard, but the response is fast. The centres of the rf and dc temperature distributions are brought into agreement by appropriately choosing the position of the electrode for dc impression.

3.3 References

1 Sucher, M., and Carlin, H.J., 'Broadband calorimeters for the measurement of low and medium level microwave power, I: Analysis and design', *IRE Trans.*, **MTT-6**, 2, pp. 188–194, April 1958.

2 James, A.V., and Sweet, L.O., 'Broadband calorimeters for the measurement of low and medium level microwave power, II: Construction and performance', *IRE Trans.*, **MTT-6**, 2, pp. 195–202, April 1958.

3 Jurkus, A., 'A coaxial radio-frequency power standard', *IEEE Trans.*, **IM-15**, 4, pp. 338–342, 1966.

4 Clark, R.F., and Jurkus, A., 'Ten watt coaxial calorimeter for RF power measurement', *Rev. Sci. Instrum.*, **39**, pp. 660–665, 1968.

5 Clark, R.F., 'A coaxial calorimeter for use as a microwave power standard', *IEEE Trans.*, **IM-14**, 1-2, pp. 59–63, Mar.–June 1965.

6 Harris, I.A., 'The theory and design of coaxial resistor mounts for the frequency band 0-4000 Mc/s', *Proc. IEE*, **103C**, 3, pp. 1–10, Mar. 1956.

7 Hudson, P.A., and Allred, C.M., 'A dry, static calorimeter for RF power measurement', *IRE Trans.*, **I–7**, 3-4, pp. 292–296, Dec. 1958.

8 Jurkus, A., 'A coaxial calorimeter and its use as a reference standard in an automated microwave power calibration system', *IEEE Trans.*, **IM-35**, 4, pp. 576–579, Dec. 1986.

9 Fantom, A.E., 'Development of coaxial calorimeter for frequency range 0–6 GHz', NPL Report DES 15, Nov. 1972.

10 Inglis, B.D., 'AC-DC transfer standards – present status and future directions', *IEEE Trans.*, **IM-34,** 2, pp. 285–290, June 1985.

11 Fantom, A.E., 'The design of precision coaxial power meters', NPL Report DES 33, July 1976.

12 Fantom, A.E., and Yokoshima, I., 'Two-thermistor technique for calorimeter type waveguide power meters', Fourth European Microwave Conference, Montreux, Sept. 1974.

13 Fantom, A.E., 'An improved coaxial calorimeter rf power meter for use as a primary standard', *Proc. IEE*, **126**, 9, pp. 849–854, Sept. 1979.

14 Post, F.J., 'Phase, attenuation and impedance characteristics of coaxial transmission lines with thin tubular conductors', *IEEE Trans.*, **MTT-11**, 2, pp. 129–136, March 1963.
15 Somlo, P.I., and Hunter, J.D., 'Microwave impedance measurement', Peter Peregrinus, 1985.
16 Boström, R., 'Coaxial microcalorimeter', Proc. IMEKO Symp. Microwave Meas., Budapest, 1972 (FOA Report B 3030-E1(E2)).
17 Boström, R., 'Coaxial microcalorimeter', Report FOA 3 A 3749-E1(E2), Research Institute of National Defence, Sweden, Oct. 1971 (in Swedish).
18 Crawford, M.L., 'A new RF-DC substitution calorimeter with automatically controlled reference power', *IEEE Trans.*, **IM-17**, 4, pp. 378–384, Dec. 1968.
19 Fantom, A.E., 'Development of Mk III calorimeter rf power meter (triple feedback version)', NPL Report DES 34, July 1976.
20 Fantom, A.E., 'Mk III calorimeter rf power meter (linear feedback version)', NPL Report DES 45, July 1978.
21 Fantom, A.E., 'A simple linear feedback method for twin load calorimeter rf power meters', *Metrologia*, **14**, 1, pp. 15–16, 1978.
22 Hollway, D.L., and Muller, W., 'The precise measurement of rf power and voltage using coaxial calorimeters and microcalorimeters', *IEEE Trans.*, **IM-33**, 4, pp. 269–275, Dec. 1984.
23 Somlo, P.I., 'The computation of surface impedance of multilayer cylindrical conductors', CSIRO (Australia) Nat. Meas. Lab. Rep. No. APR12, 1966.
24 'Type NRS calorimeter power meter', Catalogue, Rohde and Schwartz, 1986.
25 Sharpless, W.M., 'A calorimeter for power measurements at millimeter wavelengths', *IRE Trans.*, **MTT-2**, 3, pp. 45–47, Sept. 1954.
26 Yokoshima, I., 'Development and design of calorimeter type microwave power meter for the frequency range 60–90 GHZ', NPL Report DES 22, Nov. 1973.
27 Vowinkel, B., 'Broad band calorimeter for precision measurement of millimeter- and submillimeter wave power', *IEEE Trans.*, **IM-29**, 3, pp. 183–189, Sept. 1980.
28 Macklen, E.D., 'Thermistors', Electrotechnical Publications Ltd., 1979.
29 Keen, N.J., 'Milliwatt calorimeter for the 90–140 GHz waveguide band', *Electron. Lett.*, **10**, 18, pp. 384–385, Sept. 1974.
30 Nemoto, T., Inoue, T., Sugiura, I., and Toda, H., 'A modified thermal balancing calorimeter for precise power measurements in the short millimetre wave region', Conference on Precision Electromagnetic Measurements, Ottawa, 1978, pp. 57–59.
31 Aoki, T., Ohi, K., and Ueki, K., 'Some studies in behaviour of microwave dry calorimeter', *Hitachi Rev.*, **18**, 6, pp. 235–240, 1969.

Chapter 4

The microcalorimeter

The term 'microcalorimeter', although sometimes used as a general description for any calorimeter capable of measuring milliwatt power levels, is most commonly associated with the type of instrument shown in Fig. 4.1 [1]. The load is a bolometer mount and the purpose of the measurement is to determine the effective efficiency of this mount. After the measurement, the mount is removed from the calorimeter and used as a secondary standard. The technique can give excellent results, with uncertainties for waveguide bolometers as low as 0.1 percent at 10 GHz, 0.3 percent at 35 GHz and 0.5 percent up to 100 GHz. Coaxial mounts may also be calibrated by the microcalorimeter technique, but they pose an additional problem because of the difficulty of estimating uncertainties arising from heat losses through the inner conductor.

The basic design of the microcalorimeter is due to MacPherson and Kerns [2], with later refinements by Engen [3] and further improvements by workers in various national standards laboratories. The instrument was originally introduced as an improved means of evaluating waveguide barretter mounts at a time when the impedance method (see Chapter 6) was still in use, but its application was subsequently extended to commercial thermistor mounts [4]. Film bolometers may also be calibrated by the microcalorimeter method. These generally have a higher effective efficiency than thermistors, with figures greater than 99 percent at frequencies up to 94 GHz (see Chapter 6).

A broad outline of the operation of the microcalorimeteris as follows [5]. The total heat, that is the heat generated both in the bolometer element and in the walls of the mount, is registered by a thermopile. Originally this thermopile was located on the outside of the mount itself [2], but it was later moved to the waveguide flange with which the mount mates [3]. This latter position facilitates removal of the mount without disturbing the thermopile and also reduces errors. A dc self-balancing bridge is connected to the bolometer and this bridge is used in the conventional way as for bolometric power meters (see Chapter 6). The bridge may be of the Wheatstone type, as is shown in Fig. 4.1, or alternatively of the self-adjusting current loop type [6], which enables the resistance of the dc leads to the bolometer to be eliminated by means of a four-terminal resistance measurement. Audio frequency bridges, although once common, are not normally used because of errors which can arise if the thermal time constant of the element is short enough for its temperature to follow the audio waveform [7]. The basic principle of the microcalorimeter relies on the fact that the rf power can be measured simultaneously in two different ways, that is bolometrically and calorimetrically. The bolometric measurement includes only that part of the rf power which is dissipated in the bolometer element, whereas the calorimetric measurement includes the whole power. The ratio of the two power readings is the required effective efficiency of the mount.

A significant advantage of the microcalorimeter in relation to other types is that, to a first approximation, neither the power reflected by the bolometer mount nor the power lost in the feed line influences the result. This is because both the bolometric and calorimetric power readings are reduced in the same proportion. One consequence of this fact is that the feed line can be made as long as is required for good isolation, so that the compromises which are made for the thermally isolating lines in dry load calorimeters are not necessary. Microcalorimeters are frequently operated in a water bath or oil bath. They can be left unattended for long periods during operation, since it is not necessary to connect and disconnect the bolometer during the measurement. Nevertheless, there are a number of aspects which make the microcalorimeter inconvenient to use and which have resulted in its being confined to national standards laboratories. Amongst these is the fact that the calorimeter must be disassembled and reassembled when a new bolometer is inserted and several hours must then be allowed for the instrument to settle down thermally again. Consequently it is advisable to calibrate at least two bolometers at each frequency so that, by intercomparing these, changes with time can be detected without having to repeat the whole calibration procedure. A knowledge of the thermal properties of the microcalorimeter and the mount to be calibrated is required, since without this one cannot estimate the uncertainty or the correction factors. This means that one has to assess the suitability of each type of mount which it is proposed to calibrate in the instrument. Bolometers are in general less well matched than many other types of power meter, especially when wide bandwidth is required. Although this does not necessarily matter for the calibration process itself, it is important when the calibrated mount is later used as a secondary standard. For waveguide barretters the traditional solution to the matching problem is to incorporate screw tuners in the mount [3], but this approach means that each mount can be used only at one frequency, since resetting of tuners is not reliable, and several mounts are needed to cover one waveguide band. This, together with the need to calibrate more than one mount at each frequency, can result in as many as fifteen mounts being used to cover one waveguide band. Tuners are not generally used for thermistor mounts and therefore for accurate results it is necessary to apply corrections for mismatch. In principle the microcalorimeter technique could also be applied to the direct calibration of other types of power meter, but in practice the bolometer is chosen because of its good stability and relatively compact size.

Some of the more general design aspects which microcalorimeters share with the dry load type have already been discussed in Chapter 3 and attention will be focussed here on features which apply specifically to the microcalorimeter.

4.1 Theory of the microcalorimeter

Referring to Fig. 4.1, the self-balancing bridge supplies a dc bias so as to maintain the resistance of the bolometer element at the specified value R. Before the commencement of the rf measurement the sensitivity G_1 of the thermopile (in volts/watt) is determined by noting the rise in the thermopile output voltage when the dc bias power is first applied. When the rf power is switched on, the power dissipated in the element (rf plus dc) is held constant by the bridge, but power is dissipated additionally in the walls and elsewhere in the mount, causing an increase in the thermopile voltage. A simplified version of the theory will be given first before considering additional effects which

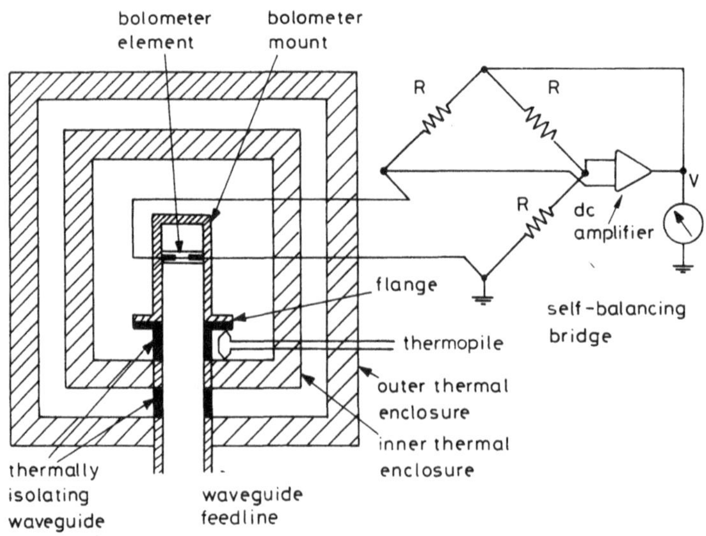

Fig. 4.1 *Simplified version of waveguide microcalorimeter. The bolometer element may be a metal wire barretter as shown, a thermistor, or a thin film bolometer. The effective efficiency of the bolometer mount is found by comparing the thermopile reading with the voltmeter reading V*

must be taken into account in a more realistic situation. Neglecting errors and corrections, the effective efficiency of the mount is given by:

$$\text{effective efficiency} = \frac{\text{retracted power}}{(\text{retracted power} + \text{wall loss})} \tag{4.1}$$

where the retracted power, that is the amount by which the dc power to the bolometer is reduced by the bridge when the rf power is applied, is given in terms of the bridge voltage V by:

$$\text{retracted power} = \frac{V^2_{rf\,off} - V^2_{rf\,on}}{4R} \tag{4.2}$$

and

$$\text{wall loss} = \frac{e_2 - e_1}{G_1} \tag{4.3}$$

Voltages e_2 and e_1 are the thermopile voltages when the rf is on and off respectively, and G_1 is the sensitivity in volts/watt. In a practical case there are three main causes of error. The first arises from the fact that the sensitivity of the thermopile to heat generated in the walls of the mount G_2 will usually be one or two percent higher than the sensitivity for heat generated in the bolometer element itself G_1. The difference is due to the fact that some of the heat generated in the element is lost down the inside of the waveguide and consequently is not detected by the thermopile. As an example, if the effective efficiency is 99 percent and the difference between G_1 and G_2 is 2 percent, then this error is approximately two percent of one percent, or 0.02 percent. On taking this effect into account equation 4.1 becomes:

$$\text{effective efficiency} = \frac{\text{retracted power}}{\text{retracted power} + \left(\dfrac{G_1}{G_2}\right)(\text{apparent wall loss})} \tag{4.4}$$

and

$$\text{apparent wall loss} = \frac{(e_2 - e_1)}{e_1} \cdot \frac{V^2_{rf\,off}}{4R} \tag{4.5}$$

The ratio G_1/G_2 is usually determined experimentally. This is done by attaching an auxiliary dc heater to the outside of the mount in such a way as to simulate the expected distribution of wall loss and then comparing the thermopile output for a given dc power dissipated in this heater with that for the same power dissipated in the bolometer element [5]. It is possible to incorporate the auxiliary heater into a feedback loop which keeps the thermopile output constant. The wall loss is then given directly by the reduction in dc auxiliary heater power when the rf power is applied. A small amount of rf power may also be dissipated in other locations inside the mount, such as tuners, the rf choke, the dielectric sheath of a barretter or the leads of a thermistor. The sensitivity to power dissipated in these regions is normally assumed equal to G_2, so that there may be a small residual error associated with this assumption.

The second potential source of error is the difference in rf and dc heat distributions within the element itself. Taking the barretter as an example, the sensitivity G_1 applies only for uniform heating along the length of the wire; its value will begreater than G_1 for heat generated near the ends of the wire and less than G_1 for the heat generated near its mid-point. To calculate with any confidence the effect of non-uniform dissipation in the element requires a knowledge of the rf and dccurrent distributions in the wire plus an analysis of the thermal circuit of the mount and calorimeter. Calculations on the current distribution have been carried out for specific geometries of waveguide mount by Jarvis and Adams [8,9] although their object was to estimate the equivalence error of the element when used as a bolometer, which is not the same as the calorimeter error under discussion here. Prior to the work of Jarvis and Adams the best estimate of equivalence error in bolometer elements was by Carlin and Sucher [10]. They obtained a value of 2 percent maximum by considering the worst possible current distribution which could arise, corresponding to infinite standing wave ratio along the barretter wire. However, neither the figures of Carlin and Sucher nor those of Jarvis and Adams relate directly to the error when the mount is calibrated by the microcalorimeter technique. It is often assumed that the error for the microcalorimeter will be smaller than that calculated by Jarvis and Adams because of additional temperature smoothing, but this assumption is difficult to justify without a detailed analysis of the thermal circuit, including the possible effects of convection. The alternative to a theoretical calculation is an experimental approach, but this is not easy. One cannot easily apply the movable heater approach used for waveguide dry load calorimeters (see Chapter 3), although one can replace the bolometric element by one of a different form, for which the heat distribution may be presumed to be different, and then seek to detect any changes in sensitivity to dc power. For example, a wire bolometer may be replaced by a thermistor. This requires great care in order not to disturb the arrangement, producing thereby a change in sensitivity due to other causes. The result is valid only for a particular design of mount, although it is usually assumed that it is permissible to scale the results to different waveguide sizes of the same design. Experiments such as these have shown that convection is important and have demonstrated the need for a convection barrier consisting of tapes or foam [3]. A check on the overall accuracy of the calorimeter can be obtained at a single frequency by carrying out a calibration of a barretter and a thermistor in the same microcalorimeter and intercomparing these.

The third source of error is the contribution to the thermopile response from heat generated in the short section of thermally isolating thin-walled waveguide between the bolometer mount and the inner thermal enclosure (Fig. 4.1). Loss here will affect the result, because it will influence the thermopile reading without significantly affecting the bolometric reading. There are two approaches to the determination of this correction factor. The first is to measure the attenuation of the isolating section and deduce the effect on the thermopile theoretically by making assumptions about the thermal circuit [3]. For this approach one would expect, as a rough guide, that the sensitivity would be slightly more than half of G_2. The second is to place a shortcircuit in front of the load and to deduce the correction from the thermopile reading, which is then due to the combined effect of the forward and reflected waves [5]. An allowance must be made for dissipation in the short circuit and also for the fact that the heating in the thermally isolating line when the short circuit is in position is no longer uniform owing to the standing waves. This necessitates assuming uniform surface resistance. In theory the effect of standing waves could be cancelled by repeating the measurement with an open circuit replacing the short circuit and taking the mean. However, an open circuit is difficult to arrange in waveguide. In equation 4.5 the thermopile voltage e_2 is the voltage after the effect of this heating has been corrected for.

In addition to the above three sources of error, a correction must also be applied for the resistance of the dc leads of the bolometer if a bridge other than of the self-adjusting current loop type is used. A dc block is required in the earth circuit of these leads in order to prevent earth loops. The possibility of leakage from the mount must also be considered. Since power lost by leakage will not be measured by the calorimeter, it is essential that the rf choke in the mount should be effective.

4.2 Some practical microcalorimeter designs

The microcalorimeter developed originally by MacPherson and Kerns [2] for the waveguide band 8.2 to 12.4 GHz (Fig. 4.2) was used to calibrate barretter mounts made from 0.25 mm electroformed copper waveguide and fitted with tuning screws. The mounts were thermally isolated by a 0.25 mm air gap, on either side of which plastic foam plugs were inserted into a waveguide in order to reduce heat convection. A power of 10 mW gave a temperature rise of 0.2 deg C, which was measured by a 20 junction constantan-to-chromel-p thermopile having a sensitivity of 800 μV/deg C. The housing was a cylindrical brass jacket immersed in an oil bath. The calorimeterhad a response time of between 10 and 30 minutes.

In evaluating their calorimeter MacPherson and Kerns considered two sources of error, namely the effect of the difference in sensitivities G_1 and G_2 and the effect of the difference in the rf and dc temperature distributions in the bolometer wire. Because of the air gap, there was no thermally isolating line

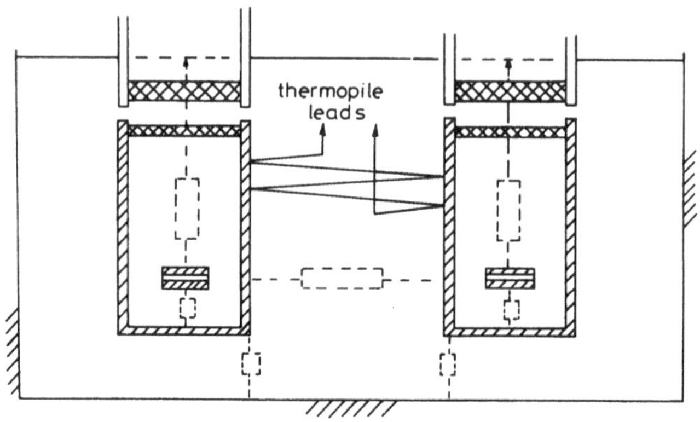

Fig. 4.2 *Arrangement of the original microcalorimeter of MacPherson and Kerns. The thermal resistances of the main heat paths are shown by dashed resistors*

dissipation to take into account. The use of air gaps was discontinued later. The error due to the different sensitivities G_1 and G_2 was examined theoretically by means of a simplified thermal circuit which assumed that transfer by convection was proportional to surface areas. These considerations gave a figure of 0.2 percent for the maximum error. Some attempts were made to check this by replacing the plastic foam plugs by thin mica sheets and noting any change in sensitivity values but apparently without conclusive results. For the second error, that due to the difference in the rf and dc distributions in the element, the authors referred to the bolometer analysis by Carlin and Sucher [10] and suggested that the effect on the calorimetric reading was probably small compared with this case, owing to the additional temperature smoothing. Attempts to demonstrate this by replacing the element by a thermistor and looking for any change in sensitivity were inconclusive because disassembly itself caused changes of up to 1 percent in the dc sensitivity.

Omori and Sakurai [11], in their microcalorimeter, also examined the effect of replacing the bolometer element by a thermistor. They found changes of 2.03 percent in the sensitivity to dc power, which were reduced to 0.35 percent when part of the thermopile was removed. They also investigated the power lost down the inside of the waveguide. This was done by observing the effect of inserting an aluminium plug. The conclusions reached were that the error due to the difference between G_1 and G_2, if not corrected for, was 0.06 percent maximum. They suggested that further investigation of the positioning of the thermopile could reduce errors.

Engen [3] incorporated several refinements in his 8–12 GHz microcalorimeter design, which is shown in Fig. 4.3. Amongst the more significant changes was the relocation of the thermopile. The thermally isolating section between the bolometer mount and theinner housing was made from 0.025 mm copper shim with the addition of plastic for mechanical rigidity. The mounts,

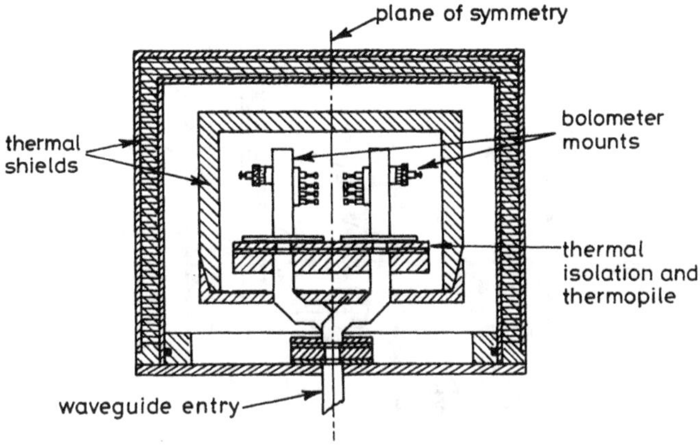

Fig. 4.3 *Engen's microcalorimeter*

which had effective efficiencies of 98–99 percent, were machined in two halves from tellurium copper, hard soldered together along the centre of the wider side, and fitted with five tuning stubs. A temperature stability of 5–10 millionths of a degree C was obtained using a triple thermal shield with a water bath [12,13] with regulation of 0.005 deg C. The difference between G_1 and G_2 was measured as lying in the range 0–2 percent. The effect of temperature distribution in the bolometer element was checked by the same method used by earlier investigators, that is by replacing the bolometer by a thermistor. At first a difference was observed due to convection and this difference was not eliminated by inverting the bolometer. When 8 PTFE membranes were inserted, agreement of 0.05 percent was obtained. For an estimation of the heating from the thermally isolating waveguide, the loss of this guide was determined as 0.05 percent. It was assumed that slightly more than half of this would be measured by the thermopile and a figure of 0.04 percent ± 0.04 percent was taken. The resulting combined uncertainty was 0.1 percent plus a further 0.1 percent for instrumentation error.

Since the time of Engen's work many microcalorimeters of broadly similar design have been built for use as primary standards in various standards laboratories. At the Royal Signals and Radar Establishment in Britain a set of microcalorimeters [5,14,15] (Fig. 4.4) was built with uncertainties of 0.1–0.3 percent between 8.2 and 35 GHz, 0.5 percent up to 110 GHz. These were designed for the calibration of waveguide barretter mounts using 2 μm diameter wire and fitted with three screw tuners. Commercial thermistors were later also calibrated directly in them. The calorimeters were of cylindrical form, the outer diameter and height being 250 mm and the total assembly weighing about 32 kg. A very stable thermal environment was provided by a double-walled enclosure, which was itself completely immersed in a large water bath of 170 litre capacity whose temperature was maintained to within

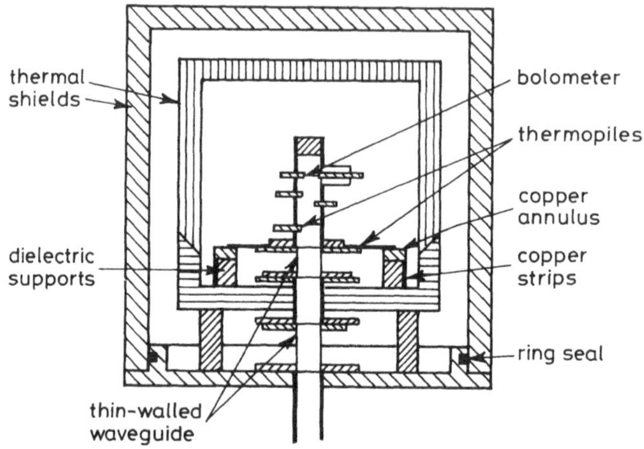

Fig. 4.4 *Microcalorimeter built at the Royal Signals and Radar Establishment. The copper annulus serves as a temperature reference*

50 μdeg C per hour. The rise in temperature of the bolometer was measured by a series of thermocouples between the bolometer input flange and a copper ring reference body connected to it thermally by copper strips. The copper strip geometry was adjusted so that the copper ring had a thermal time constant identical to that of the bolometer mount. The output of the thermopiles was amplified using a low noise chopper dc amplifier which was in good thermal contact with the inner calorimeter jacket. A mercury wetted latching relay was used to short circuit the amplifier input, enabling remote adjustment of the amplifier offset voltage. A dissipation of 15mW gave a temperature rise of approximately 0.05 deg C and a thermopile output voltage of about 120 μV. At 15 GHz the thermopile output voltage due to input waveguide heating was 0.029 ± 0.002 volt per watt of incident power on the mount. For a 98.8 percent efficient mount this resulted in a correction of +0.067 percent ± 0.005 percent. This heating effect was determined by placing a short circuit consisting of a piece of polished copper foil across the flanged joint between the bolometer mount and the microcalorimeter. When used in an automated system [16] the spread of measurements was typically 0.1 percent with a standard deviation of 0.03 percent. The mean of a set of 10 to 12 measurements repeated at 0.01 percent if the mount was removed from the calorimeter and then replaced.

An automated microcalorimeter used with a feedback circuit was described by Clark [4]. The feedback power was applied to an auxiliary heater placed next to the thermopile. This instrument was a single-load microcalorimeter used for calibrating commercial waveguide and coaxial thermistors rather than wire barretters. The frequency was stepped automatically, covering an octave band or waveguide size in 100 minutes. Inserts for the higher waveguide frequencies were inserted through the lowest frequency waveguide, which was permanently fixed. The instrument operated in a massive thermal shield in an air oven. For single load calorimeters, as for many other devices, it is important to have a good air seal in the enclosure to prevent changes in air pressure from affecting the readings [1,17].

A number of microcalorimeters have been built employing thermoelectric cooling as part of a feedback circuit, rather than using an auxiliary heater. Sakurai and Maruyama [18] used this principle in a microcalorimeter for the calibration of thermistors at 34.5 GHz. Both the thermopile and Peltier cooling element were of semiconductor (BiTe P–N alloy). In this instrument the estimated effect of the loss in the isolating waveguide was 0.1 percent, based on the assumption that half of the loss contributes to the heating. The effect of the difference in G_1 and G_2 was assumed to be the same as at 10 GHz (0.06 percent maximum) (see [11] and [3]). The effect of heat loss through the foam surrounding the loads was calculated from the leakage conductance and thermal resistance of the waveguide per unit length. A total uncertainty of 0.5 percent was quoted.

Inoue et al. [19,20] described an automated microcalorimeter system for the effective efficiency measurement of a thin film bolometer mount in the 35 GHz band. This instrument used both Peltier cooling and an auxiliary heater (Fig. 4.5). The thin film bolometer element had a width of 0.06 mm and length 1 mm. Constant current was supplied to the Peltier cooling element, while the auxiliary heater power was controlled by a feedback loop. The effect of the difference in G_1 and G_2 was estimated theoretically to be less than 0.08 percent, based on earlier work. Ref. [21] describes a similar microcalorimeter for 10 GHz. The thin film elements were nickel and gold deposited on high grade mica by vacuum evaporation techniques. The resistance was stabilised by annealing and was kept about 185 Ω. Gold plated acrylic resin waveguide was used between the bolometer mount and heat sink. The feed was of stainless steel. Other microcalorimeters are described in refs. [22–31].

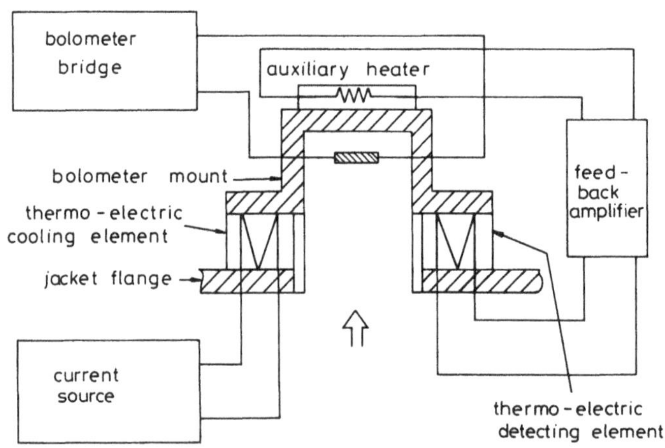

Fig. 4.5 *Thin film microcalorimeter using thermoelectric cooling and auxiliary heater*

A number of microcalorimeters for coaxial mounts are described in the literature. Fujisawa [32] described a coaxial microcalorimeter using a Peltier element (Bi_2Te_3). For this instrument the isolating section was acrylic resin coaxial tubing. A quartz thermometer with resolving power of 1/10000 deg C was used. An uncertainty of 0.48 percent was claimed. A coaxial microcalorimeter in 7 mm line was described by Hollway and Muller [33]. For this instrument a complex thermal equivalent circuit was derived. The reading time allowed was three hours and the time needed for the instrument to settle down after insertion of the thermistor was 24 hours. In all coaxial calorimeters the inner conductor provides a heat path and is also heated by thermal losses. In order to investigate the uncertainties, a model having 94 segments conducting heat between 49 nodes was derived. Since the inner and outer conductor must be considered, the rf segments had up to three conducting layers. A computer program was used to compute the rf resistance of each conducting segment, taking into account the skin effect. Account was taken of the fact that there is a delay before heat generated in the isolating line affects the thermopile, so that if readings are taken before thermal equilibrium is reached there may be an error. This error was calculated and a correction applied. Other coaxial microcalorimeters are described in refs. [4,34–41]. Most are 50 Ω instruments, but some 75 Ω versions have been used for frequencies below 1GHz [42].

4.3 References

1 Clark, R.F., 'The microcalorimeter as a national standard', *Proc. IEEE*, **74**, 1, pp. 102–104, Jan. 1986.

2 MacPherson, A.C., and Kerns, D.M., 'A microwave calorimeter', *Rev. Sci. Instrum.*, **26**, 1, pp. 27–33, Jan. 1955.

3 Engen, G.F., 'A refined X-band microwave microcalorimeter', *J. Res.* NBS, **63C**, pp. 77–82, July–Sept. 1959.

4 Clark, R.F., 'A semiautomatic calorimeter for measurement of effective efficiency of thermistor mounts', *IEEE Trans.*, **IM-23**, 4, pp. 403–408, Dec. 1974.

5 Skilton, P.J., 'Developments in United Kingdom waveguide power standards', RSRE Report 80006, April 1980.

6 Larsen, N.T., 'A new self-balancing d.c.-substitution r.f. power meter', *IEEE Trans.*, **IM-25**, 4, pp. 343–347, Dec. 1976.

7 Raff, S.J., and Sorger, G.U., 'A subtle error in RF power measurements', *IRE Trans.*, **I-9**, 1, pp. 284–291, Sept. 1960.

8 Adams, J.W., 'Current distribution in barretters and its application to microwave power measurements', *IEEE Trans.*, **MTT-17**, 10, pp. 778–785, Oct. 1969.

9 Jarvis, S., and Adams, J.W., 'Calculation of substitution error in barretters', *J. Res.* NBS-C, **72C**, 2, pp. 127–137, June 1968.

10 Carlin, H.J., and Sucher, M., 'Accuracy of bolometric power measurements', *Proc. IRE*, **40**, 9, pp. 1042–1048, Sept. 1952.

11 Omori, S., and Sakurai, K., 'A new estimating method of equivalence error in microwave microcalorimeter', *IRE Trans.*, **I-7**, pp. 307–309, Dec. 1958.

12 Larsen, N.T., '50 microdegree temperature controlled water bath', *Rev. of Sci. Instrum.*, **39**, 1, pp. 1–12, Jan. 1968.

13 Harvey, M.E., 'Precision temperature controlled water bath', *Rev. Sci. Instrum.*, **39**, 1, pp. 13–18, Jan. 1968.

14 Bagnall, I.A., 'The UK standard of microwave power measurement in WG16 and international intercomparisons', RRE Technical Note No. 772, Oct. 1972.

15 Skilton, P.J., Gordon, D.R., and Hill, L.D., 'United Kingdom waveguide power standards – recent developments', IEE Colloquium on the Measurement of Power at Higher Microwave Frequencies, London, Jan. 1979, pp. 1/1-5.

16 Bagnall, I.A., Roberts, J., and Clark, S., 'The automated system used in the calibration of the UK waveguide power standards', Conference on Precision Electromagnetic Measurements, London, July 1974, pp. 185–187.

17 Clark, R.F., Jurkus, A., McLaren, G.D., and Wood, B.M., 'The importance of an air pressure seal on standard cell enclosures with high temperature stability', *IEEE Trans.*, **IM-34**, 3, Sept. 1985.

18 Sakurai, K., and Maruyama, T., 'Millimeter-wave microcalorimeter', *IRE Trans.*, **I-11**, 3–4, pp. 270–276, Dec. 1962.

19 Inoue, T., Yamamura, K., and Nemoto, T., 'Automatic calorimeter system for the effective efficiency measurement of a bolometer mount in 35 GHz band', *IEEE Trans.*, **IM-27**, 3, pp. 205–209, Sept. 1978.
20 Inoue, T., Yamamura, K., and Nemoto, T., 'Correction to "Automatic calorimeter for the effective efficiency measurement of a bolometer mount in 35 GHz band"', *IEEE Trans.*, **IM-28**, 2, p. 171, June 1979.
21 Kothari, P.C., Bhatnagar, H.M., Aggarwal, R., Agrawal, V.K., and Chandra, K., 'Measurement of effective efficiency of thin-film barretter mount at 10.0 GHz using microcalorimetric technique established at NPL, New Delhi', *J. Inst. Electron. and Telecommun. Eng.* (India), **29**, 10, pp. 501–506, Oct. 1983.
22 Brodskii, W.I., and Pronenko, V.I., 'Microcalorimeters for measuring superhigh frequency power in the 3 cm range', *Izmer. Tekh*, **5**, p. 65, 1957.
23 Harvey, M.E., 'WR15 microwave calorimeter and bolometer unit', NBS Tech. Note 618, May 1972.
24 Chuiko, V.G., and Frumkin, V.D., 'Special state standard for the power unit of electromagnetic oscillations in waveguide channels at 25.9–37.5 GHz', *Meas. Tech.*, **17**, 7, pp. 978–980, July 1974.
25 Akhiezer, A.N., Sen'ko, A.P., Terekhov, M.V., Ostroumova, L.G., and Sidorov, V.M., 'Special state standards for the unit of power of electromagnetic vibrations in waveguide tracts in the frequency range 37.5–53.57 GHz', *Meas. Tech.*,**17**, 5, pp. 645–648, May 1974.
26 Serenko, B.G., 'A modernized calorimeter', *Meas. Tech.*, **19**, 6, p. 926, June 1976.
27 Inoue, T., Toda, H., and Sugiura, I., 'An automated precision power measurement and power meter calibration system in the millimeter wave region', Conference on Precision Electromagnetic Measurements, Boulder, June 1976, pp. 172–174.
28 Inoue, T., Yamamura, K., and Nemoto, T., 'A calorimeter for measurement of effective efficiency of 100 GHz bolometer mounts', *Trans. Soc. Instrum. and Control Engineers*, **13**, 6, pp. 609–614, Dec. 1977 (in Japanese).
29 Inoue, T., Yamamura, K., and Nemoto, T., 'Precision power measuring system for 100 GHz', *Bull. Electrotech. Lab.*, **41**, 12, pp. 928–937, Dec. 1977 (in Japanese).
30 Akhiezer, A.N., Danil'chenko, V.P., Sen'ko, A.P., Terekhov, M.V., Gordeev, K.K., Kaliberda, L.G., and Terzhova, V.P., 'Special state standard for unit power by electromagnetic oscillations in waveguide channels in the frequency range 53.57–78.33 GHz', *Meas. Tech.*, **22**, 7, pp. 767–771, July 1979.
31 Weidman, M.P., and Hudson, P.A., 'A WR 10 millimeter wave microcalorimeter', NBS Tech. Note 1044, June 1981.
32 Fujisawa, K., 'Microwave power standard at 3 GHz', *Bull. Electrotech. Lab.*, **33**, 4, pp. 100–108, 1969 (in Japanese).
33 Hollway, D.L., and Muller, W., 'The precise measurement of rf power and voltage using coaxial calorimeters and microcalorimeters', *IEEE Trans.*, **IM-33**, 4, pp. 269–275, Dec. 1984.
34 Krzhimovskii, V.I., Kshimovskii, V.V., and Sergeev, I.A., 'Equipment for certifying coaxial bolometric heads for their conversion factor at 0.03–10 GHz', *Izmer. Tekh.*, **6**, pp. 58–59, June 1967.
35 Blouet, J., 'Centimetre wave power standards', Pub. 18-B-4/1-18-B-4/7, Laboratoire Central des Industries Electriques, Fontenay-aux-Roses (in French).
36 Bayer, H., 'Microcalorimeter precision measuring equipment for determining the effective efficiency of bolometer measuring heads in the frequency range 8.2 to 12.4 GHz', *PTB Mitteilungen*, **80**, pp. 201–269, 1970 (in German).
37 Fedorov, A.M., Grigor'ev, A.V., Krzhimovskii, V.I., Morozova, T.B., and Sergeev, I.A., 'Special state standard for the unit of a.c. power in coaxial channels at high frequencies', *Meas. Tech.*, **18**, 4, pp. 556–559, April 1975.
38 Erard, L., 'Automation of the calibration of bolometer mounts at high frequencies', Bulletin BNM, No. 29, pp. 12–15, July 1977 (in French).
39 Blouet, J., 'High Frequency power standards', Bulletin BNM, 36, pp. 87–92, April, 1979.

40 Kolotygin, S.A., Kon'kova, L.T., and Chuiko, V.G., 'Measurement of the efficiency coefficient of a coaxial thermal transducer in a microcalorimeter at 18 GHz', *Meas. Tech.* **22**, 10, pp. 1253–1256, Oct. 1979.

41 Bayer, H., 'Actual position and development tendencies in the subject of high frequency measurement technology', Proceedings of the 49th PTB Seminar, PTB-E-24, June 1983, pp. 3–26 (in German).

42 Delahaye, F., and Geneves, G., 'Standard bolometric mounts on 75 ohm coaxial lines', *IEEE Trans.*, **IM-34**, 2, pp. 214–216, June 1985.

Chapter 5

Flow calorimeters

In a flow calorimeter it is the rise in temperature of the fluid on passing through the load which is measured. The rf power is proportional to the product of this temperature rise, the flow rate, and the specific heat of the fluid. Instruments have been built operating up to 40 GHz for standards use [1] and up to higher frequencies for commercial use. However, interest in the flow calorimeter for precise measurements began to decline in the 1960s and subsequent efforts were devoted largely to refinements of the dry load and microcalorimeter types. Consequently only a small number of precision flow calorimeters with uncertainties below 1 percent have been constructed. Above 3 GHz the directly heated type is the normal one, with water as the working fluid, although greater loss can be obtained with ethylene glycol or glycerine. At lower frequencies the fluid is usually oil and the instruments are normally of the indirectly heated kind [2]. It is nevertheless possible to construct directly heated low frequency versions by using a suspension of carbon particles in oil. Air has also been employed [3]. In theory gases should enable greater sensitivity to be achieved because of their smaller thermal capacity, but an additional problem is heating due to the compressibility.

As with other calorimeters, the design of the load is of great importance. Many different types of load have been investigated over the years [4–7], particularly in the 1940s and 1950s. The flow must provide enough turbulence to ensure thorough mixing of the fluid and to avoid hot spots, but not so much turbulence that heat is generated. Special mixers may be inserted between the load and the temperature sensor in order to improve this mixing. The formation of bubbles in the load may be a problem if the bubbles cannot escape. They must if necessary be removed by a bubble trap. For directly heating calorimeters the large difference in the permittivities of air and water makes an effective method of matching essential. This is usually achieved by means of a gradual transition from one medium to the other. In many of the earlier loads the inlet and outlet were placed adjacent to one another, so that the flow was forced to double back on itself (Fig. 5.1), leading to the formation of undesirable stagnant regions. Two directly heated waveguide loads of this

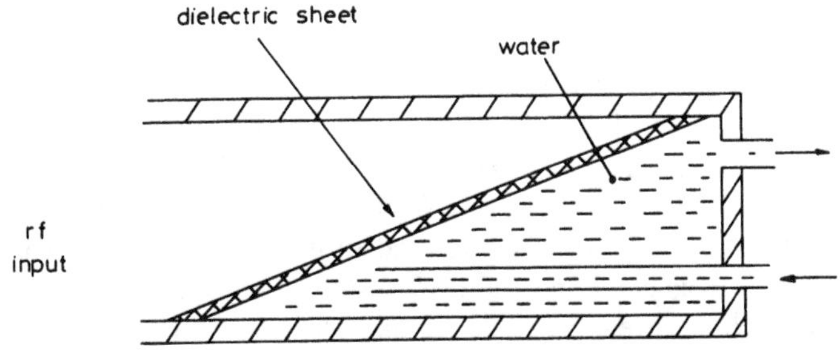

Fig. 5.1 *Early design of waveguide flow calorimeter load having adjacent inlet and outlet. This type tends to suffer from stagnant regions*

type were described by Turner [8], one based on a wedge and the other on a conical polystyrene shell. Other possibilities are a hairpin-shaped tube and a coaxial flow arrangement in which the inlet tube isplaced inside the outlet tube. In later instruments it was recognised that a design of load in which the fluid enters at one end and emerges at the other (Fig. 5.2) is most successful in avoiding the formation of stagnant regions, although it is less convenient from some other points of view, such as the prevention of rf leakage. All the instruments described below are of this type.

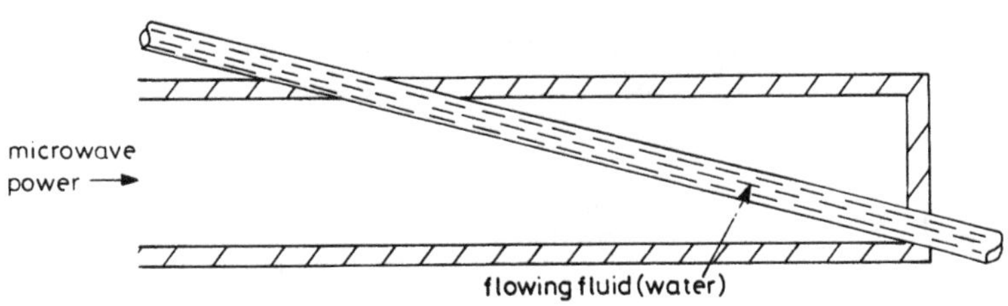

Fig. 5.2 *Usual type of waveguide flow calorimeter load*

Flow calorimeters normally operate in the power range from a few watts up to tens of kilowatts. The flow rate is chosen to suit the level of the power to be measured; a high flow rate increases the power handling capacity and speeds up the response at the expense of sensitivity. Normally the flow rate should be as great as is consistent with adequate sensitivity. This ensures maximally efficient integration of heat generated in different parts of the load and reduces heat losses by keeping the temperature difference between the load and its surroundings to a minimum. Heat losses are also minimised if the hottest part of the load is closest to the outlet.

5.1 Coaxial flow calorimeters

An indirectly heated precision coaxial flow calorimeter was described by Crawford and Hudson [2] (Fig. 5.3). This instrument, which covered a frequency range from dc to 4 GHz and was fitted with 14 mm precision connectors, operated at power levels from 2 W to 100 W. The design was a substantial refinement of an earlier flow calorimeter due to Vinding [9]. A dual load arrangement was used and the flow system, shown in Fig. 5.4, was of the parallel type, which preserves the symmetry between the two loads but requires that the ratio in which the flow divides should be stable.

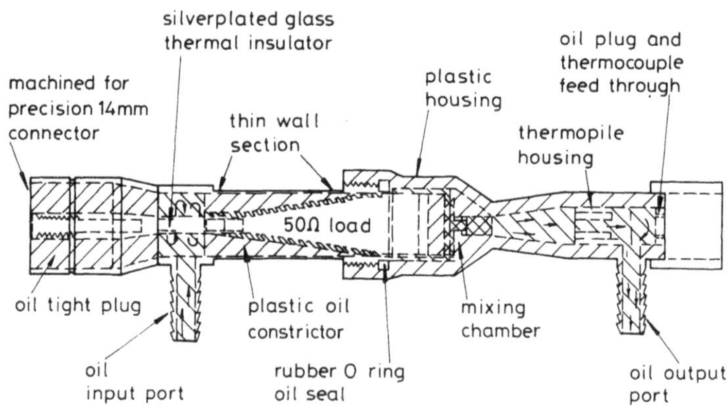

Fig. 5.3 *Details of load of precision coaxial oil flow calorimeter*

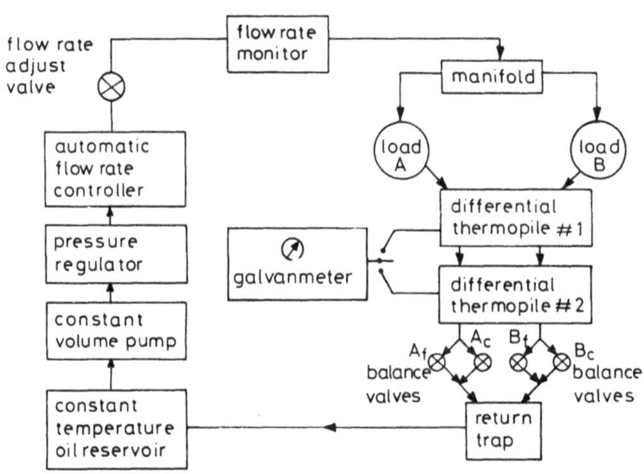

Fig. 5.4 *Parallel flow system*

The rf power was applied to one of the two identical loads, whilst dc power was applied to the other. A differential thermopile detected the temperature difference between the two streams on the downstream side of the loads. The junctions of the thermopile were located in the path of the moving oil. The oil was mixed thoroughly in a chamber downstream from the loads and then passed around and through a plastic block in which the junctions of thermopile 1 were located. Two thermopiles were used, one mounted further downstream from the other, as a check on the reading.

Prior to the measurement it is necessary to balance the system by applying equal dc powers to the two loads and adjusting the flow rate of either channel for a null at the differential thermopile output. In order to absorb 100 W without damage to the film the oil had to be circulated rapidly around and through the resistor body. At low levels of input power (2–5 W) the minimum oil flow rate was such that the temperature rise of the oil was approximately 5 deg C. The flow rate per watt was about 15 $cm^3\ min^{-1}\ W^{-1}$. At higher levels of input power the flow rate per watt was reduced to about 7.5 $cm^3\ min^{-1}\ W^{-1}$, giving a temperature rise of 10 deg C.

The load resistor was a uniform thin metal film deposited by vacuum evaporation onto a truncated conical substrate mounted inside a cylindrical outer conductor. This type of design is described in refs. [10] and [11]. The principle is similar to that of the tractorial load described in Chapter 3 [12], but it is the resistor rather than the outer conductor which is tapered. If ρ is the sheet resistivity of the film, then the resistance R is given by:

$$R = \frac{\rho}{2\pi \sin \theta} \ln \left(\frac{b}{a}\right) \tag{5.1}$$

where θ is the cone semiangle, and b/a is the ratio of outer to inner diameter of the coaxial line. The characteristic impedance Z_0 of the section of coaxial line containing the conical resistor is

$$Z_0 = \frac{Z_m}{2\pi} \ln \left(\frac{b}{a}\right) \tag{5.2}$$

where Z_m is the wave impedance of the medium ($=\sqrt{(\mu/\varepsilon)}$). By making $\rho/\sin\theta$ equal to Z_m, the values of Z_0 and R become equal at any point along the length of the load. This condition makes for low reflection coefficient and gives practically the same current distribution along the resistor for both rf and dc. The uniformity of the load resistor was checked experimentally by means of a time domain reflectometer (TDR) and the observed variations were of the order of 0.1 Ω or less in 50 Ω.

The oil flowing over the resistor surfaces tends to absorb all the heat regardless of distribution, resulting in a low rf/dc equivalence error. The main cause of error was considered to be the possibility of heat conducted away at the grounded end of the resistor where it contacts the outer conductor. In order to check for a temperature difference in this area a 10-junction thermopile was cemented to the outer conductor of one of the loads and referenced to an ice bath. Equal levels of dc or rf power were then applied alternately to the load. At a frequency of 4 GHz a difference in temperature of 0.035 deg C was observed while at 2 GHz no measurable difference was discernible. Calculations were made which showed that at 4.0 GHz with a temperature difference of 0.035 deg C an error less than 0.1 percent would exist. On the basis of the above tests the rf/dc equivalence error was estimated to be 0.1 percent at a frequency of 1 GHz and above and 0.05 percent at frequencies below 1 GHz. Negligible heat conduction to the mount was considered to occur at the input end of the load resistor, where the centre conductor is immersed in the moving oil stream for a distance of 15 mm. The resolution of the measurement due to noise and ambient temperature variations was approximately 0.05 percent. The correction for power loss in the input lines was calculated theoretically to give a figure of 0.0084 dB ± 0.0033 dB at 4 GHz. The largest contribution to the uncertainty was the effect of instability in the flow division, together with thermal drift, and corresponded to 0.15 percent of the reading for a temperature rise of 10 deg C. The calorimeter was compared at 1, 3, and 4 GHz with a dry-load calorimeter which operated over the range 50 mW to 5 W and which had an uncertainty of 0.35 percent. The disagreement was reported as being no greater than 0.2 percent [2].

A commercial dual load coaxial flow calorimeter designed in the 1950s is described in ref. [13]. This operated in the range 10 mW to 10 W at frequencies from dc to 12 GHz. The fluid was silicone oil of high stability and very low viscosity, equal approximately to that of water. The thermally isolating centre conductor consisted of a glass rod coated with a thin film of silver. The load resistor was a metallic film and in the region of this termination the tapered outer conductor was made with a very thin metal wall for thermal isolation. The temperature sensors were resistance thermometers and they were used in an ac bridge with a synchronous detector. A series flow system was chosen, avoiding the problem of unequal division from which the parallel flow type suffers. A heat exchanger equalised the temperatures of the oil at the inlets to the rf and reference loads. By applying a feedback signal to the reference load, a reading time of less than 5 seconds was achieved. No rigorous investigation of the effective efficiency was claimed, although it was believed that at 10 GHz this was approximately 97.5 percent. The uncertainty, if the uncertainty in effective efficiency is excluded, was considered to be approximately 3 percent.

5.2 Waveguide flow calorimeters

For waveguide flow calorimeters the form of load which has become established consists of a thin-walled glass or polystyrene tube fitted obliquely across a short-circuited length of waveguide. Lane [14,15] built a calorimeter of this type operating at 9.4 GHz and at a power level of 10 W. The microwave power was measured in terms of the ac power at 50 Hz necessary to produce an equal rise in temperature, as indicated by a pair of thermocouples. An ac supply was preferred to dc for calibration since, in the latter case, electrical leakage through the water would cause a current in the thermocouple circuit comparable with that being measured. The flow rate was adjusted so that the temperature rise was kept to a minimum compatible with an adequate response in the thermocouple circuit (about 0.5 deg C). The uncertainty of this instrument was believed to be in the region of 1 percent.

Abbott [1,16,17] and others describe some more sophisticated calorimeters which were developed following Lane's general design. Their design, for power levels of between 1 and 20 W and with a response time of 5 minutes, was realised in nine different waveguide sizes covering a frequency range up to 40 GHz. The version for the frequency range 8.2–12.4 GHz will be described here. The load tube, through which distilled water was circulated, was formed from high-purity fused quartz precision drawn tubing with a nominal wall thickness of 0.5 mm. At 10 GHz this material has tan δ as low as 0.0001 and a permittivity of approximately 4.0. The construction of the waveguide loads can be seen from Figs. 5.5 and 5.6. All except the lowest frequency load are

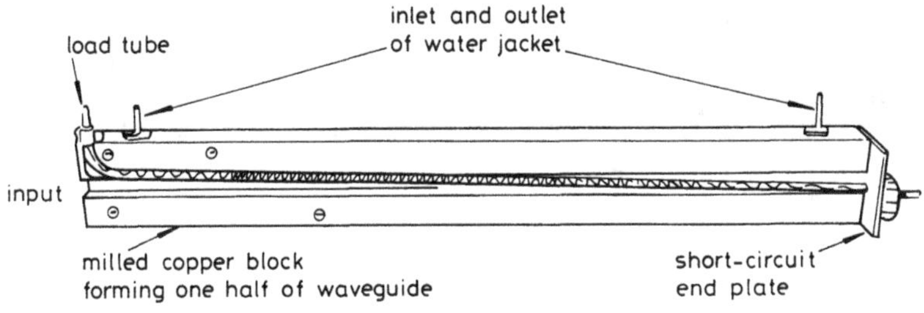

Fig. 5.5 *Details of waveguide flow calorimeter load*

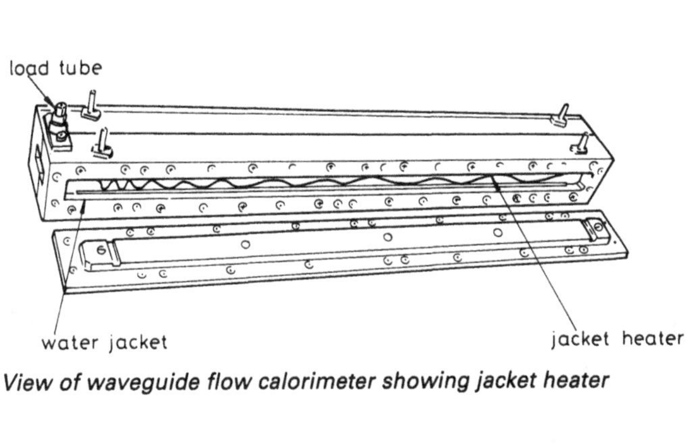

Fig. 5.6 *View of waveguide flow calorimeter showing jacket heater*

machined from solid copper in two mirror-image halves joined at the centre line of the broad walls. They are located by dowels and are demountable. A U-section longitudinal channel is machined in each half to form, with the cover plates in position, an integralwater jacket 1.5 mm wide almost surrounding the waveguide and leaving a wall thickness of 2 mm. In the inner mating surfaces of the body halves corresponding grooves are machined to form a circular channel of 4.5 mm radius which locates the load tube at an angle of 1.5° to the waveguide. For most of its length the channel is straight and of constant cross-section but bends sharply to 90° before the exit point. This bend allows the tube to be placed as close as possible to the waveguide input flange to minimize wall losses ahead of the load tube and also to form an rf choke. With the load tube in position and filled with water the channel represents a lossy coaxial line with more than 60 dB attenuation. To achieve a good wide-band VSWR it was found that the load tube must be nominally straight for two guide wavelengths within the block and at least to the mid-plane of the waveguide cavity. The load incorporated a dc heater wound from epoxy-insulated manganin-type wire. This heater was inserted inside the load tube in direct contact with the water and distributed along the tube in conformance with the distribution of rf heating of the water (Fig. 5.7). This rf distribution was obtained experimentally using thermocouples inside the tube to measure the rise in temperature over a fixed period of time when the fluid was static.

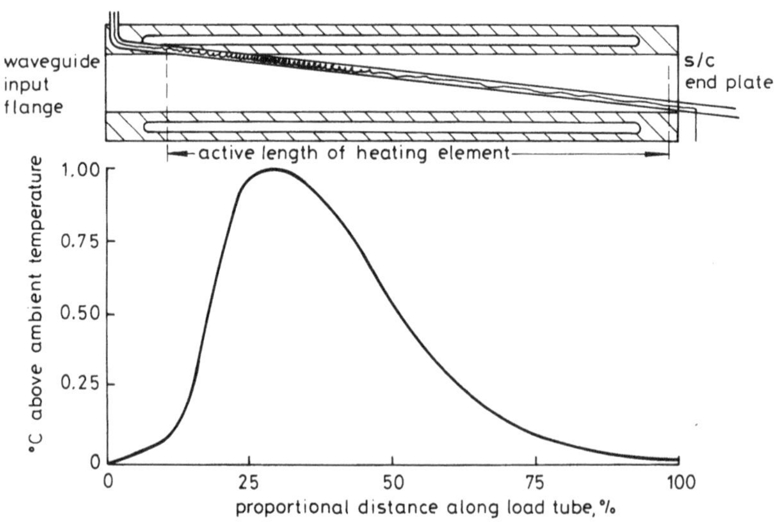

Fig. 5.7 *Heat distribution in load tube of waveguide flow calorimeter*

As in the Crawford and Hudson coaxial calorimeter, a parallel flow system was used and at 10 W the flow rate was approximately 250 ml/min through each load tube. Water was circulated through a jacket surrounding the waveguide in a separate calorimetric circuit. This made possible the measurement of the rf loss in the waveguide walls, so that the effective efficiency of the whole calorimeter could be determined.

The differential temperature of the outlets of the rf load and the reference load was measured. The outlet water temperature of the load tubes was about 1 deg C above the inlet. The thermal balance stability was 0.05 percent standard deviation. The thermopiles were of the multijunction plated type described in Chapter 2. One hundred double thermo-junctions were formed in series by overplating with copper half of the circumference of a solenoid wound with Eureka wire. Water channels in the coil former enclosed the two diametrically-opposite exposed interfaces of copper and Eureka. The load thermopiles gave a discrimination after amplification of 0.1 millidegree, which is equivalent to 0.01 percent at 1 W level. In some of the instruments the thermopile was later replaced experimentally by a pair of thermistors, which gave increased sensitivity, but the increased drift due to imperfect matching of the thermistor characteristics proved unacceptable.

The absorbed rf power is dissipated partly in the load tube and partly in the waveguide wall loss. The ratio of these two was measured by dc substitution simultaneously in the load tube heater and in a heater fitted inside the water jacket of the rf waveguide load. By successive adjustments of the dc levels heat flow balance was obtained in both circuits. The measurement takes approximately 30 minutes for each frequency point. The determined effective efficiencies were in the range 98.6 to 99.6 percent.

Most of the sources of uncertainty were considered, but the authors of the paper point out that no means of experimental verification of the uncertainties was devised once the calorimeter had been assembled. Their maximum value was estimated from experience gained during development, such as noting the effect of changes to dimension and shape. Consequently the only practical test that could be carried out was to check the instrument against a microcalorimeter. The main sources of error which were taken into account were as follows. First there are the possible rf/dc substitution errors in the load tubes. RF dissipation is almost entirely in the water itself, the quartz tube having negligible loss. Water turbulence in the tube, stimulated by the presence of the heater wire, distributes the heat and integrates it over the length and bore of the tube. DC heating is applied directly in the water, close to the tube wall where most of the rf heat is produced. DC heating is distributed along the tube in accordance with the rf heat distribution. The relative levels of rf heating at the ends of the load tube are very small and there are practical difficulties in achieving a similar distribution for the dc heating. Also, there is a small loss of dc power from the water to the lead wire which to some extent compensates. These two effects give rise to an estimated uncertainty of 0.02 percent.

Heat exchange between load tube and waveguide body takes place by conduction through the quartz tube to the channel. This is in the area of maximum waveguide wall loss and is slightly different in the rf and dc conditions. An uncertainty of ±0.03 percent is allowed for this effect.

Consider next the jacket dc substitution. The area of maximum wall heating is at the front end of the waveguide ahead of the load tube, where the rf power is greatest. It is impracticable, however, to position a heater at this point. By applying discrete sources of heat to various surfaces over the front end of the waveguide, the maximum error due to displacement was determined as ±5 percent in the differential jacket water temperature. The maximum variation is accounted for by assigning a ±0.05 percent uncertainty in the effective efficiency measurement.

Due to rf losses in the source waveguide to which the waveguide load is connected, heat is conducted through the input flange to the waveguide load, most of which will pass into the jacket water and give rise to a dc substitution error in the efficiency measurement. To eliminate this, a thin-walled heat insulating waveguide section, incorporating a water jacket on the source side, is interposed. The loss in this thin-wall section, measured as 0.2 percent, produces some heat itself. Because of its symmetrical design, approximately 50 percent of this heat passes into the waveguide load and this amount is added as a correction to the measured value of effective efficiency. An uncertainty of 0.05 percent is allowed for possible error in this assumption.

The flow calorimeters were still in use in 1987, having been incorporated into an automated system which significantly simplified their operation. Comparisons with microcalorimeters generally showed agreement to 0.2 percent, although when such comparisons were first carried out differences of up to 0.5 percent were observed. For the 8.2–12.4 GHz version the uncertainty in effective efficiency was 0.16 percent for the systematic and 0.05 percent for the standard deviation of the random component. When used to calibrate a secondary instrument the uncertainty was 0.41 percent systematic and 0.07 percent standard deviation.

5.3 References

1 Abbott, N.P., Reeves, C.J., and Orford, G.R., 'A new waveguide flow calorimeter for levels of 1–20W', *IEEE Trans.*, **IM-23**, 4, pp. 414–420, Dec. 1974.

2 Crawford, M.L., and Hudson, P.A., 'A dual flow calorimeter for RF power measurement to 4 GHz', NBS Special Publication 300, Vol. 4, pp. 47–53, 1970 (reprint of NBS Journal of Res., **71C**, 2, pp. 111–117, April/June 1967).

3 Wilchinsky, Z.W., and Kyser, R.H., 'Air flow U.H.F. wattmeter', *Electronics*, **19**, pp. 128–129, Oct. 1946.

4 Shaw, R.C., and Kircher, R.J., 'A coaxial type water load and associated power measuring apparatus', *Proc. IRE*, **35**, 1, pp. 84–87, 1947.

5 Griesheimer, R.N., 'Microwave power measurements', Chapter 3 in 'Technique of Microwave Measurements', ed. C.G. Montgomery, McGraw-Hill, 1947.

6 Penton, W.A., and Overton, I., 'A balanced water flow calorimeter for centimeter wavelengths', *New Zealand J. Sci. Tech.*, **29**, Sec. B, 4, p. 215, 1948.

7 Jaeger, T., and Schneider, M.W., 'A broad-band high-power microwave calorimeter', *Arch. Elek. Uebertragung*, **13**, pp. 21–25, 1959.

8 Turner, L.B., 'Balanced calorimeters for 3,000 and 10,000 Mc/s with tapered water loads for H-rectangular pipes', *J. IEE*, **93**, Pt. IIIA, 9, p. 1467, 1946.
9 Vinding, J.P., 'An accurate calorimeter for high microwave power', *Microwave J.*, **4**, pp. 41–46, Jan. 1961.
10 Tischer, F.J., 'Mikrowellen-Messtechnik', Springer, 1958.
11 Woods, D., 'Improvements in precision coaxial resistor design', *IRE Trans.*, **I-11**, 3–4, pp. 305–309, Dec. 1962.
12 Harris, I.A., 'The theory and design of coaxial resistor mounts for the frequency band 0–4000 Mc/s', *Proc. IEE*, **103C**, 3, pp. 1–10, Mar. 1956.
13 'Microwave power measurement', Application Note 64, Hewlett-Packard, 1965.
14 Lane, J.A., 'The measurement of power at a wavelength of 3 cm by thermistors and bolometers', *Proc. IEE*, **102B**, 6, pp. 819–824, Nov. 1955.
15 Lane, J.A., 'Microwave power measurement', Peter Peregrinus, 1972.
16 Vinding, J.P., 'Comments on "A new waveguide flow calorimeter for levels of 1–20 W"', *IEEE Trans.*, **IM-25**, 1, p. 89, Mar. 1976.
17 Abbott, N.P., Reeves, C.J., and Orford, G.R., 'Authors' reply to "Comments on 'A new waveguide flow calorimeter for levels of 1–20W'"', *IEEE Trans.*, **IM-26**, 1, pp. 79–80, Mar. 1977.

Part B

Non-calorimetric power meters

Chapter 6

Bolometers

Bolometric power meters make use of a temperature-sensitive resistor [1–5], referred to as the bolometer element, which is housed in a mount. The rise in temperature caused by the absorption of rf or microwave power produces a change in resistance, which is detected by a bridge circuit. Although the principle of the bolometer goes back to 1880 [6], present forms date from the 1940s, when the need first arose for reliable power measurements at frequencies above 1 GHz. They are of three main types*: barretters, thermistors, and film bolometers. Almost all are directed heated, that is the bolometer element functions both as the absorber of the power and as the temperature sensor. Barretters consists essentially of a thin metal wire and have a positive temperature coefficient of resistance. They have been used extensively in standards laboratories but less frequently elsewhere. Thermistors consist of small beads of semiconducting material and are the natural choice for commercial instruments. They are less fragile than barretters and possess a relatively large negative temperature coefficient of resistance. Film bolometers, which employ thin metal films as temperature-sensitive resistors, are used in some standards laboratories.

Modern bolometric power meters are built around a self-balancing dc bridge. Fig. 6.1 shows the most frequently employed form. The function of the bridge is to supply an automatically controlled dc power P_{dc} to the bolometer element so as to keep its resistance constant, thereby maintaining the bridge in balance. For the circuit shown in Fig. 6.1 the value of this dc power is given by:

$$P_{dc} = \frac{V^2}{4R} \tag{6.1}$$

where V is the dc voltage across the bridge and R is the resistance of the three fixed resistors of the bridge. When an rf signal is applied to the bolometer

* The term 'bolometer' is sometimes used in a more restricted sense to mean barretter or film bolometer, but not thermistor. It will be used to include all three.

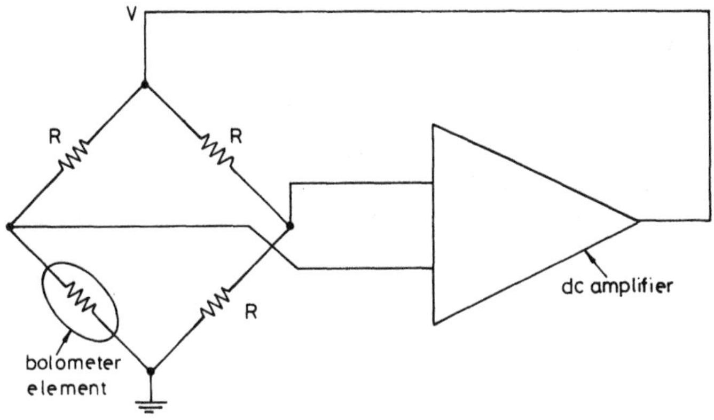

Fig. 6.1 *Bolometer with self-balancing Wheatstone bridge. The feedback loop maintains the bolometer element at a constant resistance R*

element, the dc power is automatically reduced by an amount which, in the ideal case, would be equal to the rf power being measured. This change in dc power is referred to as the retracted power, or the dc substituted power, and it is normally less than the rf power for practical instruments. The main reason for the difference isthat a proportion of the rf power is absorbed in the mount instead of in the bolometer element. Mount losses are characterised by the efficiency of the mount, which is defined as:

$$\text{mount efficiency} = \frac{\text{rf power absorbed by bolometer element}}{\text{rf power absorbed at input to mount}} \tag{6.2}$$

A second source of error arises from the fact that the current distribution in the bolometer element may not be identical for rf and dc signals. Thus there may be a difference in the corresponding temperature distributions, with the result that equal rf and dc powers do not produce equal changes in dc resistance. The error resulting from the different temperature distributions is known as the equivalence error of the bolometer element. This error is usually smaller than that due to mount losses but, unlike mount losses, it may be either positive or negative. It is not entirely a property of the element but depends partly on the field distribution in the mount. The combined effect of the equivalence error of the element and the efficiency of the mount is characterised by the effective efficiency, which is defined as:

$$\text{effective efficiency} = \frac{\text{dc substituted power}}{\text{rf power absorbed at input to mount}} \tag{6.3}$$

The process of calibrating a bolometric power meter involves two steps. The first is to check that the bridge and other dc circuitry correctly registers the dc substituted power. The second is to determine the effective efficiency of the bolometer-mount combination, either directly as described in Chapter 4 (The Microcalorimeter) and Section 6.4.1 below or by comparison with another power meter as described in Chapters 11 and 12. In addition, the reflection coefficient is normally required. Details of reflection coefficient measurements are given in Chapter 12. Sometimes calibration factor rather than effective efficiency is quoted. Calibration factor relates the dc substituted power to the power incident on the mount instead of to the absorbed power. Itis defined as:

$$\text{Calibration factor} = \frac{\text{dc substituted power}}{\text{incident power}}$$

$$= (1 - |\Gamma|^2)\,(\text{effective efficiency}) \qquad (6.4)$$

where Γ is the input reflection coefficient of the mount. Calibration factor is most useful when the source is well matched to the characteristic impedance of the transmission line, because in this case the available source power is calculated simply by dividing the power meter reading by the calibration factor, without the need to take account explicitly of the reflected power. Otherwise effective efficiency is the more useful quantity, as this generally has a smoother frequency response.

6.1 Barretters

For small equivalence error the barretter wire must be thin. There are two reasons for this. The first is that, in order to equalise the rf and dc current distributions over the cross-section, the diameter ought ideally to be less than the skin depth. This ideal is not actually achieved at the higher frequencies; for example, at 10 GHz the skin depth is only 1 μm, which is comparable with the diameter of the thinnest barretter. However, the resulting error in the measured power is not as great as might be imagined, because the high thermal conductivity of the metal tends in any case to equalise the cross-sectional temperature distribution.

The second reason for preferring a smaller diameter is that, for a given resistance, the resistive element can be made shorter, with correspondingly less opportunity for longitudinal current variations due to standing waves to arise. Some early microwaveexperiments used 10 μm diameter tungsten filaments, this being a natural development from the use of lamps as loads in the radio frequency region [7,8]. For frequencies above 1 GHz, however, smaller diameters were adopted.

The present-day barretters were developed from 5 and 10 mA instrument fuses [9]. They may be encapsulated as shown in Fig. 6.2 but can also be mounted on a substrate or etched in situ in a waveguide. Their manufacture starts from a length of Wollaston wire consisting of a core of platinum of

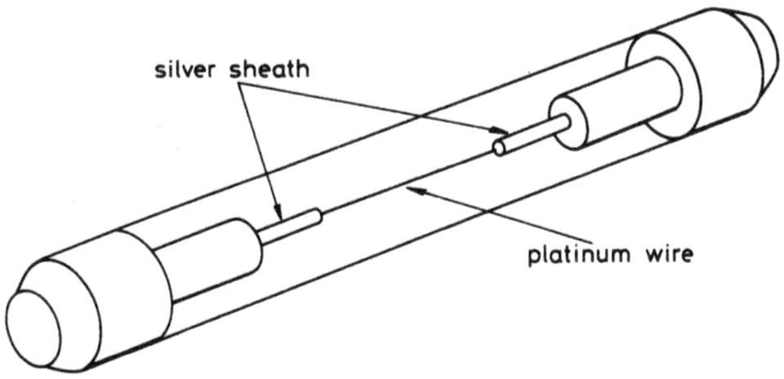

Fig. 6.2 *One form of encapsulated Wollaston wire barretter*

diameter 1–3 μm surrounded by a sheath of silver having an overall diameter of typically 25 μm. Wollaston wire is manufactured by taking a length of normal diameter platinum wire, coating it with silver, and then stretching it. To form the barretter a portion of the silver is either etched away using an aqueous nitric acid solution or removed by electrolytic deplating so as to leave the platinum exposed over a length of 2 to 3 mm. After etching the wire is spot-welded or soldered to larger diameter support leads. The manufacturing process can be made fairly reproducible, as the resistance of the barretter can be controlled by varying the length of the etched section. This resistance normally lies in the range 100 to 400 ohms at the operating temperature of between 100 and 200 °C.

Fig. 6.3 shows a typical resistance/power characteristic for a wire barretter. The relation can be represented in general by the following equation [3]:

$$R - R_0 = J\,P^n \tag{6.5}$$

where R is the resistance corresponding to dissipated power P, and J and R_0 are constants. A typical value for n is 0.9. The sensitivity, which is given by the slope of Fig. 6.3, is normally in the range 5–10 Ω/mW. Most barretters operate fairly close to their burn-out power level in order to obtain as large a change in resistance as possible, and this makes them easily damaged by overloads. The operating resistance is about twice theroom temperature resistance. The time constant is 0.1–0.4 ms. Modulated signals can be detected, provided that the modulation frequency is no more than a few kilohertz.

Fig. 6.4 shows a design for a fixed tuned waveguide barretter mount [10] suitable for the bands 8.2–12.4 GHz or 12.4–18 GHz. It is intended for standards work at a single frequency, the tuning being accomplished by means of three lockable screws of 0.5 mm diameter. The body of the mount is electroformed in copper around a polished, stainless steel mandrel. This enables good tolerances and surface finish to be achieved on the inside of the mount. It also produces good electrical continuity between the waveguide walls and the short circuit which terminates the guide at a distance of a quarter

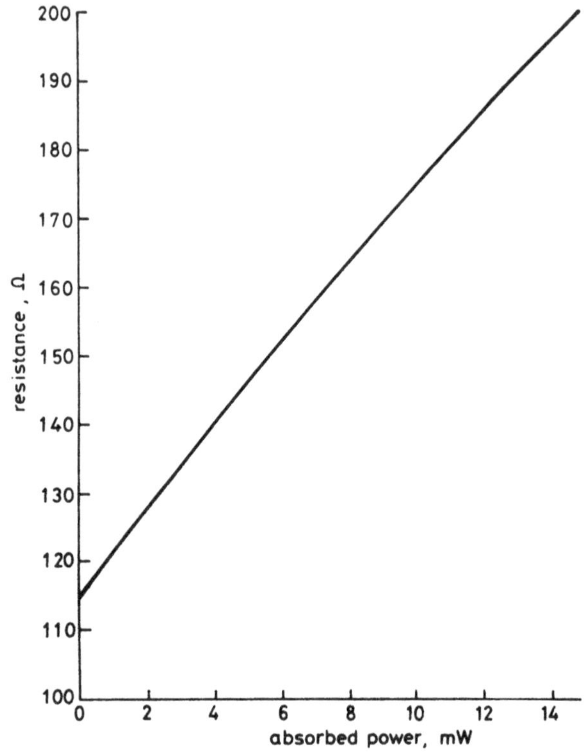

Fig. 6.3 *Typical characteristic of a Wollaston wire barretter element at an ambient temperature of 20°C*

wavelength behind the barretter element. The tuning screws and the dc lead-out are surrounded by half-wavelength, air-spaced, folded chokes, which reduce the rf leakage signal to between 55 dB and 75 dB below the incident power. When properly tuned, the voltage reflection coefficient at the input to the mount is less than 0.01 at the design frequency, corresponding to a power reflection coefficient of 0.01 percent. A convection barrier of low loss dielectric foam is inserted in the entrance of the mount to prevent circulating air currents around the element. The mount is intended to be calibrated in a microcalorimeter (Chapter 4) and is gold plated in order to minimise thermal radiation. The gold plating also prevents oxidation of the copper, which could cause a reduction in efficiency with age.

Commercial waveguide barretter mounts have been made which incorporate a sliding short circuit. The short circuit must be adjusted for minimum reflection at each frequency of operation as it is difficult to make a fixed tuned barretter mount covering a whole waveguide band. The operating resistance of the barretter is chosen so as to achieve the best average match over the tunable frequency range, the optimum value normally being in the range of 300–400 Ω.

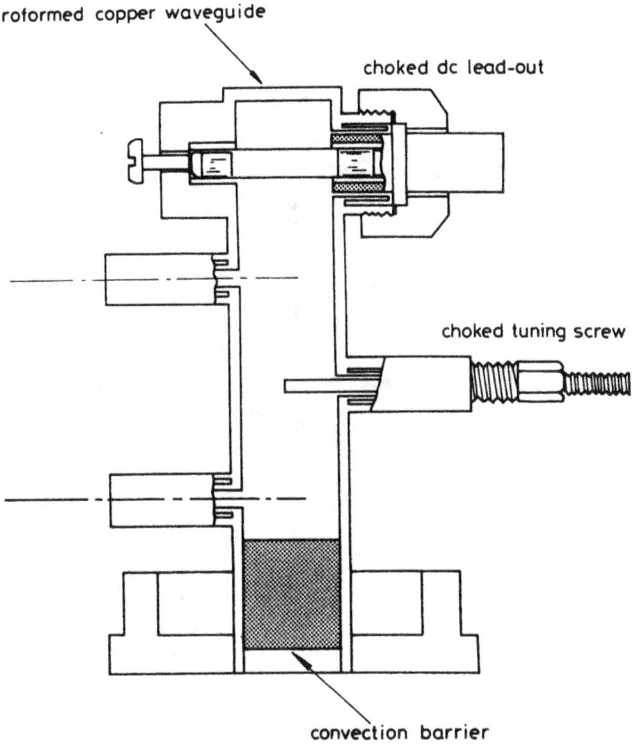

Fig. 6.4 *Tuned waveguide barretter mount designed for standards use. This design employs an encapsulated barretter and is suitable for the frequency bands 8 – 12 GHz and 12 – 18 GHz*

The short circuit can be replaced by a resistive load, enabling a broader bandwidth to be achieved, although this makes the mount very lossy [3].

Ref. [11] describes the wafer type of waveguide barretter mount, which provides for high effective efficiency of 96–98 percent at 35 GHz and low leakage. The barretter element is mounted on a 25 μm ruby mica substrate which has evaporated nichrome/gold electrodes (Fig. 6.5). The 1 μm core diameter Wollaston wire is soldered into position using a low-melting-point indium-alloy solder. This configuration gives strain-free construction and ensures that the whole of the Wollaston wire section is contained within the waveguide, thus minimising microwave leakage. It also improves the cooling mechanism for the element, resulting in a stable power handling capability of up to 25 mW. The cold resistance is approximately 110 Ω and the sensitivity of the element 7 Ω/mW, so that as bias power of about 12 mW is required to reach the operating resistance of 200 Ω. The technique can be extended for use in other waveguide bands up to at least 100 GHz.

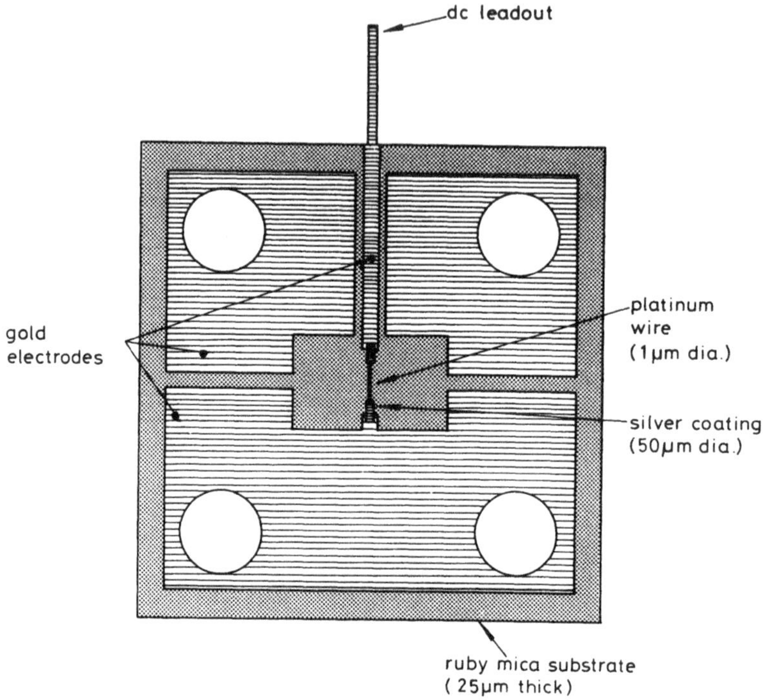

Fig. 6.5 *Unencapsulated waveguide barretter element consisting of a Wollaston wire on a mica substrate. Unencapsulated barretters are used for the higher frequencies*

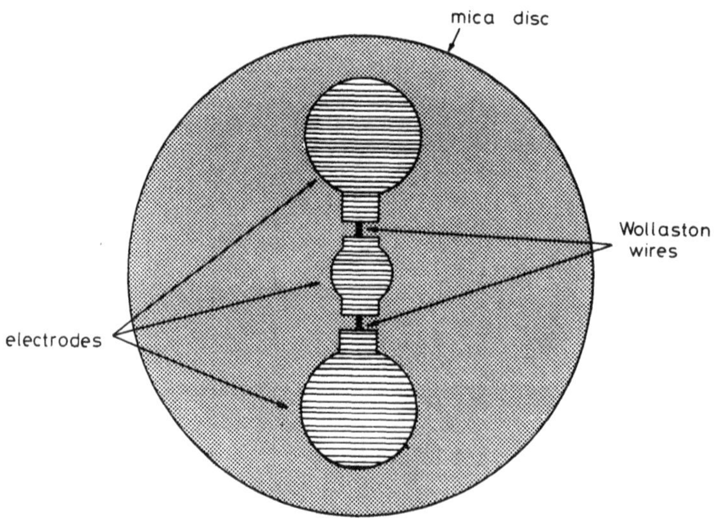

Fig. 6.6 *Coaxial bolometer element consisting of two Wollaston wires on a mica substrate*

Coaxial bolometers are usually made by mounting two Wollaston wire elements on a circular substrate [3] (Fig. 6.6). They are used in a similar way to dual element thermistor power meters (see Section 6.2).

6.2 Thermistors

Thermistors used in power meters [12–16] consist of a polycrystalline sintered ceramic made from a mixture of oxides chosen from the elements manganese, nickel, cobalt, copper and iron, and formed into a small bead. Early thermistor materials suffered from poor reproducibility because their chemical compositions were sensitive to atmospheric and temperature variations during sintering and subsequent processing, but over the years stability has improved. In general there are two important conduction mechanisms involved in oxide semiconductors [16]. The first is departure from stoichiometry (balance of elements as required by chemical formulas). For example, conduction in CuO_2 is due to an excess of oxygen, leading to p-type conductivity. The second mechanism is a 'hopping' process which can be controlled by varying the proportions of the constituent oxides. Semiconductors relying on this principle are termed controlled valency semiconductors. To take an example, Mn_3O_4 is normally an insulator in its pure form, but conduction occurs if nickel is introduced. The hopping mechanism is different from that which applies for semiconductors such as silicon, which are described by the electron band model. Instead, electrons hop from one ionic site to the next. When nickel is added to Mn_3O_4 some of the Mn^{3+} ions are induced to change to Mn^{4+}, and electrons are then able to hop between Mn^{3+} and Mn^{4+} ions, the Mn^{3+} ion becoming an Mn^{4+} ion and vice versa. The nickel itself does not take part in the conduction. The general conditions for hopping to occur are that the ions between which the electrons jump must be of the same element, must differ in valency by one unit, must occupy crystallographically equivalent sites and must not be too far apart.

The construction of a thermistor bead is shown in Fig. 6.7. The method of fabrication is as follows. The constituent oxides are first mixed and ground to a

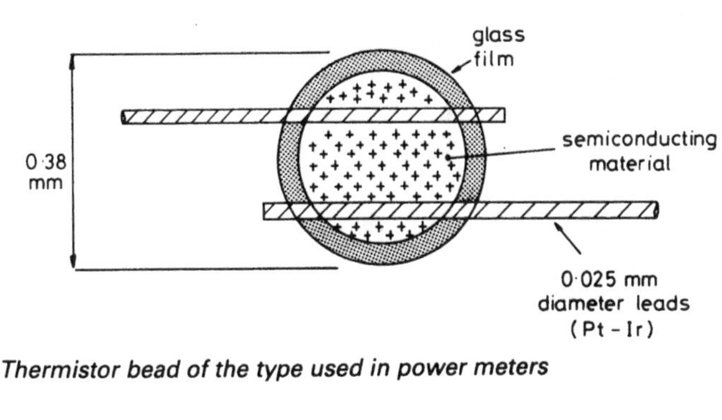

Fig. 6.7 *Thermistor bead of the type used in power meters*

homogeneous fine powder, which is then mixed with an organic binder such as polyvinyl acetate to form a slurry. A small amount of the slurry is applied so as to form a bead of diameter 50–70 μm which bridges two platinum wires of diameter 25 μm. The bead is air dried, sintered in a furnace at 1100–1200 °C so that the oxides combine chemically, and covered by a protective glass film. The two unwanted lengths of platinum wire are cut off and tungsten tipped nickel lead wires are attached to the two remaining ends. The final stage consists of encapsulation in glass. Ageing improves the stability of the resistance characteristic.

The relation between the resistance R of the thermistor and the absorbed power P can be represented approximately by the equation:

$$R = J \exp\left(\frac{B}{K + CP}\right) \tag{6.6}$$

where J, B, K and C are constants. Fig. 6.8 shows a typical characteristic. If voltage is plotted against current there is usually a region of negative slope resistance over which an increase in current produces a decrease in voltage. If the thermistor is biased in this region, care must be taken to ensure that the circuit is stable. For power measurement the operating resistance normally lies in the range 100–300 Ω, usually 100 or 200 Ω. The operating temperature is normally slightly higher than 100 °C and the maximum power which can be measured is normally 10 mW.

The large temperature coefficient of resistance of thermistors, which lies in the range 10–50 Ω/mW, enables the spread in the characteristics of different thermistors to be compensated for by adjusting the dc bias current, which happens automatically if a self-balancing bridge is used. Thermistor power meters have reasonably good ability to withstand moderate power overloads, because as the temperature approaches the burn-out value the resistance drops sharply to a value typically in the region of 5 Ω, with the result that most of the incident power is reflected.

A coaxial mount is shown in Fig. 6.9. It is of the dual element type [17,18] which avoids the quarter wave stub used in early single element coaxial mounts to provide a dc return [1]. The two elements are arranged so as to be in parallel for the rf signal and in series for the dc current. A 50 Ω mount therefore requires two thermistors operating at a resistance of 100 Ω and presenting a combined resistance of 200 Ω to the bridge. Commercial mounts of this design can cover a frequency range from 10 MHz to more than 18 GHz in a single device [18]. However, if the characteristics of the two thermistors are not identical an error known as the dual element substitution error may arise (see [19] and Section 6.5). The result is that the effective efficiency varies with power level. A way of avoiding the problem by using a single thermistor with a centre tap was proposed by Aslan [20]. Most coaxial thermistor mounts are designed for an input impedance of 50 Ω, but below 1 GHz 75 Ω mounts are also used [21].

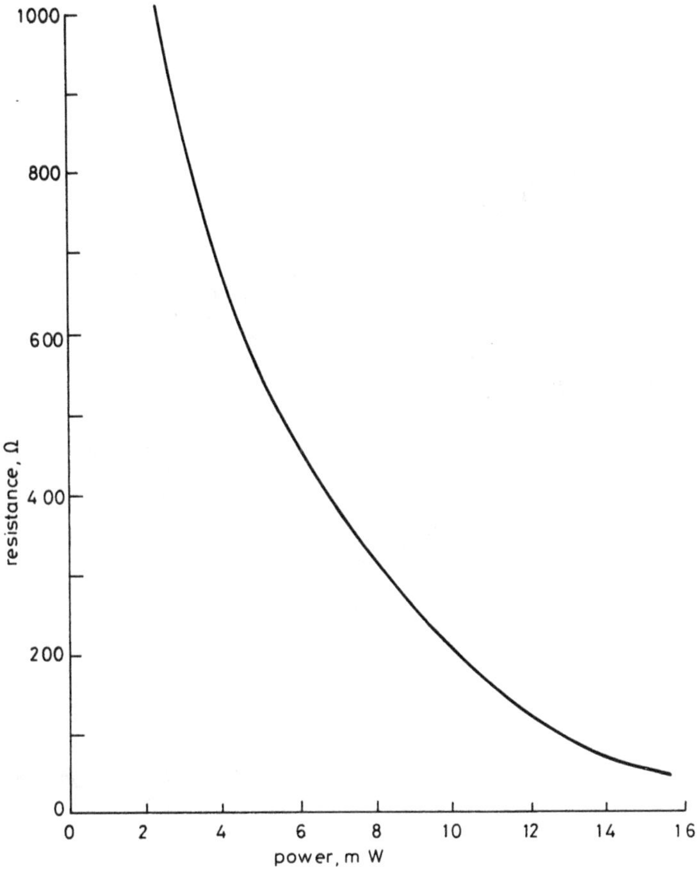

Fig. 6.8 *Typical characteristic of a thermistor bead at an ambient temperature of 20°C*

Fig. 6.10 shows the bar and post type of waveguide thermistor mount [18] which enables a whole waveguide band to be covered. An alternative form is the stepped waveguide mount [2,22] shown in Fig. 6.11. As the frequency increases, the performance of thermistor power meters usually worsens. Thermistor mounts designed for the 75–110 GHz band may have effective efficiencies as low as 40–60 percent, although figures up to 90 percent at, for example, 94 GHz are possible. The causes are discussed in ref. [23], which reports the large discrepancies which were found when a thermistor was compared with a water calorimeter and a film bolometer. The capacitance of the thermistor tends to shunt out its resistance, so that as the frequency increases a greater proportion of the power is absorbed in the leads. A theoretical curve based on an assumed equivalent circuit was fitted to the measured results, although it is not certain that all of the discrepancy reported in this work was due to the inefficiency of the thermistor.

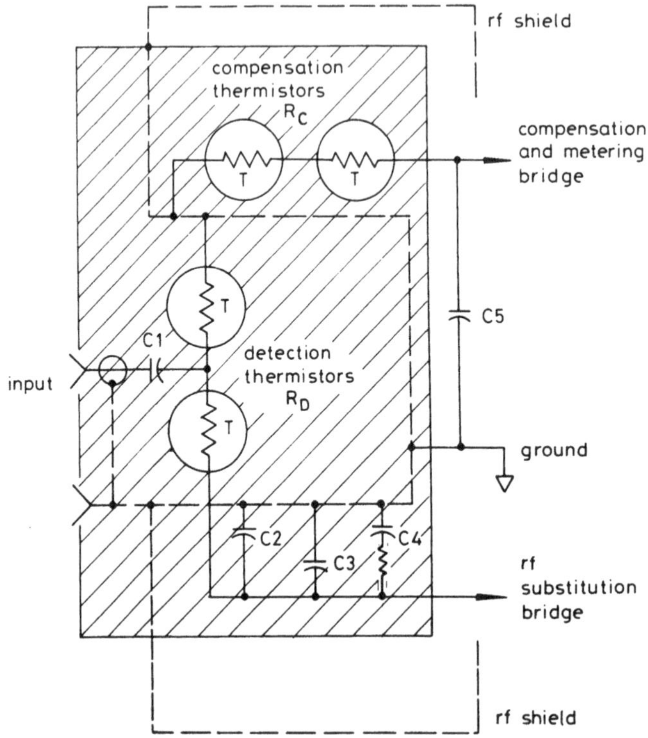

Fig. 6.9 *Coaxial thermistor mount of the dual element type. The two thermistors are in parallel for rf and in series for dc*

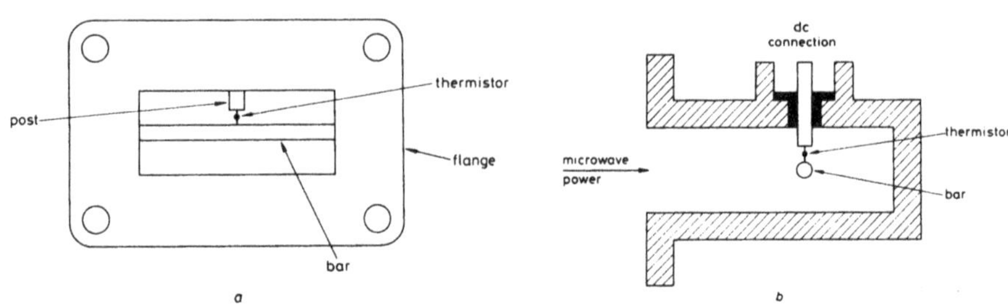

Fig. 6.10 *Bar and post type of waveguide thermistor mount*

Compensation for ambient temperature changes is usually achieved by means of a compensating thermistor bead or disc placed in thermal contact with the mount but not exposed to the rf signal [18]. Details are given in Section 6.6. Temperature-compensated thermistor power meters can measure powers down to 10 μW.

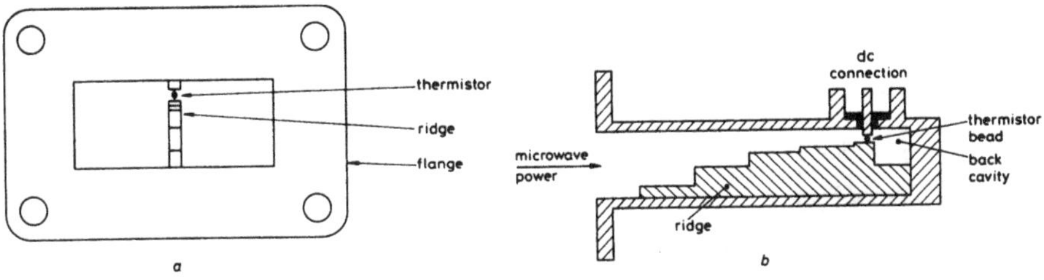

Fig. 6.11 *Stepped waveguide type of thermistor mount*

6.3 Film bolometers

Film bolometers have several useful properties. The greater surface area of the element, compared with that of barretters and thermistors, enables the power rating to be increased, so that powers of several hundred milliwatts or even a watt can be measured. The losses in the mount tend to be low, because it is not necessary to concentrate the power into a very small volume. By controlling the thickness of the film, high resistance elements can be fabricated and these can be used to make feed-through power meters which absorb only a small fraction of the power, the rest being transmitted. A disadvantage of thin film bolometers, which arises from the larger surface area, is that the equivalence error is more strongly dependent on the field distribution in the mount. Consequently theoretical analysis is difficult. The time constants of film bolometers tend to be relatively long, up to 12 seconds, although the exact value depends to a large extent on the thickness of the substrate. Practical difficulties have also been experienced in making films which are sufficiently stable with time and in ensuring reliable contacts. Materials most commonly used for thin film bolometers are gold and platinum on a glass or mica substrate.

6.3.1 Coaxial film bolometers

Coaxial film bolometers are of two main types. The first consists of a cylindrical resistor mounted in series with the inner conductor of a coaxial line. This form was first investigated by Johnson [24]. Harris [25] later developed a dual element versionusing a slit film in conjunction with a tractorially shaped outer conductor fabricated in two insulated halves (Fig. 6.12). This instrument operated over the frequency range 0.2-4 GHz and had a claimed uncertainty of 1 percent including the error due to mismatch. The reason for choosing a tractrix (see Chapter 3 and Appendix 3) for the outer conductor was to achieve uniform power absorption along the length of the load resistor, both for dc and for the rf power, so as to minimise the equivalence error of the element. The second type of coaxial film bolometer is the transverse film version, originally investigated by Norton [26]. Dual element versions using split film discs were

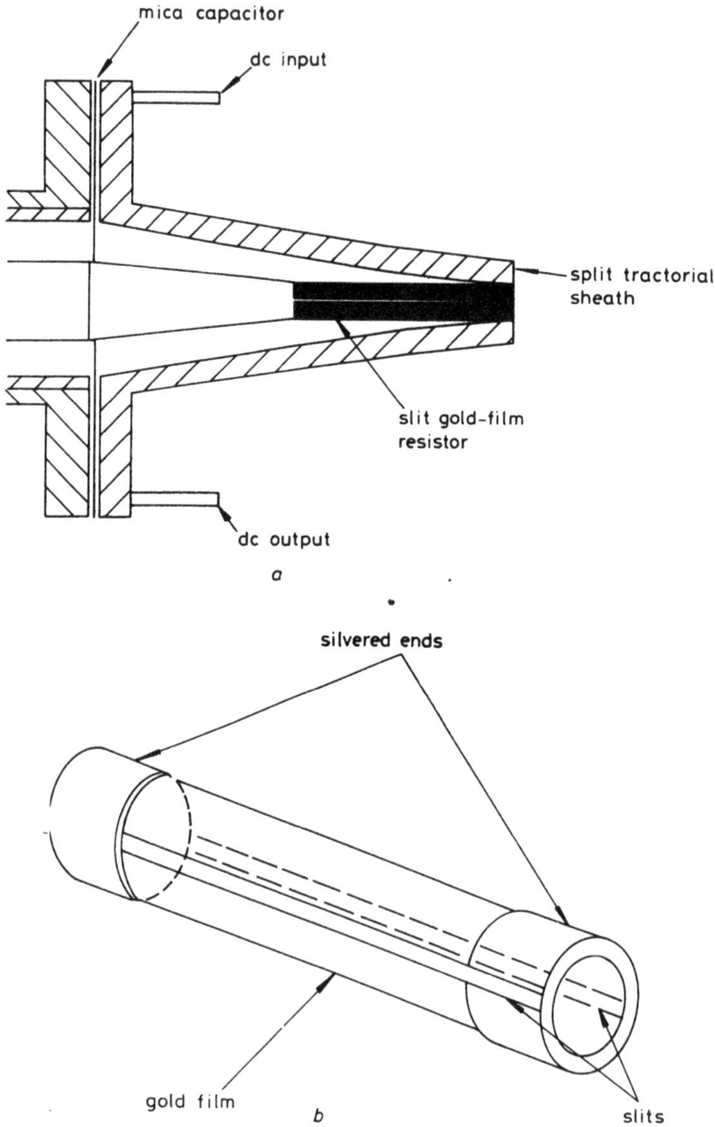

Fig. 6.12 *Coaxial film bolometer employing a slit cylindrical resistor. Each half of the resistor is in contact with one half of the longitudinally split outer conductor. The assembly is held together by an insulating housing (not shown)*

later developed by Selby [27–32] under the name of 'bolovac'. It is however more appropriate to consider this instrument as a voltmeter, since the absorbed power depends on the voltage between the inner and outer conductors, measured in the plane of the film. For further details see Chapter 15.

6.3.2 Waveguide film bolometers

Fig. 6.13 shows a waveguide transverse film bolometer mount of the general type developed by Lane [33–35]. One of the mounts used in this work had an effective efficiency of 98.8 percent when measured against the average of a flow calorimeter and a torque vane wattmeter at a frequency in the region of 9 GHz. It was suggested that this relatively high value of effective efficiency in relation to typical thermistor mounts was probably due to the uncomplicated design of the mount. Ref. [36] describes an electroformed thin film bolometer mount operating in the 74–110 GHz waveguide band. At 94 GHz it had an effective efficiency greater than 99 percent as measured by a microcalorimeter.

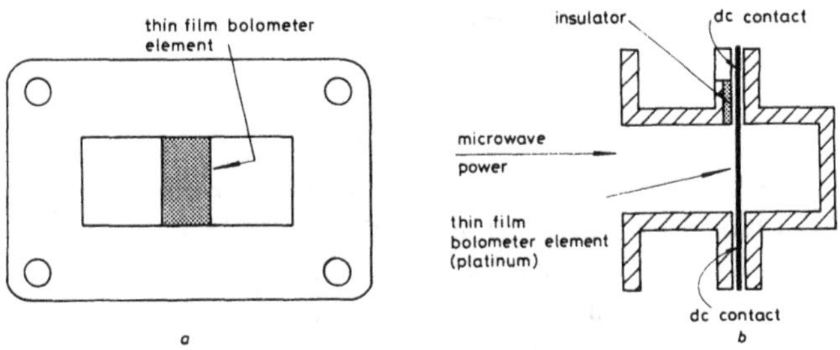

Fig. 6.13 *Waveguide film bolometer of the transverse type*

The bolometer element consisted of a thin nickel film on a 12.5 μm thick polyimide substrate. A capacitive window enabled the series inductance of the element to be tuned out, thereby increasing the bandwidth. The element had a sensitivity of 2.9 Ω/mW, a cold resistance of 143 Ω, and an operating resistance of 200 Ω. A transverse film bolometer using a wider film which fills the whole of the cross-section of the waveguide is described in ref. [37]. In this design there is no inductance associated with the film, but the difference between the rf and dc current distributions is greater. This type has not in general found favour. Other forms of thin film waveguide bolometer have employed longitudinal resistive films. An early form was the waveguide wall bolometer [38], in which the resistive film formed part of the narrow wall of the waveguide. To prevent the dc path through the film from being shorted out by the waveguide, slots were cut in such a way as not to disturb the rf fields. The instrument was used to sample the power flowing along the waveguide, absorbing only a few percent of the power. Some later thin film bolometers have used a longitudinal vane in the centre of the waveguide to present a matched load to the signal [39], as shown in Fig. 6.14.

6.4 Determination of effective efficiency

The normal method for directly determining the effective efficiency of a bolometer mount is the microcalorimeter method. As this is described in detail

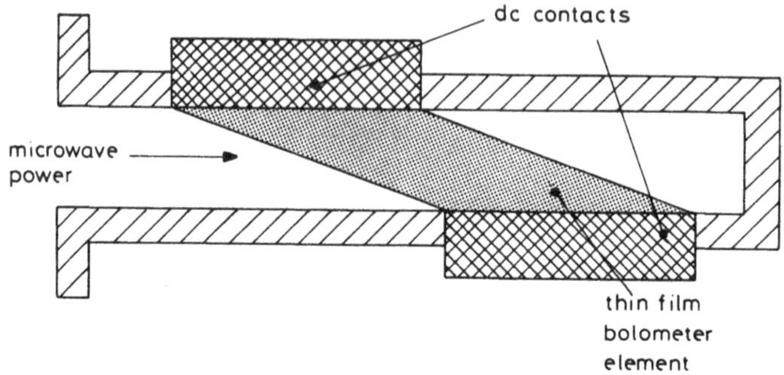

Fig. 6.14 *Waveguide film bolometer of the longitudinal type*

in Chapter 4 only a resumé need be given here. An older technique is the impedance method. This is no longer in use, as it is generally considered to be less accurate than the microcalorimeter technique and the assumptions which it makes are less reliable. Nevertheless, it is not inconceivable that the impedance method or some modification of it could return to favour at some future time. The third method of determining effective efficiency is by comparison with a known power meter whose effective efficiency has already been determined. This is dealt with in Chapters 11 and 12 and will not be discussed here.

6.4.1 The microcalorimeter method

The bolometer mount to be calibrated acts as the load of a calorimeter and its temperature rise is sensed by a thermopile. The indication of the thermopile is a measure of the total power dissipated in the combination of mount and bolometer element, that is the sum of the applied rf power, if any, and the dc bias power supplied by the self-balancing bridge. If the effective efficiency were 100 percent the thermopile indication would remain constant when the rf power was switched on, since the self-balancing bridge would reduce the dc bias power by an exactly compensating amount. For an effective efficiency of less than 100 percent there is an increase in total power and a corresponding increase in thermopile indication. The effective efficiency is deduced from the thermopile readings obtained with and without the rf power. When the measurement is complete the bolometer mount is removed from the calorimeter and its reflection coefficient measured. It can then be used as a standard against which other power meters can be compared. For further details see Chapter 4.

6.4.2 The impedance method

The impedance method makes the assumption that the equivalence error of the bolometer element is negligible and that consequently the effective efficiency is equal to the efficiency of the mount [40–44]. It is based on the fact that the efficiency of a linear reciprocal two-port network can be determined by connecting three known load impedances to the output and measuring the corresponding input reflection coefficients. Fig. 6.15 shows how this principle is applied to bolometer mounts. The reciprocal two-port represents the mount, whilst the variable load resistor represents the bolometer element, the resistance of which can be changed by altering the dc bias current. The method

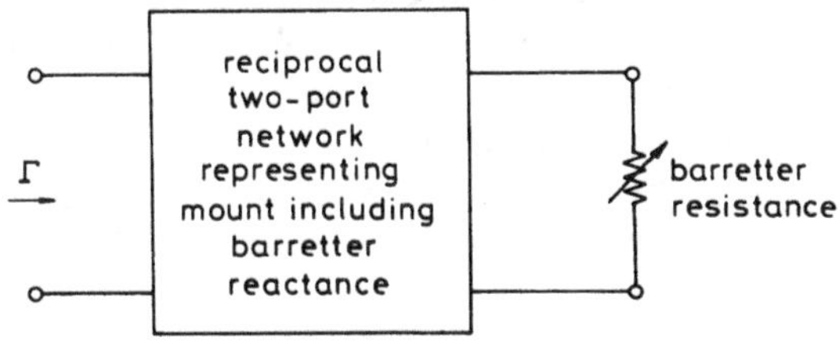

Fig. 6.15 *Determination of the efficiency of a barretter mount by the impedance method. The input reflection coefficient Γ is measured as a function of the barretter resistance, enabling the parameters of the two-port network to be deduced*

assumes that the rf impedance of the bolometer is equal or proportional to its dc resistance, an assumption which may be valid for wire barretters but not for thermistors [45]. If R_2 is the normal operating resistance of the barretter and η the efficiency of the mount (=power delivered to bolometer/power absorbed at input to mount) corresponding to this load resistance, then

$$\eta = \frac{K|\Gamma_3 - \Gamma_2||\Gamma_1 - \Gamma_2|}{|\Gamma_3 - \Gamma_1||1 - |\Gamma_2|^2|} \tag{6.7}$$

where Γ_1, Γ_2, Γ_3 are the reflection coefficients corresponding to barretter resistances R_1, R_2 and R_3 respectively, and

$$K = \frac{2R_2(R_3 - R_1)}{(R_2 - R_1)(R_3 - R_2)} \tag{6.8}$$

For good accuracy the three reflection coefficients should be well separated, since the procedure involves determining differences between them. A limit is set, however, by the fact that the range over which the barretter resistance may be varied is less than 2:1. If the mount is matched, that is, if Γ_2 is zero, equation 6.7 simplifies to:

$$\eta = K\left|\frac{\Gamma_3\Gamma_1}{\Gamma_3 - \Gamma_1}\right| \tag{6.9}$$

For matched mounts with high efficiency the phase angle between Γ_1 and Γ_3 is close to either zero or 180°, so that equation 6.9 becomes:

$$\eta = \frac{K|\Gamma_3||\Gamma_1|}{|\Gamma_3| \pm |\Gamma_1|} \tag{6.10}$$

The plus sign applies if R_2 lies between R_1 and R_3. This approximation can be used to simplify the measurement procedure, since it is then no longer necessary to measure the phases of the reflection coefficients. However, it is preferable to measure both magnitude and phase. Great care is required in carrying out the measurements, as an error of 1 percent in the VSWR can give rise to an error of up to 6 percent in the mount efficiency. The achievable uncertainty using the technique described in ref. [43] was roughly 0.5 percent excluding the error due to the assumptions about the impedance of the barretter element.

6.5 Sources of uncertainty

6.5.1 RF/dc equivalence error of bolometer elements

This error arises only if the impedence method is used. Most theoretical work has been directed towards determining an upper limit for the equivalence error of wire barretters when mounted in a waveguide. The early work is described in refs. [46–48], which are cited by Griesheimer [1], and in refs. [8,49,50]. Carlin and Sucher [51] later published a detailed study, including a summary of previous work, and it is this paper which is most frequently quoted. Ginzton [2] also devotes considerable space to this topic. The main conclusions reached are as follows. The equivalence error in a wire barretter is due to a number of effects. Firstly, the skin effect will tend to concentrate the rf current close to the surface of the wire, whereas the dc current will be more uniformly distributed over the cross section. Since power dissipated close to the surface will produce a smaller average temperature rise in the wire than the same amount of power dissipated at the centre, the result is that equal rf and dc powers do not produce the same change in barretter resistance. The error due to this cause is, however, fairly small, typically less than 0.03 percent, because the wire is too thin for large temperature variations over its cross section to occur. The major contribution to the substitution error comes from the non-uniform longitudinal distribution of the rf current. This non-uniform current can be thought of as arising from standing waves and the distribution of current is assumed to be sinusoidal by the theory. There are two components to the error, one associated with end effects and the other associated with the non-linear relation between temperature and dissipated power. Consider first the end effects. Over the middle section of the wire, heat is lost partly by convection and conduction through the air and partly by radiation. At the ends, however, there is an additional heat path through the thicker connecting wires. Heat generated near the ends of the barretter wire therefore produces a lower temperature rise than heat generated in the middle section, making the change in barretter resistance for a given power dependent on the power distribution, which depends on the current distribution. Now consider the effect of the non-linear temperature-power relation,which arises largely from the fact that the thermal conductance between the wire and its surroundings is a function of temperature. The result of this non-linearity is that a large concentration of power absorbed at one point will not produce the same average temperature rise (that is, averaged over the length of the wire) as a more uniformly distributed absorbed power. The effect is more pronounced in evacuated bolometers, for which the main cooling mechanism is radiation, because radiation results in a more non-linear temperature–power law than convection or conduction. The conclusion is therefore reached that barretters which are cooled mainly by convection, such as the normal Wollaston wire type, are to be preferred. The slightly non-linear resistance–temperature law of the barretter has a similar effect to the non-linear temperature–power relation.

The analysis of Carlin and Sucher [51] was a worst-case analysis which assumed the least favourable current distribution and gave a maximum error of 2 percent. When comparisons were carried out between the impedance method and calorimeters [43,52,53], the agreement was within 0.5 percent, which was better than expected. This suggested that the current distribution in the barretter was more uniform than had been assumed. Further work was carried out to calculate the actual distribution of the current. Jarvis and Adams [54,55] calculated the distribution to be expected in a waveguide barretter and used this to predict the substitution error. Their results indicated that the equivalence error is a function of waveguide size, frequency, and the geometry of the element and its supports. For example, in one case the calculated substitution error increased from 0.002 percent to 0.21 percent when the radius of the posts supporting the element decreased from 2.5 mm to 0.04 mm and the element dimensions remained unaltered. However, these calculations assume an abrupt transition between the plated and unplated parts of the barretter wire, which will not necessarily be the case in practice. Incorrectly etched barretter elements can have equivalence errors which are large enough to produce mount effective efficiencies greater than 100 percent [56].

6.5.2 Dual element substitution error

The dual element substitution error, first mentioned by Harris [25], refers to the variation of effective efficiency with power levels which occurs in coaxial power meters incorporating two barretter or thermistor elements. In practice, this means that it can arise in the majority of broadband coaxial bolometric instruments. Ideally, each of the two elements should absorb exactly half of the rf or dc power, but in practice this is only approximately true. The unequal division of power would be unimportant if the sensitivities of the two elements (measured in ohms per milliwatt) were identical, because a given power would then produce the same change in the total dc series resistance, irrespective of which element it was dissipated in. The error arises from the combined effect of unequal sensitivities and unequal power division.

The unequal division of power may arise in two ways. The first way is through a difference in the operating resistances of the two elements. Since the elements are connected in series for dc but in parallel for rf, the element with the larger resistance will receive the greater dc power, whilst the element with the smaller resistance will receive the greater rf power. Assuming that the total series dc resistance is held constant by a self-balancing bridge, the substitution error (in mW) may be expressed in terms of the sensitivities γ_a and γ_b (in Ω/mW) of the two elements and the change Δr (in Ω) in the dc resistance of one element when the rf signal is applied:

$$\text{error in power} = \left(\frac{1}{\gamma_a} - \frac{1}{\gamma_b}\right) \Delta r \tag{6.11}$$

The second way in which an unequal division of power may occur is through asymmetries in the mount reactances. This affects the power division for the rf signal only. Equation 6.11 also applies to this case.

The dual element substitution error was investigated theoretically and experimentally by Engen [19]. The theory, based on certain simplifying assumptions, predicts that the percentage error in the measured power will decrease with power level for barretters, but will increase for thermistors. This indeed was found to be the case. When measurements were made on typical barretter mounts, using a special bridge circuit, errors of up to 1.7 percent were found. Thermistors operating at a power level of 10 mW gave errors of up to 1 percent, but at higher power levels the error increased dramatically, reaching over 10 percent at 20 mW. The error may be minimised by selecting elements with matched characteristics. Some measurements on commercial dual element thermistor power meters have shown that in practice the matching of the characteristics of the two elements can be very good, resulting in negligibly small errors [57], but this cannot be taken for granted without calibrating the instrument against accurate attenuation standards.

6.5.3 Thermoelectric effect in bolometers

A dc current flowing across a junction between dissimilar conducting materials will produce thermoelectric heating or cooling of the junction in addition to the ohmic heating associated with the resistance of the materials. This will affect the dc heat distribution, leading to a difference in the rf and dc temperature distributions in the bolometer element and so contributing to the substitution error. The thermoelectric heating or cooling power is proportional to the current, whereas the ohmic heating power varies as the square of the current, assuming that the resistance of the bolometer element is held constant. Consequently, the percentage error produced by the thermoelectric effect increases for decreasing bias current. For this reason, thermoelectric errors are most troublesome in power meters using a combination of dc and ac biasing currents, in which the dc may supply only a small part of the total bias power (Section 6.6). Fortunately, most present day thermistor power meters operate with all-dc bridges. The thermoelectric error usually arises because of some asymmetry in the bolometer mount with respect to the two bolometer leads and its sign depends on the sign of the dc current. It can be eliminated to a first approximation by reversing the current and taking the means of the two readings.

6.5.4 Additional error in circuits using both ac and dc

For power meters which supply simultaneous dc and audio frequency bias power (see Section 6.6) there is an additional small substitution error. The af resistance of the bolometer is different from the dc resistance and it varies over the audio cycle. The size of the resulting error depends on the ratio of ac to dc bias power. Ref. [58] gives the theory of this effect together with the results of measurements. The error can be reduced by the choice of a suitably high audio frequency, but all-dc bridges are to be preferred.

6.6 Bridges and electronic circuitry

Early bridge circuits were of the Wheatstone type [1], as in Fig. 6.1, but without feedback. The out-of-balance signal, which was displayed on a meter, indicated the measured rf power. Before application of the rf power the meter was zeroed by adjusting the bias current for bridge balance. This circuit is simple to use, but it has several drawbacks which resulted in the development of a large number of more complicated circuits. The first problem is that ambient temperature variations cause not only a zero drift, which can if necessary be eliminated by frequent rebalancing of the bridge, but also a change in the bridge sensitivity. This arises because the meter reading for a given rf power depends on the bridge bias current, which changes whenever the bridge is rebalanced. Secondly, the response of the instrument does not follow a square law: that is, the relation between the meter reading and the rf power is non-linear, even if the bolometer has a linear resistance/power law. This relation also depends on the dc source resistance seen by the bridge. Thirdly, application of the rf power causes not only a change in the dc resistance of the bolometer but also a change in its rf impedance, so that the input impedance of the power meter varies with power level.

Attempts to solve the problem of the variation of sensitivity with ambient temperature took two directions [1]. One solution was to add a compensating thermistor in series with the meter, with the object of increasing the meter sensitivity at high ambient temperatures, for which the bias current required to balance the bridge was lowest. At the same time the zero drift was reduced by connecting a second compensating thermistor across the bridge. Fig. 6.16 shows the circuit. The other approach was to bias the bridge using a combination of dc and audio frequency signals supplied by two separate sources (Fig. 6.17). The meter was arranged to respond to only one of these sources, so that the other could be used for bridge balancing without affecting the sensitivity. Neither approach, however, solved the problems of the non-linear response of the meter and the variation of rf input impedance with level. These are overcome by the balanced bridge approach, that is the dc bias is re-adjusted after application of rf power so as to restore the meter reading. The decrease in dc power is then measured. Manually balanced bridges [1,3] were used first. In the days before the 1960s, when accurate digital voltmeters first became available, it was necessary to go to considerable lengths in order to obtain even moderate accuracy in the dc power measurement. A manually operated potentiometer was normally used together with an accurately calibrated switched attenuator and a means of backing off the signal to be measured. A multitude of circuits is described in the literature [3].

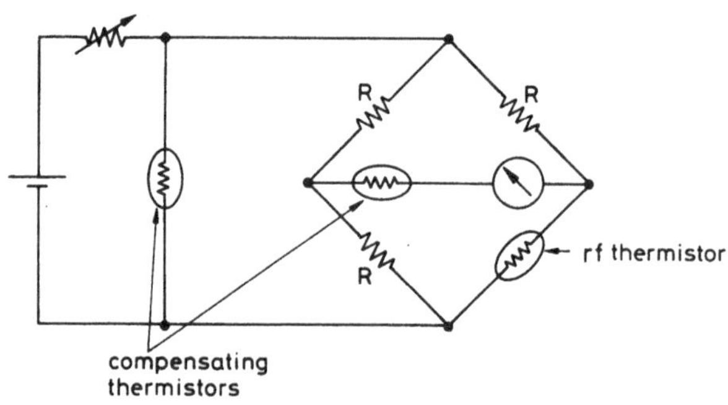

Fig. 6.16 *Early thermistor bridge employing two compensating thermistors to cancel out the effects of variations in ambient temperature*

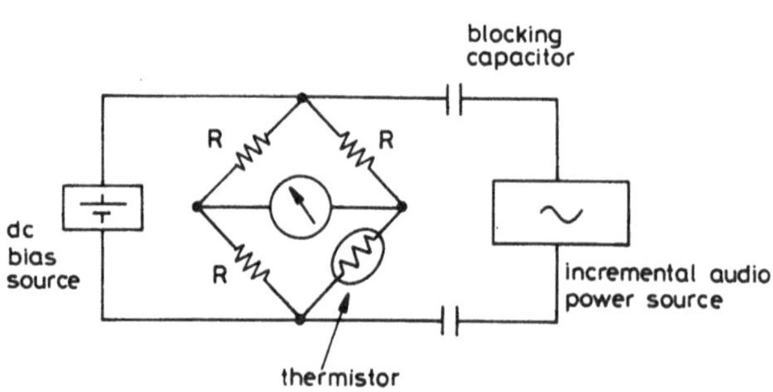

Fig. 6.17 *Thermistor bridge employing two separate sources (dc and audio-frequency). If the meter is arranged to respond only to one source, then the other may be adjusted to compensate for ambient temperature variations*

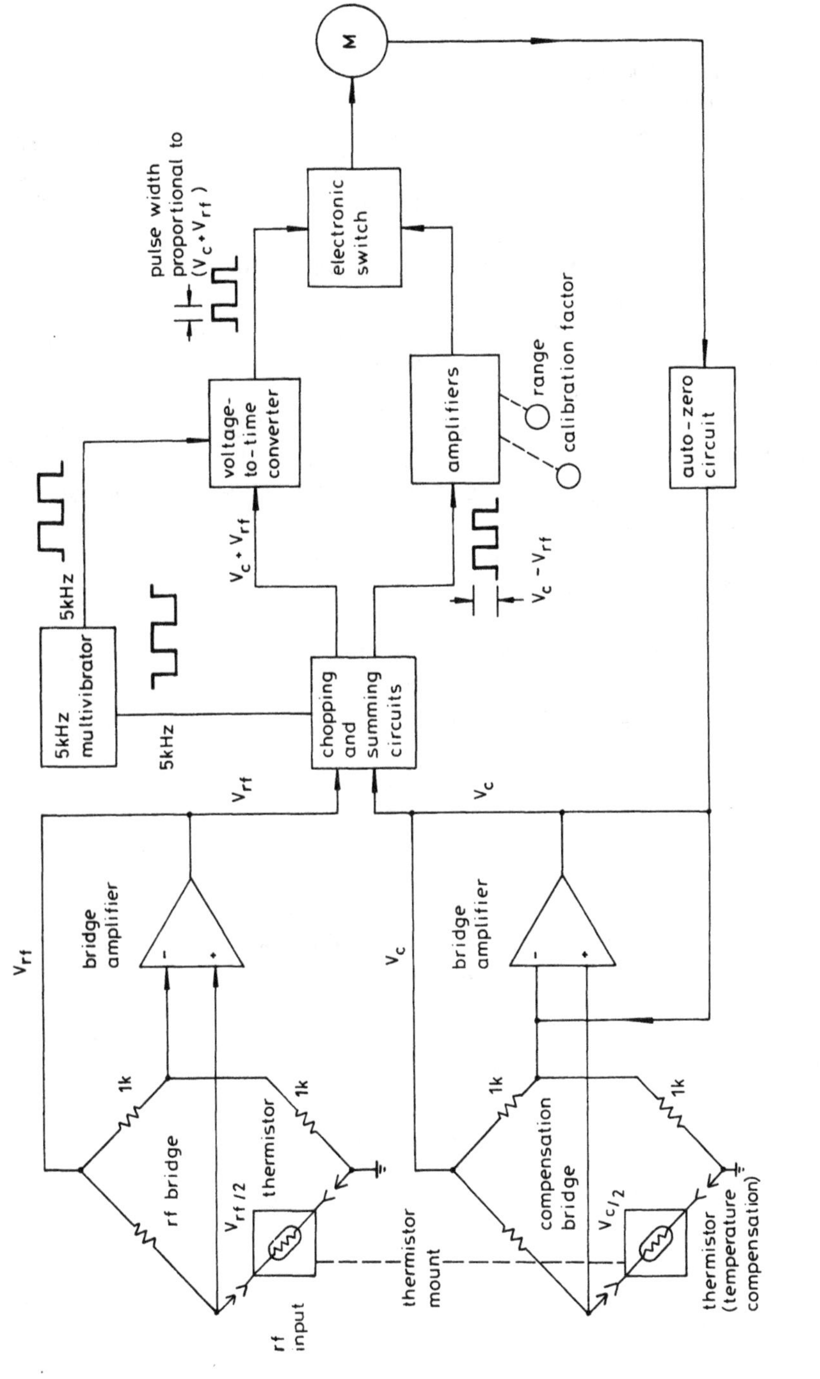

Fig. 6.18 *Circuit of a commercial thermistor power meter incorporating a second bridge for temperature compensation. Analogue multiplying circuits process the signals from the bridges, which are dc operated, so as to obtain a meter indication which is proportional to the measured rf power*

The modern form is the self-balancing bridge [59]. In this the unbalance signal of the bridge is amplified and fed back to the bridge (Fig. 6.1). If the loop gain is sufficiently high the level of the signal automatically adjusts itself until the bridge is close to balance. The polarity of the feedback is chosen so that the feedback is positive when the bolometer is at room temperature. Both dc and audio frequency self-balancing bridges have been used. For the audio frequency version a narrow band filter must be incorporated into the feedback amplifier in order to make the circuit oscillate at the required frequency.

Compensation for the zero drift caused by ambient temperature variations is achieved by using a dual bridge circuit [18,60,61]. One bridge contains the operating thermistor or barretter whilst the other contains the compensating thermistor which is attached to the outside of the mount. Each bridge is regarded as a separate power meter and the rf power is obtained from the difference in the power readings of the two bridges. For accurate measurements the calculation of the rf power from the bridge voltages (or sometimes currents as in refs. [62,63]) is done digitally, but in the past some sophisticated analogue methods were developed. One commercial thermistor power meter used a self-balancing audio frequency bridge for the rf thermistor [18]. This audio frequency signal was then measured using a self-balancing dc bridge which incorporated the compensating thermistor. The dc signal was then squared and a backing off signal added to derive an output proportional to the measured power. Later versions of this instrument used all-dc bridges together with analogue multiplier circuits using pulse length modulation, as shown in Fig. 6.18.

A significant development of the 1970s was the introduction of the self-adjusting current loop [64,65] as an alternative to the Wheatstone bridge. This circuit, which is based on a four-terminal resistance measurement, eliminates errors due to the dc resistance of the bolometer leads. The underlying principle is illustrated by the arrangement shown in Fig. 6.19, consisting of an operational amplifier with associated power supply, two four-terminal resistors, a variable voltage source and a voltmeter. Because no currents flow in the amplifier input leads the currents in the two resistors R_1 and R_2 must be equal. The amplifier maintains the current at a level which keeps the potential difference between its input leads equal to zero. The voltmeter connected between the other two potential terminals of R_1 and R_2 displays a reading which is proportional to the difference in R_1 and R_2. If R_2 is a thermistor, its resistance can be controlled by adjustment of the variable voltage source. Adjusting the source so that the voltmeter reads zero makes R_2 equal to R_1.

Replacing the voltmeter and the source by a second operational amplifier together with its power supplies results in the circuit shown in Fig. 6.20. The two amplifiers maintain a zero potential difference across their respective inputs, and the circuit forces the thermistor resistance into equality with R_1. There are two stable states, as the current can flow in either direction. The

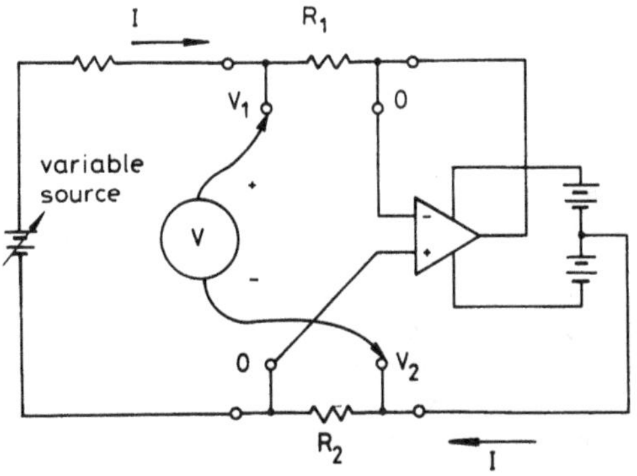

Fig. 6.19 *A method of intercomparing two four-terminal resistors. This method forms the basis of the self-adjusting current loop*

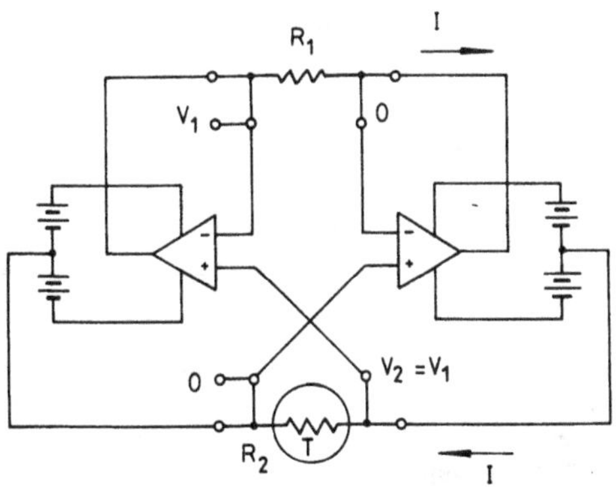

Fig. 6.20 *The self-adjusting current loop. This circuit represents an improvement over the older self-balancing Wheatstone bridge, in that the resistance of the thermistor leads cancels. In addition, only one standard resistor is required (R_1)*

system works equally well with positive temperature coefficient bolometers such as barretters provided that the bolometer and R_1 are interchanged. Further benefits claimed for the self-adjusting current loop are that, compared with the self-balancing Wheatstone bridge, it is less susceptible to the effects of noise and poor common mode rejection in the amplifiers. In addition, only a single standard resistor is required, which can if desired by switched between 100 Ω and 200 Ω operation without the usual errors due to switch contact resistances. A disadvantage is the need for a highly isolated power supply.

6.7 References

1 Griesheimer, R.N., 'Microwave power measurements', Chapter 3 in 'Technique of microwave measurements', ed. C.G. Montgomery, McGraw-Hill, 1947.

2 Ginzton, E.L., 'Microwave measurements', McGraw-Hill, 1957.
3 Carlin, J.J., 'Measurement of power', Vol. 1, Chapter 3 in 'Handbook of microwave measurements', ed. M. Sucher and J. Fox, Polytechnic Press, 1963.
4 Casper, S., 'Bolometer characteristics', *Microwaves*, **1**, pp. 56–63, Nov. 1963.
5 Adam, S.F., 'Microwave theory and applications', Prentice Hall, 1969.
6 'McGraw-Hill Encyclopedia of Science and Technology' 5th edition, McGraw-Hill 1982.
7 Linder, E.G., 'The use of gas-filled lamps as high dissipation, high-frequency resistors, especially for power measurements', *RCA Rev.*, **4**, 1, p. 83, 1939.
8 Bleaney, B., 'Radio-frequency power measurement by bolometer lamps at centimetre wavelengths', *J. IEE*, **93**, Pt. IIIA, 9, Mar.–May 1946.
9 Eberle, E.E., 'The Sperry type 821 barretter for microwave power measurement and detection', Sperry Gyroscope Company Report No. 5224-1042, 24 July 1945.
10 Skilton, P.J., 'Developments in United Kingdom waveguide power standards', RSRE Report 80006, April 1980.
11 Hill, L.D., 'New method for producing and mounting barretter elements for power measurement in waveguides', *Electron. Lett.*, **19**, 9, pp. 352–354, 28 April 1983.
12 Becker, J.A., Green, C.B., and Pearson, G.L., 'Properties and uses of thermistors – thermally sensitive resistors', *Bell Syst. Tech. J.*, **26**, 1, p. 170, 1947.
13 Scarr, R.W.A., 'Thermistors, their theory, manufacture and application', *Proc. IEE*, **107**, Pt. B, 35, pp. 395–409, Sept. 1960.
14 Hyde, F.J., Thermistors', *Proc. IEE*, **115**, 1, pp. 100–102, Jan. 1968.
15 Hyde, F.J., 'Thermistors', Butterworth, 1971.
16 Macklen, E.D., 'Thermistors', Electrotechnical Publications Ltd., 1979.
17 Harris, I.A., and Fowler, C.S., 'Some comparisons of power measuring standards at 300 and 400 Mc/s', *Proc. IEE*, **111**, 9, pp. 1597–1601, 1964.
18 'Microwave power measurement', Application Note 64, Hewlett-Packard, 1965.
19 Engen, G.F., 'A d.c.–r.f. substitution error in dual-element bolometer mounts', *IEEE Trans.*, **IM-13**, 2–3, pp. 58–64, 1964.
20 Aslan, E.E., 'A single bead broadband coaxial thermistor mount', *IEEE Trans.*, **IM-14**, 3, pp. 102–106, Sept. 1965.
21 Delahaye, F., and Geneves, G., 'Standard bolometric mounts on 75 ohm coaxial lines', *IEEE Trans.*, **IM-34**, 2, pp. 214–216, June 1985.
22 Aslan, E.E., 'Temperature-compensated microwatt power meter', *IRE Trans.*, **I-9**, 2, pp. 291-297, Sept. 1960.
23 Collard, J., Nicoll, G.R., and Lines, A.W., 'Discrepancies in the measurement of microwave power at wavelengths below 3 cm', *Proc. Phys. Soc. B*, LXIII, p. 215, 1950.
24 Johnson, S.A., 'Metallized-Glass Bolometers', NDRC Report 14-524, Polytechnic Institute of Brooklyn, Oct. 1945.
25 Harris, I.A., 'A coaxial film bolometer for the measurement of power in the UHF band', *Proc. IEE*, **107**, Pt. B, 31, pp. 67–72, Jan. 1960.
26 Norton, L.E., 'Broadband power measuring methods at microwave frequencies', *Proc. IRE*, **37**, 7, pp. 759–766, July 1949.
27 Selby, M.C., 'Voltage measurement at high and microwave frequencies in coaxial systems', *Proc. IEEE*, **55**, 6, pp. 877–882, June 1967.
28 Selby, M.C., 'Bolometric voltage and current (BOLOVAC) standard for high and microwave frequencies', *J. Res. NBS*, **72C**, 1, pp. 61–79, Jan.–Mar. 1968.
29 Selby, M.C., 'Bolovac application for HF and microwave power measurement and standardization', *J. Res. NBS*, **74C**, 3 and 4, pp. 123–133, July–Dec. 1970.
30 Selby, M.C., 'BOLOVAC systems for measuring electrical quantities from 0.5 MHz through microwaves', NBS Monograph 123, Jan. 1972.
31 'Bolovac reassessed', NBS Tech. Bull., p. 227, Sept. 1972.
32 Selby, M.C., 'Microwave two-port coaxial power and voltage mounts', *IEEE Trans.*, **IM-22**, 2, pp. 166–173, June 1973.

33 Lane, J.A., 'Transverse film bolometers for the measurement of power in rectangular waveguides', *Proc. IEE*, **105B**, 19, pp. 77–80, 1958.

34 Lane, J.A., and Evans, D.M., 'The design and performance of transverse-film bolometers in rectangular waveguides', *Proc. IEE*, **108B**, 37, pp. 133–135, Jan. 1961.

35 Lane, J.A., 'Microwave power measurement', Peter Peregrinus, 1972.

36 Inoue, T., Yokoshima, I., and Sasaki, M., 'High-performance thin-film barretter mount for power measurement in W-band', *Electron. Lett.*, **21**, 5, pp. 170–172, 28 Feb. 1985.

37 Hinton, L.J.T., and Burry, L.F., 'An enthrakometer for the band 26.0–40 Gc/s', EMI Electronics Ltd., Document CP 198, 1959.

38 Collard, J., 'The enthrakometer, an instrument for measurement in rectangular waveguides', *J. IEE*, **93**, Pt. IIIA, 9, pp. 1399–1402, 1946.

39 Nemoto, T., Fujisawa, K., and Inoue, T., 'Newly developed bolometer mounts for the short millimetre wave region', *IEE Trans.*, **IM-21**, 4, pp. 480–483, Nov. 1972.

40 Kerns, D.M., 'Determination of efficiency of microwave bolometer mounts from impedance data', *J. Res. NBS*, **42**, 579, RP1995, 1949.

41 Beatty, R.W., and Reggia, F., 'An approved method of measuring efficiencies of ultra-high frequency and microwave bolometer mounts', *J. Res. NBS*, **54**, pp. 321–327, RP2594, June 1955.

42 Lane, J.A., 'Measurements of efficiency of bolometer and thermistor mounts by impedance methods', *Proc. IEE*, **104**, Pt. B, 17, p. 485, Sept. 1957.

43 Engen, G.F., 'A bolometer mount efficiency measurement technique', *J. Res. NBS*, **65C**, pp. 113–124, April–June 1960.

44 Tamaru, T., 'A note on bolometer mount efficiency measurement technique by impedance method in Japan', *IEEE Trans.*, **MTT-14**, 9, p. 437, Sept. 1966.

45 Lane, J.A., 'Application of impedance measurements in an investigation of a waveguide thermistor mount', *Electron. Lett.*, **4**, pp. 186–188, 1968.

46 Bleaney, B., 'Power measurements by bolometer lamps at 3-cm wavelengths', Report CD Misc. 7, CVD Research Group, New Clarendon Lab., Oxford, England, Sept. 1942.

47 Feenberg, E., 'The frequency dependence of the power–resistance relation in hot wire wattmeters', Sperry Gyroscope Company, Report No. 5220-108, Mar. 1943.

48 Webber, H.E., 'Power measurement in the microwave region', Sperry Gyroscope Company, Report No. 5220-109, Mar. 1943.

49 Moreno, T., and Lundstrom, O.C., 'Microwave power measurement', *Proc. IRE*, **35**, 5, p. 514, May 1947.

50 Gainsborough, G.F., 'Some sources of error in microwave milliwattmeters', *J. IEE*, **95**, Pt. III, 36, pp. 229–238, July 1948.

51 Carlin, H.J., and Sucher, M., 'Accuracy of bolometric power measurements', *Proc. IRE*, **40**, 9, pp. 1042–1048, Sept. 1952.

52 Engen, G.F., 'Recent developments in the field of microwave power measurements at the National Bureau of Standards', *IRE Trans.*, **I-7**, 3–4, pp. 304–306, Dec. 1958.

53 Adams, J.W., and Desch, R.F., 'Experimental confirmation of barretter substitution error', *IEEE Trans.*, **MTT-16**, 3, pp. 201–202, Mar. 1968.

54 Jarvis, S., and Adams, J.W., 'Calculation of substitution error in barretters', *J. Res. NBS-C, Eng. and Instrum.*, **72C**, 2, pp. 127–137, June 1968.

55 Adams, J.W., 'Current distribution in barretters and its application to microwave power measurements', *IEEE Trans.*, **MTT-17**, 10, pp. 778–785, Oct. 1969.

56 Oldfield, L.C., Chap. 7, p. 109, in 'Microwave measurements', ed. A.E. Bailey, Peter Peregrinus, 1985.

57 Orford, G.R., and Abbott, N.P., 'Some recent measurements of linearity of thermistor power meters', IEE Colloquium on 'Microwave Measurement', London, May 1981, pp. 5/1-4.

58 Raff, S.J., and Sorger, G.U., 'A subtle error in RF power measurements', *IRE Trans.*, **I-9**, 1, pp. 284–291, Sept. 1960.

59 'NBS Type II Power Measurement System', *Microwave J.*, p. 70, Feb. 1977.
60 Aslan, E.E., 'Temperature-compensated microwatt power meter', *IRE Trans.*, **I-9**, 2, pp. 291–297, Sept. 1960.
61 Aslan, E.E., 'Accuracy of a temperature compensated r.f. power bridge', *IEEE Trans.*, **IM-18**, 3, pp. 232–236, Sept. 1969.
62 Engen, G.F., 'A self-balancing d.c. bolometer bridge', Joint URSI-IRE Meeting, National Bureau of Standards, Washington, D.C., May 1955.
63 Engen, G.F., 'A self-balancing direct-current bridge for accurate bolometric power measurements', *J. Res. NBS*, **59**, pp. 101–105, Aug. 1957.
64 Larsen, N.T., 'A new self-balancing d.c.-substitution r.f. power meter', *IEEE Trans.*, **IM-25**, 4, pp. 343–347, Dec. 1976.
65 Larsen, N.T., 'NBS Type IV rf power meter operation and maintenance', NBSIR (Interagency report) 77-866, Oct. 1977.

Chapter 7

Thermoelectric power meters

Thermoelectric power meters take several forms. At frequencies below 1 GHz instruments based on the traditional thermoelement may be used. Thermoelements are indirectly heated evacuated wire thermocouples which form the basis of many rf/dc substitution instruments (see Chapter 15). Their rf/dc substitution error is usually small below 30 MHz but becomes appreciable above this frequency, so that it is necessary to apply correction factors. Above 300 MHz these corrections become large as a consequence of resonances. Directly heated wire thermocouples are potentially capable of operation up to higher frequencies than thermoelements because of their comparative simplicity, but their development was cut short by the relative success of wire barretters, which are even simpler and can be made with smaller diameters if the Wollaston wire technique is used (Chapter 6). Consequently thermoelectric power meters did not become common for frequencies above 1 GHz until the development and application of thin film technology, which resulted in the appearance of commercial thin film instruments in the 1960s. A further advance was made in the 1970s, when instruments employing a combination of thin films and microelectronics techniques first appeared. Since that time thin film thermoelectric power meters have shown a steady increase in popularity, largely at the expense of the thermistor. Nevertheless they possess a number of disadvantages which have only partly been overcome. The main problems are that the thermocouple output voltage is not exactly proportional to the rf power level, which can lead to errors of several percent if not corrected for, and the sensitivity is dependent on ambient temperature. In addition, the variation of sensitivity with frequency is larger for thermoelectric devices than for thermistors. The difficulties are reduced by the use of microprocessor-based instruments which store and automatically apply correction factors, but such corrections are usually only approximate and, although they add greatly to the convenience of the instruments, the necessity for them makes precision measurements more difficult. Both indirectly heated and directly heated thin film thermoelectric power meters exist, but most instruments in common use are of the directly heated type. These cannot be used for rf/dc substitution measurements, as they incorporate a dc blocking capacitor which separates the input signal from the thermocouple output voltage. Instead, they often incorporate a reference oscillator which supplies a known rf power at a fixed frequency and which enables changes in sensitivity with time to be detected and corrected for.

Experimental work has been carried out on the use of hot carrier thermoelectric diodes in power meters (Section 7.6). These diodes, which aroused attention in the 1960s, are somewhat similar in construction to point contact rectifying diodes but operate thermoelectrically. They differ from conventional thermocouples in that only the charge carriers, not the crystal lattice, are heated and therefore the response is very rapid. They are not to be confused with Schottky barrier hot carrier diodes (Chapter 8).

7.1 Seebeck, Peltier and Thomson effects

The Seebeck effect, on which the action of a thermocouple depends, is one of three interrelated effects, the other two being the Peltier and Thomson effects. The latter are important in thermal rf/dc substitution instruments, because they modify the heat distribution when a dc input power is applied, and an outline of all three will be given here.

The Seebeck effect was discovered in 1821 [1,2]. In a circuit containing two dissimilar conductors an emf will arise if the two junctions between the dissimilar conductors are at different temperatures (Fig. 7.1). The emf is

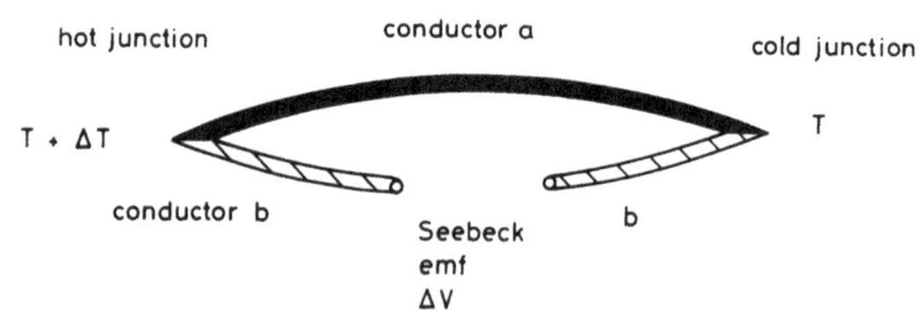

Fig. 7.1 *The Seebeck effect. A temperature difference ΔT between the two junctions formed by the dissimilar conductors a and b gives rise to an emf ΔV*

related to the temperature difference between the junctions by the differential Seebeck coefficient α_{ab} which is defined as

$$\alpha_{ab} = \lim_{\Delta T \to 0} \frac{\Delta V}{\Delta T} \tag{7.1}$$

where ΔV is the emf produced by a temperature difference ΔT. The differential Seebeck coefficient depends on the temperature and on the particular combination of materials employed. It is approximately 40 μV/deg C for a copper-constantan thermocouple at room temperature, whilst for bismuth–antimony (Bi–Sb) its value is around 110 μV/deg C. Thermocouples employing semiconductor materials can have higher differential Seebeck coefficients of over 1 mV/deg C.

The Peltier effect, discovered in 1834, occurs when a current passes through a junction formed by two dissimilar conductors. Heat is either absorbed or generated at the junction, depending on the direction of current flow. This heat is in addition to the normal Joule heating due to the resistance of the conductors. The Peltier coefficient Π_{ab} is defined as

$$\Pi_{ab} = \frac{Q}{I} \tag{7.2}$$

where Q is the rate of reversible heat generation due to current I.

It was realised by Thomson, later Lord Kelvin, that there must be a relation between Seebeck and Peltier effects. He also concluded from thermodynamical arguments that there is a third effect, the Thomson effect, which causes heating or cooling of a current-carrying conductor in which there is a temperature gradient. This effect is characterised by the Thomson coefficient γ, defined as

$$\gamma = \lim_{\Delta T \to 0} \frac{\Delta Q}{I \Delta T} \tag{7.3}$$

where ΔQ is the rate of reversible heat generation due to a temperature difference ΔT in the direction of the flow of the current I.

The three coefficients α_{ab}, Π_{ab} and γ are linked by the Kelvin relations, which can be derived by applying the law of conservation of energy and the second law of thermodynamics. They are:

$$\alpha_{ab} = \frac{\Pi_{ab}}{T} \tag{7.4}$$

$$\frac{d\alpha_{ab}}{dT} = \frac{\gamma_a - \gamma_b}{T} \tag{7.5}$$

The differential Seebeck coefficient of a thermocouple junction can be expressed as the difference between the absolute Seebeck coefficients, otherwise known as thermoelectric powers, of the two materials comprising the junction. The definition of thermoelectric power is derived from the second of the Kelvin relations (equation 7.5):

$$\frac{d\alpha}{dT} = \frac{\gamma}{T} \tag{7.6}$$

or

$$\alpha = \int_0^T \frac{\gamma}{T} dT \tag{7.7}$$

For semiconductors the thermoelectric power is a function of the carrier concentration and therefore depends on the electrical conductivity [3]. *P*-type semiconductor materials have positive thermoelectric powers whilst *n*-type materials have negative thermoelectric powers.

7.2 Thermoelements

A thermoelement consists of a metal heater wire and a single thermocouple junction which is electrically insulated from the heater (Fig. 7.2). The junction materials are usually copper and constantan and the heater is made of nickel chromium or of some other alloy with a low temperature coefficient of resistance. Both heater and thermocouple are mounted in an evacuated glass

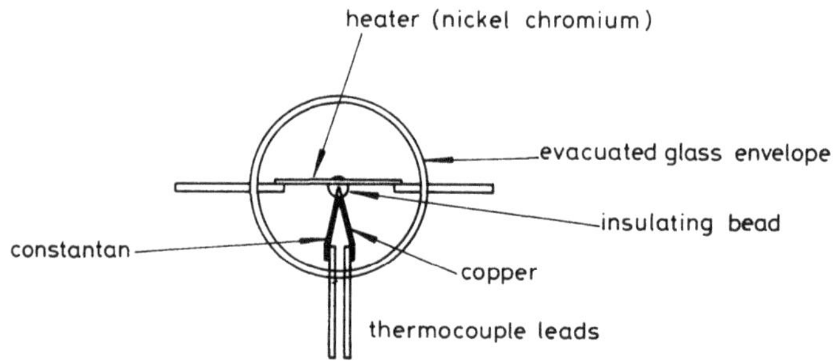

Fig. 7.2 *Construction of an ultra-high-frequency thermoelement*

envelope. The dc output voltage of a typical thermoelement is around 12 mV, with a response time of the order of 6 seconds from zero to full output. Corrections for the rf/dc substitution error are necessary above 30 MHz and the highest usable frequency is determined by the stray reactances, which produce resonances and lead to very large errors and difficulties in rf matching. A practical upper limit for thermoelement power meters is 300 MHz, although thermoelements are used up to 1 GHz for voltage and current measurements. Other disadvantages of thermoelements are that the heater is easily burnt out and that the dynamic range is limited to approximately 10 to 1 in power. Breakdown of the insulation is also a frequent cause of failure. For a dc input signal, the thermocouple output voltage usually suffers a change if the polarity of the input is reversed. This change, which is due to Peltier and Thomson heating, is known as the dc reversal error and it is eliminated by taking the mean of the readings for the two polarities. However, reversing the polarity does not eliminate second order errors due to the Peltier and Thomson effects, although for rf measurements the remaining errors are usually negligibly small compared with the total uncertainties. Further information on these errors and on thermoelements in general is given in Chapter 15 and ref. [4].

A commercially produced milliwatt test set using a thermoelement is described in ref. [5]. The instrument enables a standard 1 mW power level to be reproduced over a frequency range from 10 Hz to 100 MHz, or with degraded performance at frequencies up to 300 MHz. The equivalent rf circuit of the probe containing the thermoelement is shown in Fig. 7.3. The components to the right of the switch (a reed relay contact) represent the thermoelement itself, whilst the compensating circuit to the left of the switch

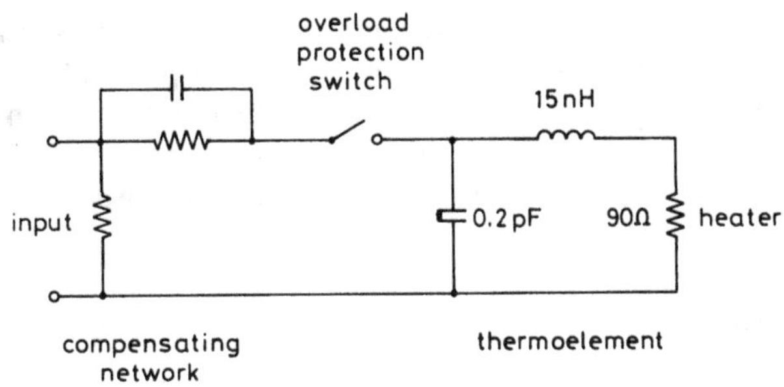

Fig. 7.3 *Equivalent circuit of a thermoelement type power sensor usable up to 300 MHz. The compensating network to the left of the switch provides the correct input impedance*

provides the correct input impedance (either 50 Ω or 75 Ω) over the operating bandwidth. The function of the switch is to provide protection against overload, a necessary precaution owing to the fact that the heater operates close to burn-out. The switch is controlled by a photo-transistor which senses the temperature of the heater and causes the switch contacts to open when the temperature exceeds a certain value. A feedback circuit supplies a reference current, of magnitude equal to the rf current, to a reference thermocouple. This reference current serves as a driver current for the panel meter, which indicates the measured level. A dc calibration source is also incorporated in the instrument. The voltage reflection coefficient of the probe is less than 0.025 up to 300 MHz. The rf/dc substitution error may be expressed as the effective efficiency or as the calibration factor, both of which are defined in the same way as for bolometers:

$$\text{effective efficiency} = \frac{\text{dc substituted power}}{\text{absorbed rf power}}$$

$$\text{calibration factor} = \frac{\text{dc substituted power}}{\text{incident rf power}} \tag{7.8}$$

It is possible for the effective efficiency of this type of sensor to be greater than 100 percent at some frequencies because of the frequency response of the resistive matching network.

Thermoelements have been used in feedthrough power meters. Ref. [6] describes a series of such instruments which were designed as power transfer standards. They operated at fixed frequencies in the range 10 to 1000 MHz and at power levels from 1 to 1000 W. Each unit consisted of a coaxial directional coupler and a 10 mA thermoelement, which was connected to the side arm of the coupler via a step attenuator. The resistance of the heater of the thermoelement was 25 Ω, which was matched to 50 Ω at a single frequency by a π network or, in the case of the highest frequency, by a stub tuner. The instrument was calibrated by separately measuring the power sensitivity of the thermoelement and the coupling factor of the coupler. The uncertainties totalled approximately 1 percent, of which the thermoelement sensitivity accounted for 0.5 percent. A series of small, portable, feed-through wattmeters was also constructed for general laboratory use.

7.3 Directly heated wire thermocouples

At first sight the relatively simple structure of a directly heated wire thermocouple would appear to make it attractive for use at the higher frequencies, where the stray reactances of thermoelements become troublesome. This possibility was investigated in the 1940s as a potential means of extending the range of thermoelectric instruments into the microwave region [7]. However, a number of problems were encountered and efforts in this direction were discontinued in favour of the Wollastone wire barretter (see Chapter 6). The nature of these problems can be seen most clearly from a comparison of the two devices. Firstly, a thermocouple wire cannot be made as thin as that of the barretter, since the use of the Wollaston wire technique is not practicable. Consequently the thermocouple resistance is often lower than would be desirable and broadband matching is made more difficult. A similar problem led to the demise of the tungsten filament barretter. Increasing the length of the wires does not help in this respect, because long wires lead to increased reactance and large substitution errors. Several alternative methods of increasing the resistance were investigated experimentally [7–10]. For example, a carbon filament or a metal coated glass bead was connected between the two thermocouple wires. Such ideas were not, however, pursued very far. An alternative approach of embedding the thermocouple in lossy material is described in ref. [11].

The second important difference between the directly heated thermocouple and the barretter is the way in which the substitution error depends on the current distributions for the rf and dc (or audio frequency) signals. The thermocouple senses the temperature ideally at a single point (the junction), whereas the barretter responds to the average temperature along the length of the wire. Consequently the behaviour of the two devices is very different. Gainsborough [12] published an approximate analysis of the two cases. If one considers a barretter cooled only by convection – that is, no heat is lost by conduction through the ends of the wire – then the substitution error can be shown to be zero if the heat loss from any point on the surface is proportional to the temperature rise at that point. The thermocouple, on the other hand, has a greater substitution error under such conditions than if it is cooled only by conduction through its ends. An additional complication from which the thermocouple suffers is the fact that the two halves are made from different materials. The difference in resistivity and skin depth in the two wires causes the ratio of their rf resistances to differ from the ratio of their dc resistances, with the result that the absorbed rf and dc powers divide in different proportions between the two halves. This gives rise to an additional substitution error unless the cooling is entirely by conduction through the ends of the wires. Thus the conclusion is reached that the directly heated wire thermocouple is inferior to the Wollaston wire barretter as regards substitution error.

7.4 Thin film thermoelectric power meters

7.4.1 General properties

Attractive features of thin film instruments are the potentially very wide dynamic range and good power handling capacity. The wide dynamic range, which can be 40 or 50 dB, results from the inherently low zero drift of thermoelectric devices, whilst the power handling capacity is due to the ability of thin film loads to dissipate up to several hundred milliwatts. By way of comparison, a typical temperature-compensated thermistor power meter is limited to about 30 dB in dynamic range and has a maximum rf input power of 10 mW. The low zero drift is of particular advantage in situations where the instrument cannot be zeroed immediately before the measurement. Both the indirectly and directly heated types have been the subject of an appreciable amount of development work. Indirectly heated instruments can be used for rf/dc substitution measurements, although thin film loads are not ideal for this purpose because their relatively large surface area permits differences in the rf and dc current distributions to occur unless the amount is very carefully designed. This makes the validity of the substitution process questionable and such instruments would normally be calibrated by determining the effective efficiency against a calorimeter. Substitution errors can also be caused by dissipation of power directly in the thermocouple instead of in the thin film load. Relatively little effort has been devoted, however, to an examination of the substitution process in thermoelectric instruments. Most commercial instruments cannot be used for rf/dc substitution measurements, because they are of the directly heated type and the dc blocking capacitor prevents application of a dc input to the sensor. Many instruments incorporate an rf oscillator giving a known output power, typically 1 mW at 50 MHz, by means of which variations in sensitivity with temperature or due to ageing can be calibrated out. This is done under automatic control in microprocessor-based instruments. A further problem in directly heated instruments is the dependence of the rf input impedance on temperature and power level, which arises because of the difficulty of choosing materials which combine good thermoelectric properties with low temperature coefficients of resistance. It is also necessary to correct for the non-linear relation between the thermocouple output voltage and the rf power, unless the power is restricted to low levels.

7.4.2 Indirectly heated instruments

A series of indirectly heated waveguide instruments was developed in the 1950s by Lane and others [13–15]. The absorbing resistor was a nichrome film having a resistance in the range 400 to 500 Ω and mounted transversely in a short circuited length of waveguide (Fig. 7.4). Nichrome has a low

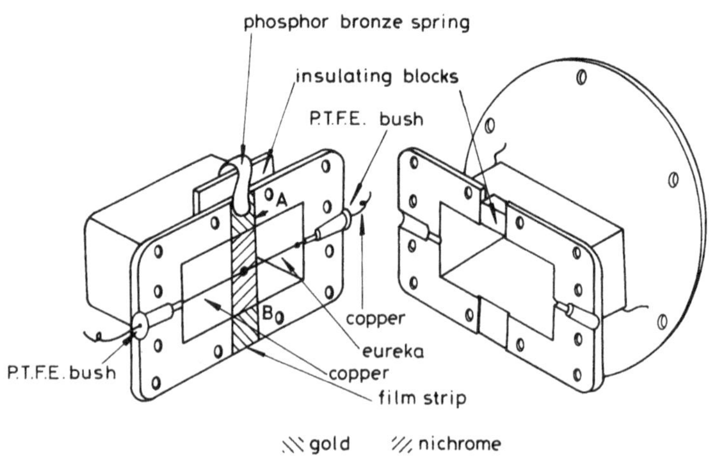

Fig. 7.4 *Indirectly heated waveguide thermoelectric power meter using transverse thin film load. The thermocouple leads are perpendicular to the electric field*

temperature coefficient of resistance and tests showed that a temperature change from 15°C to 70°C produced a change of less than 1 percent in the film resistance. One end of the nichrome film was insulated from the waveguide in order to provide an input for the substituted dc power. During the course of this work an investigation was carried out to determine the correct value of waveguide characteristic impedance Z_0 for use in the equivalent circuit. This was necessary because the characteristic impedance of a waveguide can be defined in more than one way. It was discovered that the value giving agreement with experimental results was

$$Z_0 = 2\left(\frac{b}{a}\right)z\left(\frac{\lambda_g}{\lambda}\right) \tag{7.9}$$

where a is the broad dimension, b the narrow dimension, λ_g the guide wavelength, λ the free space wavelength and z is the impedance of free space [16]. The thermocouple was of the metal wire type, consisting of a single copper Eureka junction attached to the centre of the thin film load. The thermocouple wires were arranged at right angles to the electric field in order to avoid direct pick-up. The sensitivity of this particular instrument was in the region of 10–15 μV/mW, which is not nearly as high as can be achieved with semiconductor thermocouples. Consequently only powers in excess of 1 mW could be measured with reasonable accuracy. The same basic design was also used for a series of thin film bolometers, with the exception that in this case platinum was used as the thin film material (see Chapter 6). Commercial versions of these instruments were produced.

Coaxial thin film thermoelectric power meters of the indirectly heated type present an additional problem not encountered in waveguide versions, because the thermocouple cannot be arranged to be everywhere perpendicular to the electric field. This absorption of power directly in the thermocouple is inevitable and this can only be minimised by making the thermocouple of high resistance. Work on coaxial instruments was carried out by Barlow and others [17–19], using bismuth–antimony telluride and later p-type and n-type germanium. Germanium gave a higher differential Seebeck coefficient of 1.3 mV/deg C and its higher melting point also resulted in a wider dynamic range. The substrate was plastic-resin of 15 μm thickness. Two nichrome 100 Ω film resistors were deposited on one side of the substrate, whilst two germanium thermocouples were evaporated onto the other side.

7.4.3 Directly heated instruments

Directly heated thin film thermoelectric power meters have been manufactured commercially in coaxial line and waveguide, using both transversely mounted and longitudinally mounted thermocouples. Refs. [20–23] describe the transverse type employing bismuth–antimony thin film thermocouples. Differences in temperature between the inner and outer conductors of the coaxial line can cause zero drifts, but this can be balanced out in the thermal design as described by Aslan in ref. [24]. Fig. 7.5 shows a longitudinally mounted thermocouple waveguide power meter, also using bismuth-antimony [25,26]. The advantage of the longitudinal thermocouple arrangement is that a better match to the rf signal can be obtained. A slit in the thermocouple film concentrates the absorbed power in the region of the junction. Commercial versions based on this design have sensitivities of at least 60 μV/mW at frequencies up to 260 GHz. Above 110 GHz the thermocouple is mounted in

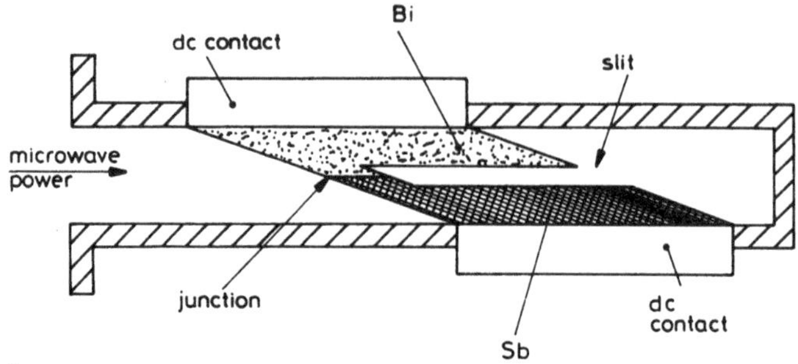

Fig. 7.5 *Directly heated waveguide thermoelectric power meter employing longitudinal film. The slit concentrates the absorbed power in the region of the junction*

square waveguide with a suitable transition to rectangular waveguide. The extra waveguide height results in increased sensitivity [26].

7.5 Application of microelectronics techniques

The application of microelectronics techniques to directly heated thin film coaxial thermoelectric sensors is described in references [27–30] (see Figs 7.6 and 7.7). The two thermocouple materials are tantalum nitride (Ta_2N) and *n*-type silicon. The tantalum nitride film provides a stable, low temperature coefficient load for the rf power, whilst the *n*-type silicon provides most of the thermoelectric power. The doping of the silicon is controlled in such a way as to achieve a compromise between high thermoelectric power and low resistivity, since the silicon is not intended to absorb a significant proportion of the rf power. The design figure for the thermoelectric power of the

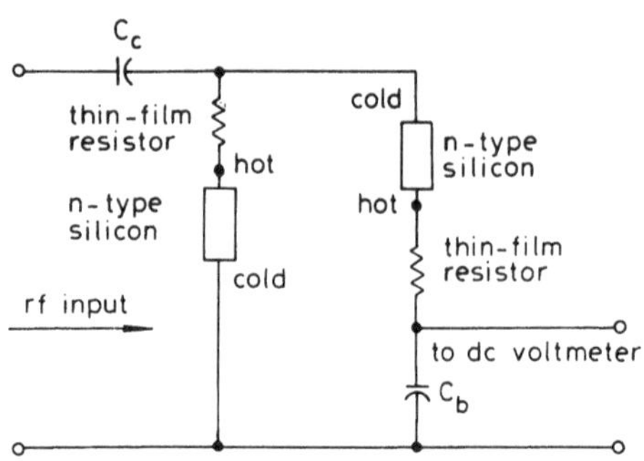

Fig. 7.6 *Schematic diagram of directly heated thermoelectric power sensor using microelectronics techniques. The thin film resistor absorbs most of the rf or microwave power and the n-type silicon provides most of the thermoelectric emf*

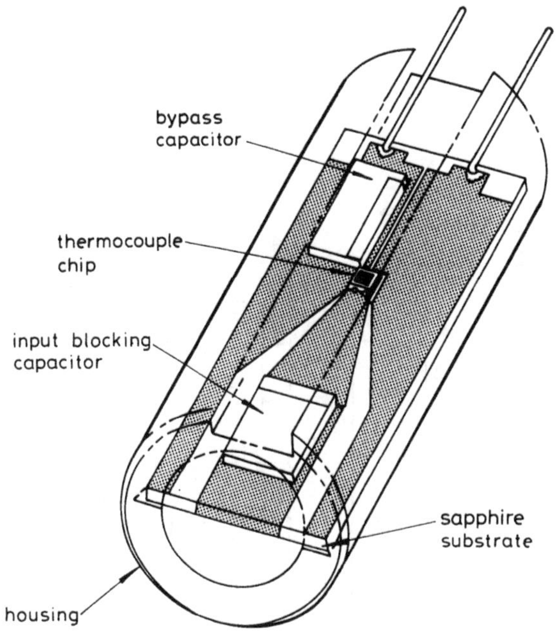

Fig. 7.7 *Sketch of thermoelectric power sensor based on the circuit of Fig. 7.6*

thermocouple is 250 μV/deg C, giving a substantial improvement over the 110 μV/deg C of a bismuth–antimony thermocouple. The thermal resistance of the element is 0.4 deg C/mW, resulting in a power sensitivity of 100 μV/mW. Two elements are used in a single mount to give a theoretical overall sensitivity of 200 μV/mW, although thermal interaction between the elements reduces this to 160 μV/mW. The whole element is manufactured on a silicon chip 0.8 mm square and the two elements are mounted on a coplanar waveguide structure on a sapphire substrate. The thermocouple output voltage is amplified by a chopper amplifier mounted in the sensor head. At the high power end of the range there is considerable departure from linearity (10 percent at 100 mW), and it is necessary to compensate electronically both for this and for the temperature dependence of the sensitivity. A 50 MHz 1 mW reference oscillator is built into the instrument. Sensors have been built for frequencies up to 26 GHz in coaxial line and up to 50 GHz in waveguide. The waveguide versions have separate inputs for the microwave power and the 50 MHz calibrating signal. Another manufacturer [31–33] has produced a similar instrument in which the sensor is characterised not only by its calibration factor and effective efficiency, but also by a linearity factor which varies from sensor to sensor. A resolution of 0.001 μW can be obtained by averaging, although this slows the response. Fig. 7.8 shows a simplified schematic of the electronic circuitry.

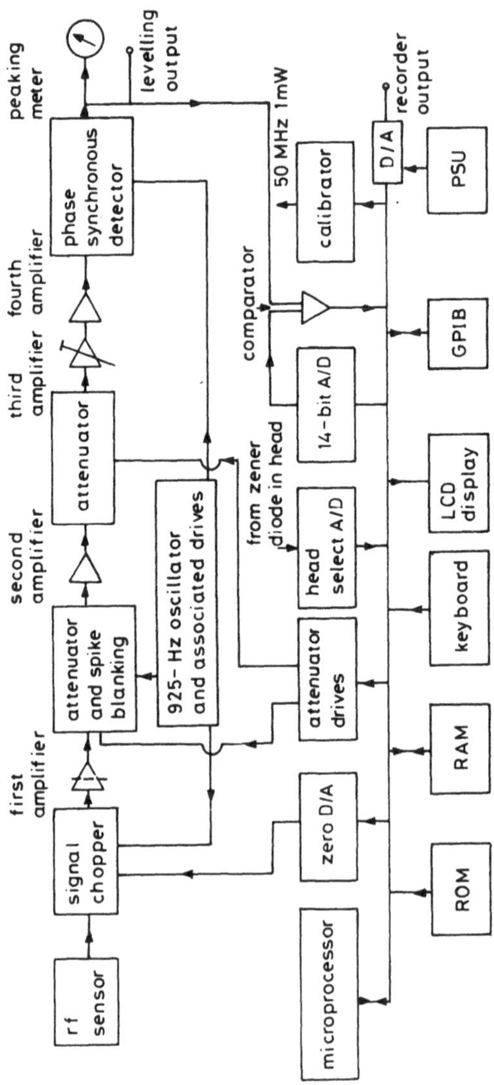

Fig. 7.8 *Simplified block diagram of a commercial power meter employing a thermoelectric sensor*

7.6 Hot carrier thermoelectric diodes

The hot carrier thermoelectric diode [34] (Fig. 7.9) is a thermoelectric device with a construction resembling that of a point contact diode. It has two near-ohmic contacts and relies for its operation on the asymmetry due to the large difference in the areas of these contacts. The semiconductor is usually p-type germanium with a resistivity in the range 0.03 to 35 Ω cm. The whisker

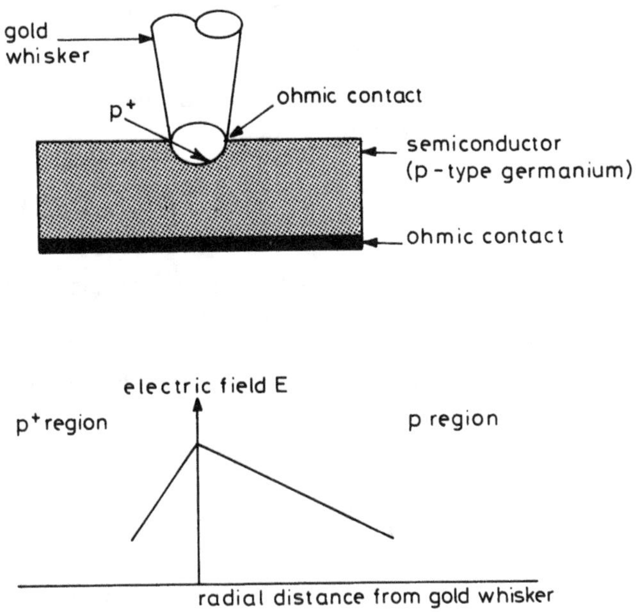

Fig. 7.9 *Hot carrier thermoelectric diode and corresponding distribution of electric field E with radial distance from the gold whisker. Maximum electric field occurs at the p^+-p junction. Carriers in this region are heated without an appreciable rise in temperature of the crystal lattice*

is a gallium doped gold wire of diameter 0.1 to 0.3 mm which is pulse bonded to the semiconductor to form a hemispherical metal–p^+–p junction. As a result of the geometry, the rf electric field is concentrated mainly in the region around the point contact and it has its maximum value at the p^+–p junction. The carriers on either side of this junction are heated by the field without any appreciable rise in the temperature of the crystal lattice. The effective temperature T_e reached by the carriers is given by the formula [35]:

$$\frac{T_e}{T_0} = \tfrac{1}{2}[1 + \{1 + 8\beta E^2\}^{\frac{1}{2}}] \tag{7.10}$$

where E is the electric field, T_0 the lattice temperature and β depends on the mobility of the carriers. From equation 7.10 the rise in temperature is therefore

$$T_e - T_0 = \frac{T_0}{2}[\{1 + 8\beta E^2\}^{\frac{1}{2}} - 1] \tag{7.11}$$

Expressing the field E in terms of the diode voltage V and the radius a of the point contact enables equation 7.11 to be re-written as:

$$T_e - T_0 = \frac{T_0}{2}\left[\left(1 + \frac{8\beta V^2}{a^2}\right)^{\frac{1}{2}} - 1\right] \tag{7.12}$$

The thermoelectric power Q of the semiconductor material depends on the carrier concentration and is given by:

$$Q = \frac{k}{e}\left[(\tfrac{5}{2} - S) - \ln\left(\frac{p}{N_v}\right)\right] \tag{7.13}$$

where p is the carrier concentration, k the Boltzman constant, e the charge of the carriers, N_v the effective density of states in the valence band and S is a constant which depends on the scattering process involved. Therefore the p^+ and p regions have different values of Q and this gives rise to a thermoelectric voltage V_{th} equal to

$$\begin{aligned} V_{th} &= (Q - Q^+)(T_e - T_0) \\ &= \frac{k}{e}\ln\left(\frac{p^+}{p}\right)(T_e - T_0) \end{aligned} \tag{7.14}$$

The sense of V_{th} is such that the sign of the dc potential appearing on the metal whisker is opposite to that of the carrier. This is the reverse of the rectifying junction. The radius a can be controlled during fabrication of the diode by a suitable choice of voltage during the bonding operation. Equation 7.12 shows that for a fixed rf voltage a small value of a results in a large temperature rise. A small value of radius a also increases the diode resistance and reduces the absorbed rf power for a given voltage. The cut-off frequency of the thermoelectric diode, if parasitic reactances are excluded, is well above the microwave region, making the device attractive as a potential wide band power meter or voltmeter. The voltage responsivity (output voltage/rf input power) is greater than for conventional thermoelectric sensors but less than that of a Schottky barrier diode, unless the resistance is made at least 100 kΩ [36]. Equation 7.12 shows that the temperature rise is proportional to V^2 for low voltages and V for higher levels.

Experimental voltmeters and power meters [35,37–41] have been built using thermoelectric diodes. Ref. [35] describes voltage measurements in stripline. Several diodes with resistance between 100 Ω and 10 kΩ were tested. The rf voltage required for an output of 1.5 μV was between 2 mV and 30 mV, depending on diode resistance and the resistivity of the material. The measurements were carried out at a frequency of 9.5 GHz. Ref. [41] describes an experimental power meter which used a diode with a resistance of 50 Ω and operated over the frequency range 1 to 12 GHz. It was found that optimum noise performance was obtained by using a low value of resistivity equal to 0.03 Ω cm. The voltage responsivity was 20 V/W. Measurements of response speed showed that the rise time was no greater than 20 ns, a figure which is roughly comparable with conventional microwave detector diodes and less than one millionth of the thermal time constant.

7.7 References

1 Goldsmid, H.J., 'Applications of the thermoelectricity', Methuen and Co, London, 1960.

2 Kinzie, P.A., 'Thermocouple temperature measurement', Wiley, 1973.

3 Johnson, V.A., 'Theory of the Seebeck effect in semiconductors', *Progress in Semiconductors*, Heywood, 1, 65, 1956.

4 Inglis, B.D., 'AC-DC transfer standards–present status and future directions', *IEEE Trans.*, **IM-34**, 2, pp. 285–290, June 1985.

5 Schuon, E., 'Milliwatt test set 10 Hz to 100 MHz', *IEEE Trans.*, **IM-20**, 4, pp. 338–341, Nov. 1971.

6 Hudson, P.A., 'A precision RF power transfer standard', *IRE Trans.*, **I-9**, 2, pp. 280–283, Sept. 1960.

7 Griesheimer, R.N., 'Microwave power measurements', Chapter 3 in 'Technique of microwave measurements', ed. C.G. Montgomery, McGraw-Hill, 1947.

8 Becker, J.A., and Shive, J.N., 'Thermocouples for use as high frequency wattmeters', BTL Report MM-43-110-4, Mar. 1943.

9 Shive, J.N., 'Temperature coefficients of sensitivity and resistance, overload and burnout data for several types of ultra-high frequency thermocouples', BTL Report MM-43-110-18, June 1943.

10 Carlin, J.J., 'Measurement of power', Vol. 1, Chap. 3 in 'Handbook of microwave measurements', ed. M. Sucher and J. Fox, Polytechnic Press, 1963.

11 Burry, L.F., 'A thermocouple milliwattmeter for the frequency band 26.5 Gc/s to 40 Gc/s', Report No. DMP 1164, Feb. 1962.

12 Gainsborough, G.F., 'Some sources of error in microwave milliwattmeters', *J. IEE*, **95**, Pt. III, 36, pp. 229–238, July 1948.

13 Lane, J.A., 'Transverse film bolometers for the measurement of power in rectangular waveguides', *Proc. IEE*, **105B**, 19, pp. 77–80, 1958.

14 Lemco, I., and Rogal, B., 'Resistive film milliwattmeters for the frequency bands 8.2–12.4 Gc/s, 12.4–18 Gc/s, and 26.5–40 Gc/s', *Proc. IEE*, **107**, Pt. B, 35, pp. 427–430, Sept. 1960.

15 Lane, J.A., and Evans, D.M., 'The design and performance of transverse-film bolometers in rectangular waveguides', *Proc. IEE*, **108B**, 37, pp. 133–135, Jan. 1961.

16 Lane, J.A., 'Microwave power measurement', Peter Peregrinus, 1972.

17 Tajima, I., Fukuda, Y., Koike, R., and Iijima, Y., 'The design of thin-film thermoelements for high-frequency-power measurements', *Trans. Inst. Electron. Commun. Eng. Jap. A, B, C*, **53**, 7 pp. 24–25, July 1970 (reproduced in *Electron, Commun. Jap.*).

18 Barlow, H.E.M., Fukuda, Y., Koike, R., and Tajima, I., 'Thin film thermoelectric transducer for accurate measurements of power between 0 and 6 GHz', *Proc. IEE*, **118**, 3/4, pp. 626–629, Mar./April 1971.

19 Koike, R., and Tajima, I., 'Thin film thermoelectric transducer for measurements of low level power between 0 and 11 GHz', *Proc. IEE*, **121**, 1, pp. 28–32, Jan. 1974.

20 Luskow, A.A., 'R.F. power meter uses thin film thermocouple', *Marconi Instrum.*, **12**, 6, p. 26, June 1969.

21 Luskow, A.A., 'This microwave power meter is no drifter', *Marconi Instrum.*, **13**, 2, pp. 30–34, Aug. 1971.

22 Higham, P., and Luskow, A.A., 'Tft powermeter goes automatic', *Marconi Instrum.*, **13**, 6, pp. 145–147, Dec. 1972.

23 Luskow, A.A., 'Microwave powermeter for the military environment', *Marconi Instrum.*, **15**, 6, pp. 128–133, Autumn 1977.

24 Aslan, E.E., 'Design of low zero drift thermocouple elements', *IEEE Trans.*, **IM-20**, 1, pp. 16–19, Feb. 1971.

25 Toda, H., Sasaki, K., Nakagawa, Y., and Sugiura, I., 'A matched-load type thermoelectric transducer for power measurements in the millimeter wave region', *IEEE Trans.*, **IM-23**, 4, pp. 408–413, Dec. 1974.

26 Sugiura, I., and Toda, H., 'Thermoelectric transducers for short millimetre wave power measurement', *Int. Microwave Symp.*, 1978, pp. 165–167.

27 Jackson, W.H., 'A thin film/semiconductor thermocouple for microwave power measurements', *Hewlett-Packard J.*, **26**, pp. 16–18, Sept. 1974.

28 Lamy, J.C., 'Microelectronics enhances thermocouple power measurements', *Hewlett-Packard J.*, **26**, pp. 19–24, Sept. 1974.

29 Edwards, A.P., 'Digital power meter offers improved accuracy, hands-off operation, systems compatibility', *Hewlett-Packard J.*, **27**, pp. 2–7, Oct. 1975.

30 Pratt, R., 'Dual sensor power meter simplifies coupler power measurements', *Microwave J.*, **26**, 11, p. 141, Nov. 1983.

31 'Speed and accuracy with microprocessor instrumentation', *Microwave J.*, **26**, 4, pp. 143–144, April 1983.

32 McAllister, P., 'Accuracy improvements in power measurements', Digest, *13th European Microwave Conf.*, Nuremberg, Sept. 1983, pp. 629–634.

33 Browne, J., 'RF power meters: Measurement workhorses', *Microwaves and RF*, **23**, 1, pp. 65–70, 92, 143, Jan. 1984.

34 Harrison, R.I., and Zucker, J., 'Hot carrier microwave detector', *Proc. IEEE*, **54**, 4, pp. 588–595, April 1966.

35 Kikuchi, K., and Oshimoto, A., 'Broad band microwave voltmeter using thermoelectric effects of hot carriers', *IEEE Trans.*, **IM-26**, 4, pp. 323–328, Dec. 1977.

36 Cowley, A.M., and Sorenson, H.O., 'Quantitative comparison of solid state microwave detectors', *IEEE Trans.*, **MTT-14**, 12, pp. 588–602, Dec. 1966.

37 Oshimoto, A., and Kikuchi, K., 'Microwave power using hot electron', *Proc. European Microwave Conf.*, Vol. 2, paper B.15.4, 1973.

38 Oshimoto, A., and Kikuchi, K., 'Microwave power meter using ohmic hot diode', *Conv. Rec. IECE*, Japan, No. 845, 1973.

39 Svetlichnyi, V.M., Plaksii, V.T., and Satyukov, A.I., 'Sensitivity of thermoelectric point-contact microwave power indicators', *Izv. VUZ Radioelektron.*, **18**, 2, pp. 79–82, Feb. 1975 (in Russian).

40 Kikuchi, K., 'Performance of millimeter wave detector using hot carriers in semiconductor', *Trans. IECEJ*, **J64-C**, 10, pp. 658–665, Oct. 1981.

41 Kikuchi, K., and Oshimoto, A., 'Broad band high speed microwave power sensor using hot carriers in semiconductor', *IEEE Trans.*, **IM-30**, 4, pp. 239–242, Dec. 1981.

Chapter 8

Diode power meters

Diode power meters are essentially semiconductor diode detectors operating in their square law region. They are capable of measuring small powers – less than a nanowatt – and possess low zero drift coupled with an intrinsically fast response. In many respects, however, they compare unfavorably with other common types of power meter. Historically the main problems have been poor stability and reliability, temperature-dependence of the diode parameters, and difficulties in obtaining a flat frequency response. Until the 1960s the behaviour of diodes which could function in the microwave range was considered too poor for serious use in power measurement applications, their main uses being mixing and detection. New manufacturing techniques [1,2] resulted in devices of greater stability and ruggedness, while improved circuit construction enabled a flatter frequency response to be achieved. These advances made possible the use of semiconductor diodes for commercial coaxial power meters operating up to 18 GHz and beyond [3]. Such instruments became common in the 1970s. Nevertheless, the detector diode is still far from an ideal device for power measurement. In particular the sensitivity is inherently temperature dependent, the square law response is followed only at low power levels, and the rf impedance of the diode varies with both temperature and signal level, which creates problems in rf matching. Theoretically the sensitivity (measured by the voltage or current responsivity and expressed in volts per watt or amps per watt respectively) is calculable, but in practice the performance does not agree well with theory and calibration is therefore necessary.

The diodes used in rf and microwave power meters are of the metal–semiconductor type [4], which first came into common use shortly after 1900 [5–7]. In the very early devices the rectifying contact was made by pressing a pointed metal wire (the cat's whisker) against the surface of a crystal of silicon or lead sulphide (galena). A contact made in this way is very fragile, tending to change its characteristics when subjected to mechanical shock, and it was not long before such point contact metal–semiconductor detectors were replaced for most purposes by thermionic valves. From 1927 onwards metal–semiconductor rectifiers based on copper oxide and later selenium became available and these were used extensively both for power frequency rectification and in ac measuring instruments. They were more reliable than the point contact diodes but the usable frequency range was limited to below 1 MHz by their high capacitance. When around 1939 a low-level detector was required for microwaves, silicon point contact diodes came into their own again for lack of a suitable alternative [8,9]. Fig. 8.1 shows such a diode mounted in a common type of cartridge which dates from this period. Microwave point contact diodes were not however regarded as reliable power or voltage measuring devices and efforts were made to extend the use of thermionic diodes into the microwave region [10–12]. It seemed possible at that time that the thermionic devices would eventually replace point contact semiconductor diodes and, with this aim in mind, some miniature vacuum tubes were produced with dimensions similar to crystal diode cartridges [13]. The advantages which thermionic diodes offered as against point contact diodes were greater stability, smaller temperature dependence, and better ability to withstand overloads, although their sensitivity was much lower. In the event the semiconductor diode outlived the thermionic devices.

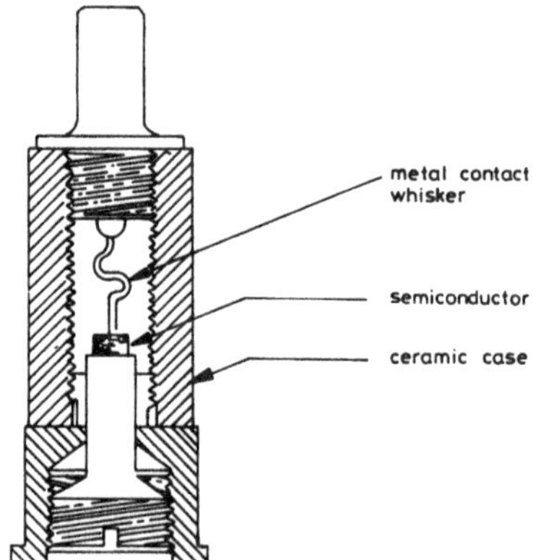

Fig. 8.1 *Microwave point contact diode mounted in a common type of cartridge*

The new manufacturing techniques introduced in the 1960s were the fabrication of metal–semiconductor diodes by evaporation or sputtering of the metal on to a thin epitaxial layer of *n*-type silicon or GaAs. Fig. 8.2 shows the construction. Diodes made in this way are known as planar Schottky barrier diodes or sometimes as hot carrier diodes (not to be confused with the thermoelectric hot carrier diodes discussed in Chapter 7). The diodes made by the new methods were more predictable then point contact diodes, were more stable, and had lower 1/f noise by about 15 dB [14]. It was also possible to control the zero bias resistance of the diode by controlling the barrier height.

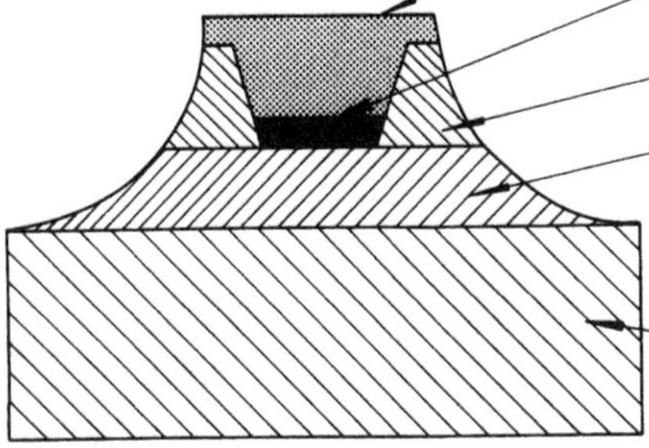

Fig. 8.2 *Structure of photometallurgically built low barrier Schottky diode. This type of diode is more stable than point contact diodes and also has better performance*

For low level measurements a low barrier height is required; otherwise the diode resistance is too high for use without forward bias, which creates additional noise. Fig. 8.3 shows a typical current-voltage characteristic of a planar low barrier height Schottky barrier diode.

The widespread adoption of metal–semiconductor diodes at high frequencies can be understood by considering the limitations of some other types of diode [1]. The ordinary *p–n* junction diode is restricted mainly to frequencies below 1 GHz because of the charge storage phenomenon, which reduces the rectification efficiency at high frequencies. This effect arises because, when an applied voltage is suddenly switched from forward to reverse, the carriers which have just crossed the junction are swept back again before recombination occurs. A reverse current then flows for a short time. A second limitation of the *p–n* junction is that the barrier height is normally too large for efficient low level detection without forward bias.

The backward diode, which is related to the tunnel diode, is made by increasing the doping of a *p–n* junction until some of the energy levels in the conduction band on one side of the junction overlap those in the valence band on the other side (Fig. 8.4). Under such circumstances electrons can tunnel through the junction without a change in energy. The tunnelling current is largest under what for a conventional diode would be the reverse bias condition and it disappears above a certain forward voltage in the conventional sense. Thus the direction of easy current flow is the opposite of that which occurs in a normal *p–n*. The backward diode is essentially a low level device, because the voltages must be kept well below the value at which conventional *p–n* conduction begins (typically 500 mV). The junction impedance is also comparatively low, which facilitates matching. Charge storage is not a problem because the mechanism does not rely on minority carriers. Consequently the backward diode makes a useful detector at microwave frequencies [15–18],

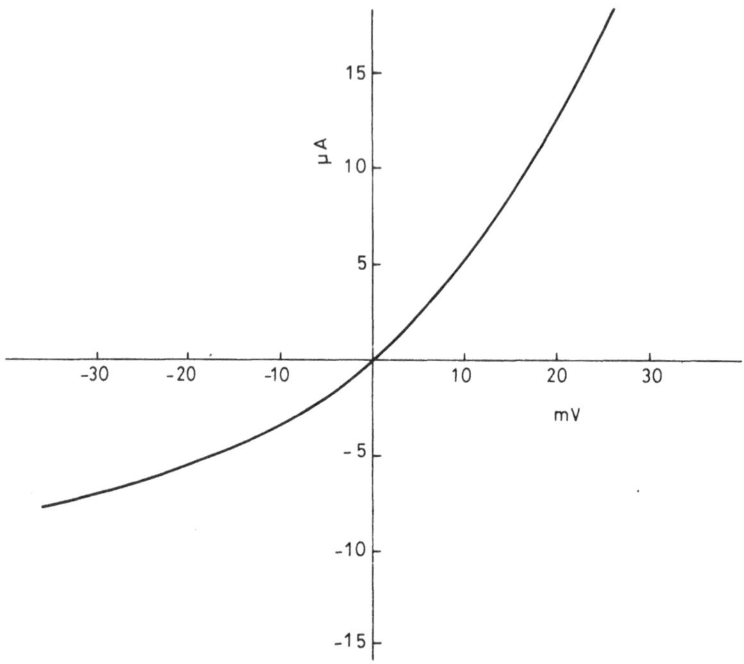

Fig. 8.3 *Square law portion of the current-voltage characteristic for a typical low-barrier Schottky diode*

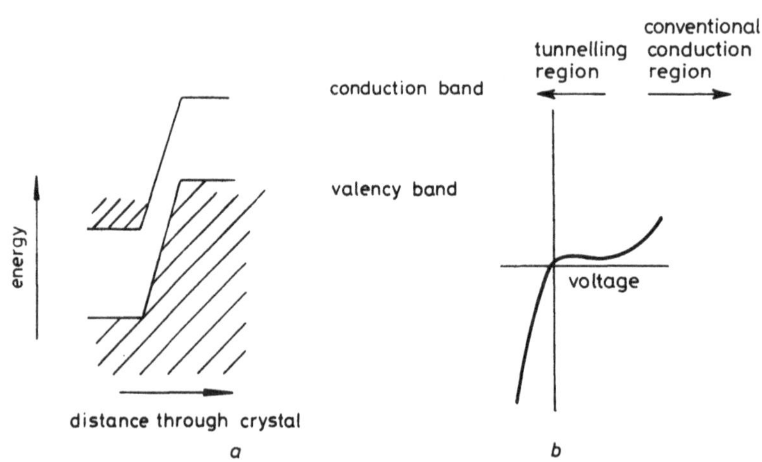

Fig. 8.4 *Energy bands for backward diode and corresponding current-voltage characteristic. Conduction is due to electrons tunnelling between the valence band on one side of the junction and the conduction band on the other. Backward diodes have rather low rectification efficiency and are limited to low level signals*

although it has not been used as widely as was thought likely at the time of its introduction in the 1960s. The current–voltage characteristic is relatively insensitive to temperature, since the tunnelling current depends mainly on the number of states available for the electrons to tunnel into. A disadvantage is that the characteristic is not very sharp and therefore the rectification efficiency is rather low.

The hot carrier thermoelectric diode (see Chapter 7) has been the subject of some experimental work for both power and voltage measurement. In this diode the charge carriers are heated directly by the rf electric field while the crystal lattice itself remains cool. The upper frequency limit lies well above the microwave range but the diode is much less sensitive than rectifying junctions unless the impedance is made extremely high.

8.1 Physics of Schottky barrier diodes

When an *n*-type semiconductor is brought into contact with a metal having a Fermi level lower than that of the semiconductor, electrons diffuse from the conduction band of the semiconductor into the metal. Some electrons also diffuse from the metal into the semiconductor, but because of the lower Fermi level of the metal there are a smaller number of these with a sufficiently high energy level. Consequently the net flow will be from semiconductor to metal. Under zero bias conditions a depletion layer will form in the semiconductor, creating a potential barrier for electrons arriving from the semiconductor side of the contact and thus reducing the net flow of electrons to zero (Fig. 8.5*a*). Application of an external voltage to the diode alters the height of the potential barrier and upsets the equilibrium. A forward bias voltage will reduce the width of the depletion layer and the height of the potential barrier, enabling a relatively large diffusion flow of electrons from semiconductor to metal (Fig. 8.5*b*). Application of a reverse bias voltage will increase the width of the depletion layer and with it the height of the potential barrier, thus suppressing the flow of electrons from semiconductor to metal (Fig. 8.5*c*). In

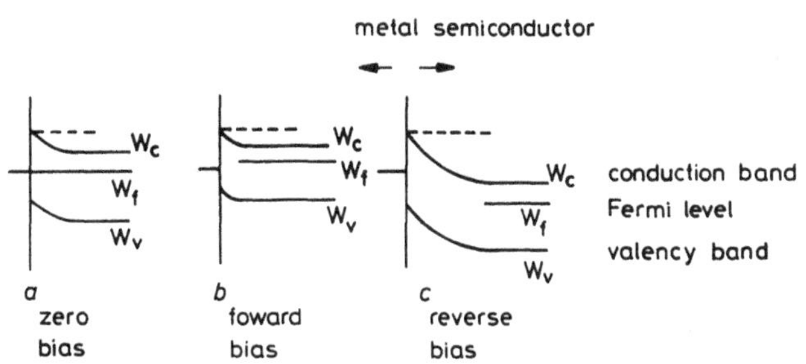

Fig. 8.5 *Energy bands for Schottky barrier diode corresponding to three different bias conditions*

contrast, the barrier height seen by electrons arriving from the metal side of the contact is independent of the external bias voltage, being equal to the difference in the Fermi level in the metal and the bottom of the conduction band in the semiconductor. Thus the contribution to the current arising from the flow of electrons from metal to semiconductor always has the same value, this value being known as the saturation current, denoted by I_s.

The barrier height as seen from wither side of the contact can be controlled by choosing a metal which has a suitable Fermi level, but it is also very dependent on surface effects at the metal-semiconductor interface. The presence of surface states results in an accumulation of charge at the interface, which reduces the barrier height. This reduction in barrier height gives rise in turn to an increase in the saturation current and a decrease in the width of the depletion layer at zero bias. The effect is usually desirable, since a relatively large saturation current of the order of 10 μA is required for appreciable current flow at low voltage. The current–voltage relation is:

$$I = I_s\,(e^{(eV/kTn)} - 1) \tag{8.1}$$

where V is the voltage, I the current, I_s the saturation current, k Boltzman's constant, T the absolute temperature, e the charge of the electron, and n is a dimensionless constant known as the ideality factor. Simple theory suggests that n should be unity, but more refined theories and practical measurements give values greater than one. For a good planar Schottky diode a typical value of n is 1.1, whilst for a point contact diode its value is nearly 2. The precise value is dependent on the diode manufacturing process and varies with batch. At room temperatures e/kT has a value of about 40 V^{-1}.

Equation 8.1 is valid until the externally applied forward voltage approaches the zero bias potential difference across the depletion layer, after which the depletion layer disappears and the resistance of the semiconductor material determines the current. Charge storage is not a problem in Schottky barrier diodes as the electrons, after being injected into the metal, rapidly drop into lower energy states and cannot be drawn back across the junction, although there is still a barrier capacitance which limits the high frequency performance. It is basically the freedom from charge storage effects together with the ability to control the barrier height which makes the Schottky barrier diode most suitable for power measurement.

8.2 Application of metal–semiconductor diodes to power measurement

8.2.1 Detection theory

Fig. 8.6 shows a simple diode detector circuit. The function of the resistor R_L is to reduce the dependence of the rf input impedance on the diode impedance, since the latter varies with frequency, temperature and power level. Thus for an input impedance of 50 Ω the diode resistance must be considerably higher

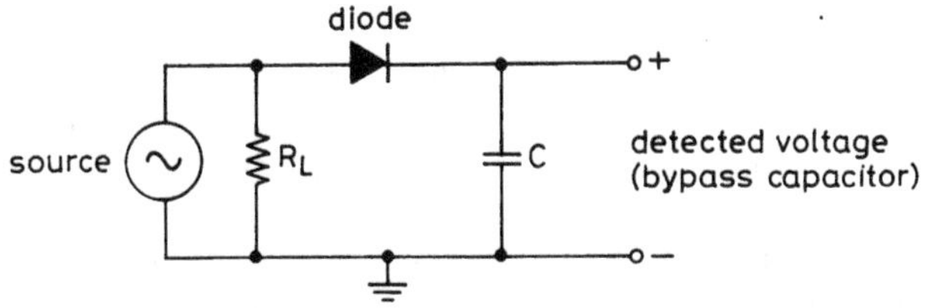

Fig. 8.6 *Single diode detector circuit. Resistor R_L makes the input impedance less dependent on the diode impedance*

than 50 Ω, typically around 2.5 kΩ. The capacitor C provides a by-pass for the rf current.

For small signals the detector operates in the square law mode and its sensitivity may be deduced from the diode characteristic given by equation 8.1. It will be assumed in the following that the ambient temperature is 300 K and the ideality factor n lies in the range 1 to 2. Using small signal approximations one can show that [19]

$$I_{dc} = I_s = \left(\frac{e}{kTn} V_{dc} + \text{higher terms in } V_{dc} + \frac{1}{4}\left(\frac{e}{kTn}\right)^2 V_{rf}^2 + \text{higher terms in } V_{rf} \right) \tag{8.2}$$

where V_{rf} and V_{dc} are the peak rf voltage and the dc voltage across the diode and I_{dc} is the dc current. The dc output voltage of the detector, when feeding into an open circuit load, is thus obtained by putting I_{dc} equal to zero:

$$V_{out} = -V_{dc} = \frac{e}{4kTn} V_{rf}^2 = \frac{e}{2kTn} V_{rf(rms)}^2 \tag{8.3}$$

In terms of the rf power absorbed in the 50 Ω impedance, equation 8.3 becomes:

$$V_{dc} = \frac{25e}{kTn} P_{rf} \tag{8.4}$$

Substituting numerical values in equation 8.4 gives a range of 0.5 to 1 mV/μW for the theoretical sensitivity, the exact value depending on the ideality factor.

As the rf power level is increased, the small signal approximations cease to be valid and there is a deviation from square law operation [20]. This happens when the exponent in equation 8.1 approaches a value of one, corresponding to a peak rf voltage of 25 to 50 mV or a peak power of 12.5 to 50 μW in 50 Ω. At these signal levels significant deviation from a square law is to be expected, but the exact law followed depends on the dc load resistance. There is usually one load resistance value, typically in the range 1 kΩ to 5 kΩ, which minimises the deviation from a square law. Fortunately this optimum load resistance is almost independent of rf frequency, so that optimisation is possible over a wide bandwidth. Measurements carried out by Hoer, Roe and Allred [21] on point contact diodes showed that a dynamic range of 50 dB could be achieved with a departure from linearity of ±0.02 dB. The point contact detectors described by Riley [22] retained their square law characteristic to within ±0.5 dB for an output of up to 50 mV, corresponding to an rf input power of about 100 μW. The sensitivity was greater than 0.4 mV/μW, which was three times that of earlier models. Fong-Tom and Cronson [23] reported an experimental comparison at 94 GHz of a number of different diodes for which a departure from square law of 0.2 dB occurred at levels ranging from 25 to 100 μW. Detectors can be operated above the square law region if the departure from a square law is corrected for and provided that no harmonics are present in the rf signal. This approach was followed by Somlo and Hunter for diodes used in a six-port [24]. By calibrating the diodes against a thermistor, operation up to 250 μW was possible. At high levels the performance of a diode approaches that of a linear detector. The short-term maximum power limit for a diode power meter is usually about 100 mW unless an attenuator is built into the sensor.

8.2.2 Frequency response

Early wideband coaxial detectors showed large variations in sensitivity of approximately ±4 dB over the frequency range 10 MHz to 12.4 GHz [25]. In the improved detectors described by Riley [22] the shunting disc resistor R_L, the point contact diode and the 10 pF by-pass capacitor were mounted together in a small replaceable capsule. This combination had a frequency response which was flat to within ±0.5 dB up to 12.4 GHz. Waveguide detectors of the same period were described by Prickett and Reniham [26] in 1965. The performance of coaxial detectors improved again in the early 1970s as a result of the use of low barrier height Schottky diodes allied to thin-film circuit technology and the frequency range was also extended to 26.5 GHz. Modern low barrier Schottky barrier diodes used in power meters have a frequency response which is flat to within ±0.25 dB to 12.4 GHz, ±0.5 dB to 15 GHz and ±0.6 dB to 18 GHz.

8.2.3 Temperature effects

The ambient temperature affects the performance of a diode detector in three ways. Firstly, it causes a variation in the rf impedance of the diode. Secondly, the optimum dc load resistance varies with temperature in order to preserve a square law response over a range of temperatures a temperature-dependent load is desirable. For the diodes measured in ref. [21] the required temperature coefficient of resistance was calculated to be −5.3 percent per degree C. Thirdly, the voltage responsivity will be temperature-dependent and this effect must be compensated for. Changes of output with temperature ranging from −0.15 percent to +2.5 percent per degree C were reported by various workers [23,27,28] and a modified law of temperature dependence for one type of detector was derived by Cullen and An [28]. In an early power meter built by Woods [29] two ganged potentiometers were used to correct simultaneously for the temperature-induced changes in optimum load resistance and in sensitivity.

8.2.4 Dual diode detectors

Diode detectors, when operating at or above the limit for square law operation, are particularly susceptible to errors caused by harmonics in the rf signal [30,31]. The dc output voltage then depends not just on the total power but also on the phase relationships between the harmonics and fundamental. One way of combating this problem is the use of a dual diode circuit, or voltage doubler, as shown in Fig. 8.7. A circuit of this type was employed by Woods

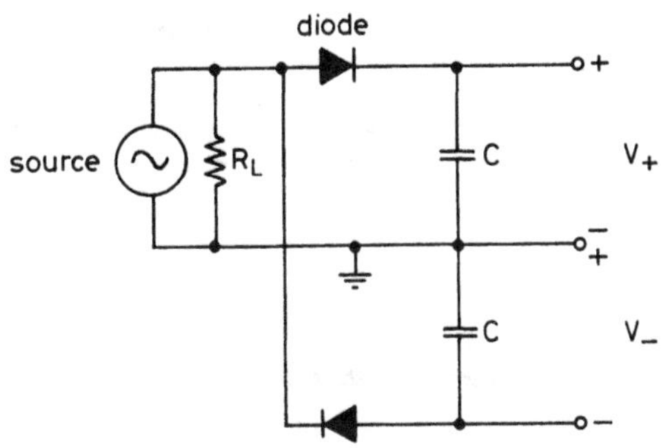

Fig. 8.7 *Dual diode detector circuit. The effect of even harmonics in the signal is greatly reduced compared with a single diode circuit*

[29] in 1962 in a diode power meter operating up to 1 GHz. It is also discussed by Wetenkamp [32] (see also [33]). Dual diode circuits tend to act as peak-to-peak detectors and this reduces the sensitivity to second and all even harmonics. For accurate calibration and standards work, however, the normal solution is to filter off the harmonics so that they are negligible.

8.2.5 Noise

The noise of a dector diode is expressed by the Noise Equivalent Power (NEP), defined as the rf power needed to produce a signal-to-noise ratio of unity for a 1 Hz bandwidth [14]. An earlier parameter was the tangential sensitivity, which can be calculated from the NEP or determined experimentally [34]. In the latter case an oscilloscope with a specified bandwidth is used to observe when the detected signal is perceptibly above the noise and then the rf input power is measured.

8.2.6 Practical details

Fig. 8.8 shows the circuitry of a commercial power sensor using a low barrier silicon Schottky diode [2,3,35]. The sensor is intended to operate in the frequency range 10 MHz to 18 GHz at power levels between 100 pW and 10 μW. Above this power level the departure from square law response becomes

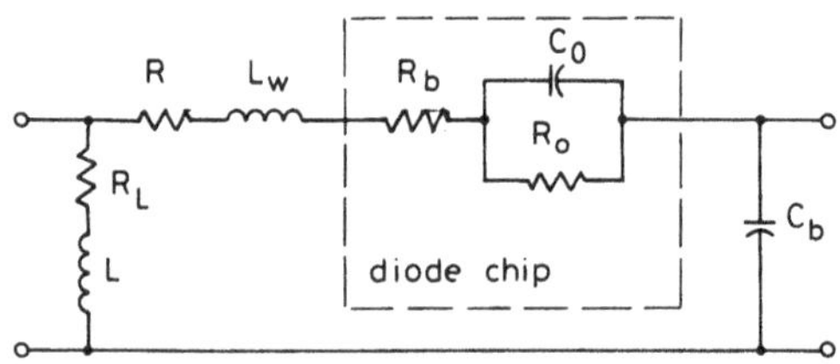

Fig. 8.8 *RF circuit associated with a low barrier Schottky diode power sensor. C_0 is the diode capacitance, R_0 the diode resistance, R_b the bulk resistance of the semiconductor, L_W the inductance associated with the bonding wires to the diode, C_b the by-pass capacitor and R_L the 50 Ω shunting resistor. R and L serve to improve the flatness of the frequency response*

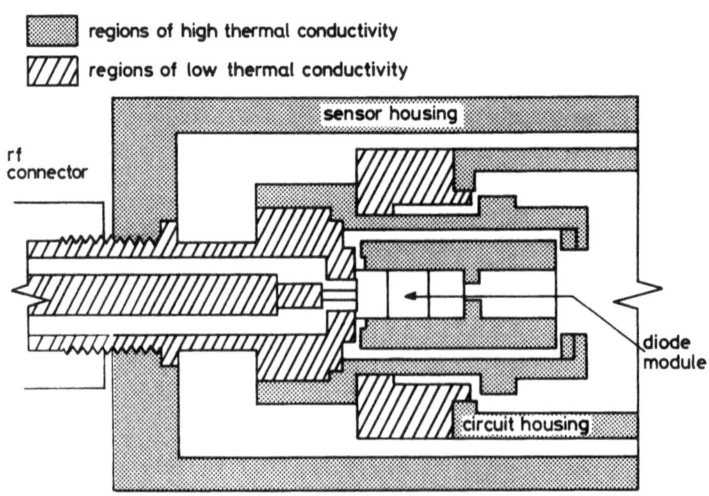

Fig. 8.9 *Details of sensor of diode power meter. To avoid thermoelectric voltages due to temperature gradients the thermal circuitry is arranged so that heat entering via the rf connector is directed to both ends of the diode. The circuit of the diode module is shown in Fig. 8.8*

appreciable, although the diode will withstand powers of up to 200 mW without damage. The saturation current of the diode is 10 μA, corresponding to a zero bias resistance of 2.5 kΩ. This resistance value is thus 50 times larger than the 50 Ω shunting resistor R_L. The diode capacitance C_0 is 0.1 pF, R_b is the bull resistance of the semiconductor, L_W the inductance associated with the bonding wires to the diode, and C_b the by-pass capacitor. The resistance R, which is 30 Ω, and the inductance L serve to improve the flatness of the frequency response. All rf circuitry, including the 50 Ω shunting resistor R_L, is constructed on a sapphire substrate using thin film techniques. The voltage responsivity is 500 V/W. Fig. 8.9 shows the diode module mounted in the sensor. The output voltage from the diode is fed via gold leads to a dc chopper amplifier having an input resistance of 5 kΩ. Since an rf power of 100 pW produces a dc voltage of only 50 nV it is extremely important to avoid temperature gradients across the diode which could give rise to thermoelectric voltages. In order to overcome this problem the sensor is well insulated thermally and the thermal circuitry is arranged so that heat entering via the rf connector is directed equally to both ends of the diode.

8.3 References

1 Watson, H.A. (ed.), 'Microwave semiconductor devices and their circuit applications', McGraw-Hill, 1969.

2 Szente, P.A., Adam, S., and Riley, R.B., 'Low barrier Schottky-diode detectors', *Microwave J.*, **19**, 2, pp. 42–44, 60, Feb. 1976.

3 Pratt, R.E., 'Very-low-level microwave power measurements', *Hewlett-Packard J.*, **27**, 2, pp. 8–10, Oct. 1975.

4 Rhoderick, E.H., 'Metal semiconductor contacts', *Proc. IEE*, **129**, Pt. I., 1, pp. 1–14, Feb. 1982.

5 Warner, F.L., 'Detection', in 'Millimetre and submillimetre waves', Benson, F.A. (ed.), Iliffe, 1969.

6 Smith, R.A., 'Semiconductors', 2nd ed., Cambridge Univ. Press, 1979.

7 Sze, S.M., 'Semiconductor devices; physics and technology', Wiley, 1985.

8 Bleaney, B., Ryde, J.W., and Kinman, T.H., 'Crystal valves', *J. IEE*, **93**, Pt. IIIA, 5, pp. 847–854, 1946.

9 Torrey, H.C., and Whitmer, C.A., 'Crystal rectifiers', McGraw-Hill, New York 1948.

10 Bronwell, A.B., Wang, T.C., Nitz, I.C., May, J., and Wachowski, H., 'Vacuum-tube detector and converter for microwaves using large electron transit angles', *Proc. IRE*, **42**, 7, pp. 1117–1123, July 1954.

11 Papp, G., 'Mechanism of rectification in vacuum tube diodes at microwave frequencies', *Elec. Commun.*, **31**, pp. 215–219, Sept. 1954.

12 Redhead, P.A., 'Microwave detection in a thermionic diode', *Proc. IRE*, **43**, 8, pp. 995–1000, Aug. 1955.

13 Carlin, H.J., 'Measurement of power', Chap. 3 in 'Handbook of microwave measurements', Sucher, M., and Fox, J. (ed.), Polytechnic Press, 1963 (p. 224, 'Thermionic diodes').

14 Cowley, A.M., and Sorenson, H.O., 'Quantitative comparison of solid state microwave detectors', *IEEE Trans.*, **MTT-14**, 12, pp. 588–602, Dec. 1966.

15 Eng, S.T., 'Low noise properties of microwave backward diodes', *IRE Trans.*, **MTT-9**, pp. 419–425, 1961.
16 Shurmer, H.V., 'Backward diodes as microwave detectors', *Proc. IEE*, **111**, 9, p. 1511, Sept. 1964.
17 Simonyan, R.G., and Avetisyan, E.V., 'Reversed diode as a measurement detector', *Meas. Tech.*, **17**, 2, pp. 165–167, Feb. 1974.
18 Blaine, C., 'A case for back diodes', *Microwave J.*, **24**, 3, p. 80, Mar. 1981.
19 Griffin, E.J., 'Detectors and detection for measurement', IEE Vacation School on Microwave Measurements, University of Kent, England, Sept. 1983.
20 Sorger, G.U., and Weinschel, B.O., 'Comparison of deviations from square law for rf detectors', *IEEE Trans.*, **I-8**, 3, pp. 103–111, Dec. 1959.
21 Hoer, C.A., Roe, K.C., and Allred, C.M., 'Measuring and minimising diode detector non-linearity', *IEEE Trans.*, **IM-25**, 4, pp. 324–329, Dec. 1976.
22 Riley, R.B., 'A new coaxial crystal detector with extremely flat frequency response', *HP Journal*, **15**, 3, Nov. 1963.
23 Fong-Tom, R.A., and Cronson, H.M., 'Diode detector characteristics for a 94 GHz sixport application', *IEEE Trans.*, **MTT-31**, 2, pp. 158–164, Feb. 1983.
24 Somlo, P.I., Hunter, J.D., and Arthur, D.C., 'Accurate six-port operation with uncalibrated nonlinear diodes', *IEEE Trans.*, **MTT-33**, 3, pp. 281–282, Mar. 1985.
25 Schrock, N.B., 'A new 10 Mc to 12 kMc coaxial crystal detector mount', *Hewlett-Packard J.*, **6**, 6, 1955.
26 Prickett, R., and Reniham, L., 'New waveguide crystal detectors with flat response', *Hewlett-Packard J.*, **16**, 6, Feb. 1965.
27 Somlo, P.I., and Hunter, J.D., 'A six-port reflectometer and its complete characterization by convenient calibration procedures', *IEEE Trans.*, **MTT-30**, 2, pp. 186–192, Feb. 1982.
28 Cullen, A.L., and An, T.Y., 'Microwave characteristics of the Schottky-barrier diode power sensor', *Proc. IEE*, **129**, Pt. H, 4, pp. 191–198, Aug. 1982.
29 Woods, D., 'A coaxial millivoltmeter/milliwattmeter for frequencies up to 1 Gc/s', *Proc. IEE*, **109**, Pt. B., suppl. 23, pp. 750–756, May 1962.
30 Oliver, B.M., 'Which RF voltmeter?', Application Note 60, Hewlett-Packard, 1967.
31 Oliver, B.M., and Cage, J.M., 'Electronic measurements and instrumentation', McGraw-Hill, 1971.
32 Wetenkamp, S., 'Comparison of single diode vs. dual diode detectors for microwave power detection', IEEE MTT-S International Microwave Symposium Digest, pp. 361–363, Boston, June 1983.
33 An, T.Y., and Cullen, A.L., 'Double Schottky-barrier diode power sensor', *Proc. IEE*, **130**, Pt. H, 2, pp. 160–165, Mar. 1983.
34 Lucas, W.J., 'Tangential sensitivity of a detector video system with r.f. preamplification', *Proc. IEE*, **113**, 81, pp. 1321–1330, Aug. 1966.
35 'Fundamentals of rf and microwave power measurements', Application Note 64-1, Hewlett-Packard, Aug. 1977.

Chapter 9

Force-operated instruments

The fact that mechanical forces can be produced by the interaction of electromagnetic fields with charges and currents forms the basis of many traditional electrical measuring instruments. For example the dc moving coil ammeter depends on the force produced by a magnetic field acting upon a current-carrying conductor, whilst the electrostatic voltmeter relies on the force produced by an electric field acting between charged plates. AC instruments based on this principle include the dynamometer wattmeter [1]. These and similar instruments have their counterparts at rf and microwave frequencies. They are generally referred to as force-operated instruments, although the term 'ponderomotive instruments' is also applied. In general the practical impact of force-operated instruments at high frequencies has been very limited, since for most applications they are neither the most accurate, nor the most convenient, nor the most sensitive type. The forces which must be detected are small, so that the instruments are easily upset by mechanical vibrations and their use is normally confined to laboratory environments. Here they are in direct competition with the very precise thermal instruments which have been developed. On the other hand, from a metrologist's point of view force-operated instruments are amongst the most interesting of the many types of power-measuring devices in existence. One important attribute which they possess is that they can be designed for use as absolute standards [2–4], enabling power to be derived directly from mass, length and time, rather than via dc measurements. Force-operated instruments of the feed-through kind can have very low power losses, because the principle of operation does not rely on conversion of power to some other form such as heat. They are also of some historical interest, because some of the earliest work on force-operated power meters for microwave frequencies was believed to represent the first reasonably accurate verification in this part of the spectrum of the formulas for calculating radiation pressure [5]. Not all force-operated instruments are of course suitable as absolute standards. Many are intended as indicators of relative level only, unless previously calibrated. A few, mainly voltage and current measuring instruments, utilise the rf/dc substitution principle. Many feed-through force-operated power meters read correctly only for one particular load impedance and in this sense they are not true power meters. In fact the term voltmeter or ammeter might often be more appropriate. However, in order to avoid being diverted by such questions it seems reasonable to choose the name which corresponds to the intended use, irrespective of other considerations. The present chapter concentrates on instruments which are intended for the measurement of power, whilst Chapter 15 treats voltage and current measuring instruments. First some formulas relating forces to fields are considered (Section 9.1). Section 9.2 then introduces the concept of radiation pressure, which enables forces to be calculated in certain circumstances without a detailed knowledge of the fields. In Section 9.3 the resonator action theorem is introduced. This greatly extends the number of situations in which a knowledge of the fields is not necessary. Some practical force-operated power meters are described in Section 9.4.

9.1 Basic formulas relating forces and fields

9.1.1. Pressure exerted on perfectly conducting surfaces

The instantaneous pressure due to an electric field acting on a perfectly conducting surface can be derived by considering the force of attraction

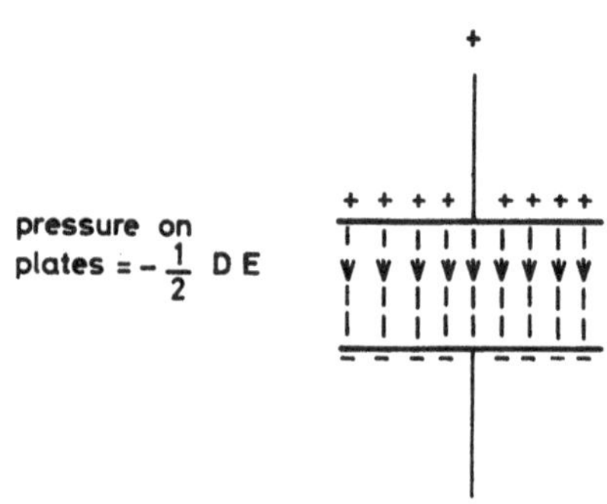

Fig. 9.1 *Pressure on a perfect conductor due to an electric field*

between the plates of an ideal parallel plate capacitor (Fig. 9.1). One can show that:

$$\text{pressure} = -\tfrac{1}{2}DE \tag{9.1}$$

where D is the instantaneous electrical displacement and E is the instantaneous electric field, both of which will be normal to the surface. If alternatively E and D are taken as the rms values of the fields, then equation 9.1 gives the mean pressure rather than its instantaneous value. The pressure is negative, indicating that it acts outwards from the surface, and the effect may thus be described as one of suction.

For a magnetic field acting on a perfectly conducting surface the corresponding formula is:

$$\text{pressure} = \tfrac{1}{2}BH \tag{9.2}$$

where B is the magnetic flux density and H is the magnetic field strength, both of which will be parallel to the surface. The pressure is positive, indicating that it acts into the conducting surface.

When both electric and magnetic fields are present, the net pressure may be either positive or negative according to whether the magnetic or electric field dominates. It is given by

$$\text{pressure} = \tfrac{1}{2}BH - \tfrac{1}{2}DE \tag{9.3}$$

Equation 9.3 is valid at any frequency and can be used to calculate, for example, the pressure on the walls of a resonating cavity or on the conductors of a transmission line. For the special case in which the ratio of E to H is equal to the impedance of free space the pressures due to the electric and magnetic fields cancel. Thus the TEM mode in a coaxial line does not exert any pressure on the walls unless standing waves are present.

The right hand side of equation 9.3 is simply the difference between the magnetic and electric stored energy densities. It can also be regarded as the density of the Lagrangian function, the Lagrangian function of an electromagnetic system being the difference between the total magnetic stored energy and the total electric stored energy.

9.1.2. Forces on dielectrics

Calculation of the forces exerted on a dielectric in terms of the fields is more complicated than for conducting surfaces. The fields can have both normal and tangential components at the surface and, in addition to the pressure at the surface, inhomogeneous dielectrics will experience a volume force which is distributed throughout the bulk of the material. Only simple examples involving electrostatic fields will be considered here. Practical cases are more likely to be dealt with by calculations based on radiation pressure (Section 9.2) or by application of the resonator action theorem (Section 9.3). The pressure due to an electrostatic field normal to the surface of a dielectric may be found by considering a partially filled capacitor as shown in Fig. 9.2. If the thickness of the dielectric is varied while the charge on the plates is held constant, then the pressure on the surfaces of the dielectric can be calculated by equating the mechanical work to the change in electrical stored energy, which is derived from the charge on the plates and the change in capacitance. This leads to the result:

$$\text{pressure} = (\tfrac{1}{2}DE)_{dielectric} - (\tfrac{1}{2}DE)_{air} \tag{9.4}$$

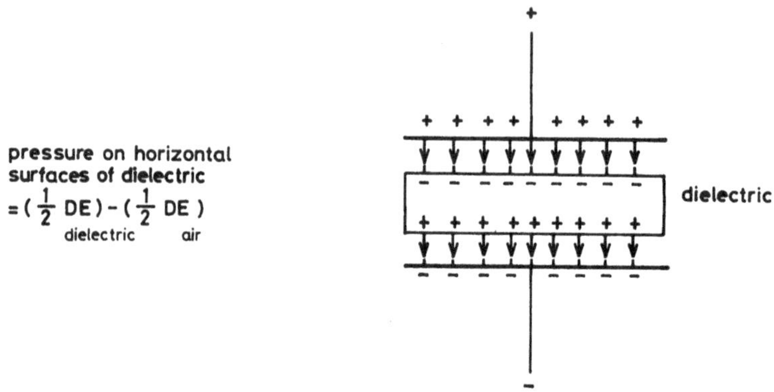

Fig. 9.2 *Pressure on the surface of a dielectric due to a normal electric field*

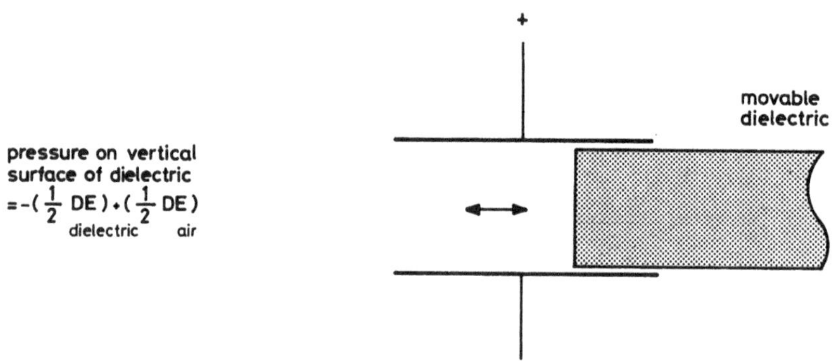

Fig. 9.3 *Pressure on the surface of a dielectric due to a tangential electric field*

The electric field E is reduced inside the dielectric, while D has the same value on both sides of the interface. The pressure is therefore negative and acts towards the less dense medium, in this case air. The formula for a tangential electric field is found by considering the partially filled capacitor shown in Fig. 9.3. In this case the mechanical work is equated to the change in electrical stored energy as the dielectric is moved into or out of the capacitor:

$$\text{pressure} = -(\tfrac{1}{2}DE)_{dielectric} + (\tfrac{1}{2}DE)_{air} \tag{9.5}$$

The value of D is reduced inside the dielectric, while E now has the same value on both sides of the interface. The pressure is once again negative, acting towards the less dense medium. The effect is to pull the dielectric material into the region of highest field strength.

The volume forces due to an electrostatic field in an inhomogeneous dielectric can be calculated by imagining the permittivity to vary in infinitessimally small steps throughout the material. The pressure on the surfaces formed by the steps is then calculated. In this way one can show that the force $\boldsymbol{\delta F}$ on a small volume δv is given in vector notation by

$$\boldsymbol{\delta F} = -\tfrac{1}{2}E^2 \text{ grad } \varepsilon\, \delta v \tag{9.6}$$

where ε is the permittivity. The force acts in the opposite direction to the gradient of the permittivity.

9.2 Radiation pressure

The concept of radiation pressure enables one to calculate forces in many cases without a detailed knowledge of the fields. A beam of radiation in free space possesses momentum. The force which the beam can exert depends on its rate of momentum, that is the momentum per unit time. This is given by

$$\text{rate of momentum} = \frac{p}{c} \tag{9.7}$$

where p is the power in the beam and c the speed of light. When the beam impinges on a surface, the force on the surface is equal to the rate of change of momentum. Radiation pressure was predicted for electromagnetic waves by Maxwell [6], who derived it from energy considerations. The same result was later shown by Bertoli [7] to apply to any stream of energy, using thermodynamic arguments. A more general discussion is given in ref. [8].

Three cases which can be calculated quite simply using the radiation pressure concept are illustrated in Fig. 9.4. These are reflection by a perfectly conducting surface, partial reflection by a dielectric sheet, and absorption by a non-reflecting surface. In the case of a perfectly conducting surface the momentum of the incident beam is totally reversed, assuming normal

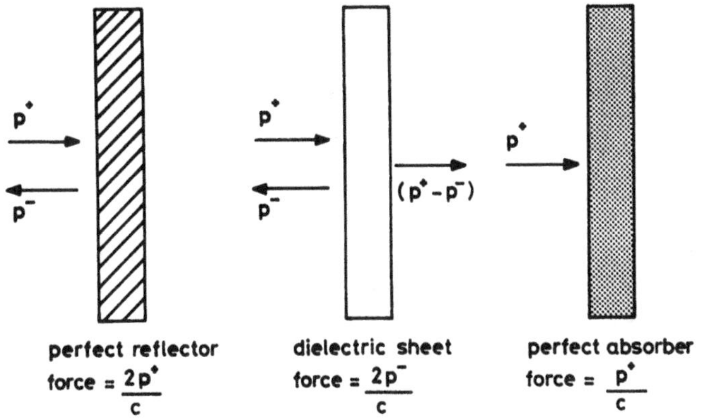

Fig. 9.4 *Force exerted by a beam of electromagnetic radiation. P+ is the incident power, P- the reflected power, and c the speed of light*

incidence, so that the change in momentum is twice the momentum in the original beam and

$$\text{force} = \frac{2p^+}{c} = \frac{2p^-}{c} \tag{9.8}$$

where p^+ and p^- are the incident and reflected powers. For a dielectric sheet at normal incidence the change in momentum is made up of the momentum in the reflected beam plus an equal amount due to the reduction in the momentum of the transmitted beam: 9.9

$$\text{force} = \frac{2p^-}{c} \tag{9.9}$$

where the reflected power p^- is due to the combined effect of reflections at both surfaces. For the case of a reflectionless absorbing surface the momentum in the incident beam is completely destroyed and the force is given by

$$\text{force} = \frac{p^+}{c} \tag{9.10}$$

To give an example of the size of the force, absorption of a power of 1 watt would result in a force of 0.333×10^{-8} newton. The mechanism by which the absorption of power in a body results in a force is very closely related to the Hall effect (Chapter 10). The force arises by interaction between the magnetic field and the currents in the absorber. Since the currents are carried by electrons (or holes in the case of *p*-type semiconductors), it is these charge carriers which primarily experience the force, which pushes them in the direction of propagation of the beam and thereby creates the Hall emf. The resulting displacement of the charge carriers sets up an electric field in the material, which acts on the fixed ionised atoms and transfers the force to the bulk of the material.

The concept of radiation pressure is also valid for guided waves as opposed to waves in free space, provided that any difference in phase velocity is taken into account [2,3,9,10]. Equation 9.8 then becomes

$$\text{force} = \frac{2p^+}{c}\,\frac{\lambda}{\lambda_g} \tag{9.11}$$

where λ and λ_g are the free space and guide wavelengths respectively. Equation 9.11 can be used to calculate the force on a movable short circuiting plunger inserted in a waveguide (Fig. 9.5).

A beam of circularly polarised radiation possesses angular momentum in addition to its linear momentum. The rate of angular momentum is given by

$$\text{rate of angular momentum of circularly polarised wave} = \frac{p^+}{\omega} \tag{9.12}$$

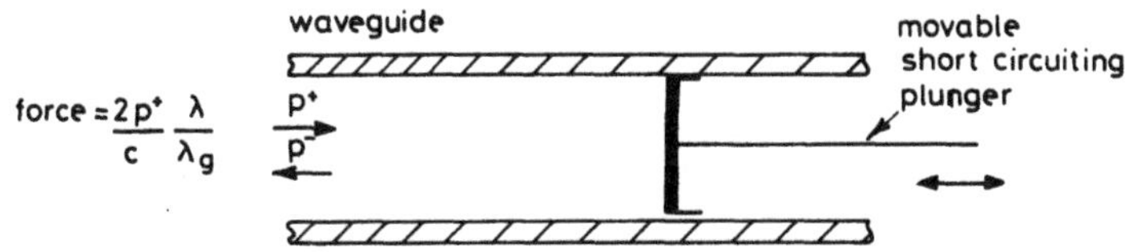

Fig. 9.5 *Force exerted on a short-circuiting movable plunger*

where ω is the angular frequency. Right-handed and left-handed polarisations have angular momentum in opposite senses. When a plane polarised beam in free space is converted to circular polarisation, the polariser experiences a torque. Absorption of a circularly polarised beam by a non-reflecting surface also produces a torque, as does reflection when accompanied by a change in polarisation. It will be seen from equation 9.12 that the angular momentum increases with decreasing frequency, so that very large torques may be produced at low frequencies.

The angular momentum possessed by guided waves is less straightforward than for waves in free space. The fields cannot be circularly polarized everywhere, since they must be linearly polarised at any metal surfaces, and the way in which the polarisation varies over the cross section will depend on the shape of the waveguide. This problem is discussed in ref. [11]. However, the use of the free space formula (equation 9.12) is valid for waveguides in which the field pattern rotates without suffering a change in form, provided that the value of ω is interpreted as the rate of rotation of the field pattern (in radians/second) rather than the angular frequency. For the field pattern to remain unchanged during rotation the waveguide must obviously be circular. For the TE_{nm} mode in circular guide the rate of rotation is ω/n, where n refers to the number of cycles of the field pattern around the circumference.

The first practical demonstration of radiation pressure was at optical frequencies [12]. Subsequent measurements [13–15] confirmed the magnitude given by equation 9.7. Similar experiments were also carried out at rf [16,17] and microwave [18] frequencies, but it was found difficult to obtain accurate quantitative results because of the difficulty of forming a narrow beam at these longer wavelengths. Thus it was not until experiments on guided waves were carried out that meaningful measurements were achieved [5]. The formula for angular momentum (equation 9.12) was confirmed experimentally by Beth [19].

9.3 The resonator action theorem and its applications

The resonator action theorem (action being the product of energy and time) is a very general theorem which applies to lossless resonators, both electrical and mechanical. An example will serve to illustrate it. Consider the resonating

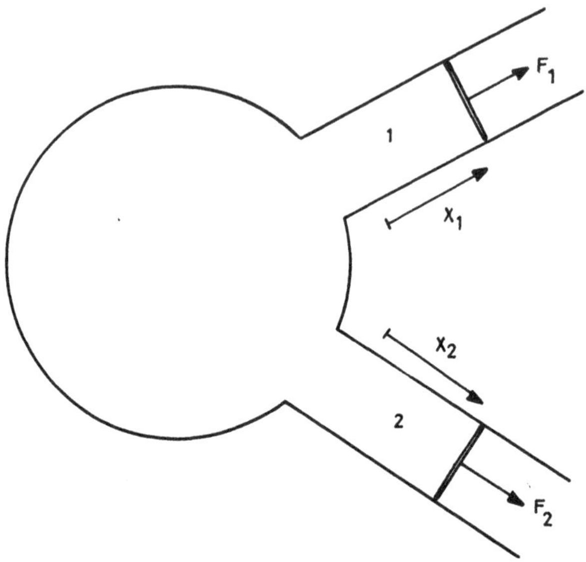

Fig. 9.6 *Comparison of forces on pistons*

cavity shown in Fig. 9.6, consisting of an arbitrarily shaped enclosure joined to two lengths of waveguide, each fitted with a short circuiting plunger. The two plungers will experience an outward force caused by the radiation pressure associated with the standing waves inside the guides. If now one of the plungers is moved against the force, this will result in an increase both in the resonant frequency of the cavity and in the stored energy. The extra stored energy is equal to the work performed in moving the plunger. The resonator action theorem states that, provided that the change is made slowly, the product of the period of oscillation and the stored energy is constant. In other words, the stored energy is proportional to the resonant frequency. In terms of the quantum theory this is equivalent to saying that the number of photons in the cavity remains constant as the plunger is moved, since each individual photon has an energy equal to hf, where f is the frequency and h is Planck's constant.

A more general statement of the theorem as applied to electrical resonators is as follows:

> In a lossless electromagnetic resonator the action of each mode, i.e. the product of total energy and period, is invariant against an adiabatic deformation.

The term adiabatic implies that the change takes place with perfect uniformity and infinite slowness, this being necessary in order to avoid the possible excitation of other resonant modes. The theorem applies both to cavities having movable walls and to cavities containing movable materials, as long as there is no absorption of power. A proof for a lossless cavity containing movable isotropic dielectric and magnetic materials is given by MacLean [20].

If the deformation is carried out in such a way that the resonant frequency is unchanged, for example by displacing both plungers simultaneously, one in an outward direction and the other in an inward direction, then it follows that the total work done is zero. Stated more formally:

In a lossless electromagnetic resonator the total energy content is invariant against any adiabatic variation which leaves the period unchanged.

A proof of this more restricted version of the theorem is given by Cullen [21] for a resonator containing no dielectric or magnetic materials.

The restricted form of the resonator action theorem forms the basis of a general theory of absolute force-operated feed-through power meters due to Cullen [21]. This theory enables one to determine the power flowing through a lossless instrument in terms of the force on an arbitrary movable element located anywhere inside it provided that certain subsidiary measurements can be carried out. It allows considerable freedom in the design of the instrument because it is not necessary to be able to calculate the exact field distribution. Consider again Fig. 9.6. If F_1 and F_2 are the forces on the plungers then the total work done by the cavity when the plungers are moved through distances δx_1 and δx_2 respectively is given by:

$$\text{work} = F_1\,\delta x_1 + F_2\,\delta x_2 \tag{9.13}$$

Assuming that the resonant frequency of the cavity remains unchanged, then by equating the total work to zero a simple relation is obtained between the forces on the two plungers:

$$F_2 = F_1\left(-\frac{\delta x_1}{\delta x_2}\right) \tag{9.14}$$

In a practical instrument one of the plungers would be replaced by a movable element of arbitrary shape. Equation 9.14 then enables the force on this element to be determined in terms of the force on the short-circuiting plunger, which can be related to the incident power on the plunger by radiation pressure. The only restriction on the form of the arbitrary movable element is that it should have only one degree of freedom.

Consider now how this result can be applied to an absolute feed-through power meter. Fig. 9.7 shows such an instrument in a generalised form. The output port would normally be connected to a matched load, but for the purpose of evaluation this load is replaced by a short-circuiting plunger. The principle of the evaluation procedure is that the movable element and the plunger are simultaneously displaced in such a way that zero work is done. The ratio of forces on them is then calculated from the ratio of the two displacements. The criterion for zero work is that the position of the voltage

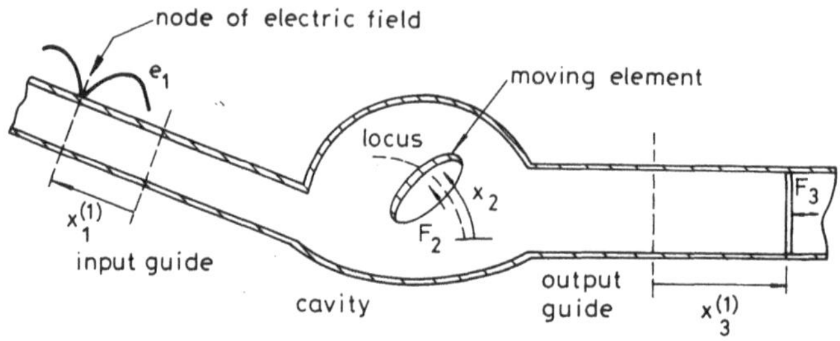

Fig. 9.7 *Generalised force-operated power measuring apparatus*

zero in the standing wave pattern at the input port should remain unchanged. One can see that this criterion follows from the resonator action theorem by imagining that a short circuit is placed in the input guide so as to coincide with the voltage zero of the standing wave pattern. This effectively converts the instrument into a cavity which has the same resonant frequency both before and after the displacements.

The above arguments lead to the following equation relating the force F_2 on the movable element to the power P_{3sc} incident on the plunger:

$$F_2 = -2\frac{dx_3}{dx_2}\frac{P_{3sc}}{c}\frac{\lambda}{\lambda_g} \tag{9.15}$$

The interpretation of F_2 and dx_2 can be generalised to include rotation, in which case F_2 represents the torque and dx_2 the angular displacement. In order to be able to calculate the force for the matched load case from measurements made with the short circuiting plunger in place, it is necessary to move the plunger by a quarter wavelength and repeat the evaluation procedure. The method relies on the fact that a travelling wave can be represented as the sum of two standing waves displaced by a quarter wavelength in space and 90° out of phase. By adding in quadrature the standing wave field distributions for the two plunger positions (assuming the same incident power P_{3sc} on the plunger in each case) one can synthesise the travelling wave field pattern corresponding to the matched load case. The power P_{out} in this travelling wave will be equal to $4P_{3sc}$. Because of the phase quadrature it is permissible to add arithmetically the individual forces on the moving element due to the two standing waves. Thus:

$$F_2 = \frac{P_{out}}{2c}\frac{\lambda}{\lambda_g}\left[-\frac{\delta x_3^{(1)}}{\delta x_2} - \frac{\delta x_3^{(2)}}{\delta x_2}\right] \tag{9.16}$$

Cullen's general theory can be applied to a large number of different types of power meter, but it is usually associated with the torque vane wattmeter, which is amongst the instruments described in Section 9.4.

For mismatched loads the theory does not give the correct value for the absorbed load power, although one can reduce the error by using two movable elements to form a double vane wattmeter (Section 9.4) [22].

Absorption-type power meters using high-Q cavities of known Q factor can also be constructed for use as absolute power meters. In such cases the relation between the force (or torque) and power is determined by first noting the change in the resonant frequency of the cavity when the moving element is displaced by a measured amount and then applying the general form of the resonator action theorem, under the assumption that the cavity losses are small. This gives a relation between the force on the moving element and the stored energy of the cavity. For example, if the frequency increases by one percent then for an adiabatic change the energy would also increase by one percent and the initial stored energy must have been one hundred times the work done on the moving element. The measured displacement relates the force to the work done and the measured Q factor then enables the relation between the stored energy and the absorbed power to be found. Thus:

$$\text{power} = 2\pi f \frac{\text{stored energy}}{Q} \qquad (9.17)$$

$$= 2\pi \,\text{force} \frac{dx}{df} \frac{f^2}{Q} \qquad (9.18)$$

where f is the resonant frequency and x the position of the moving element.

9.4 Some practical force-operated instruments

9.4.1. Radiation pressure wattmeter

Fig. 9.8 shows an experimental radiation pressure wattmeter [5] constructed by Cullen in the 1950s. It is a feed-through instrument designed around a lossless waveguide T-junction. One arm of the T-junction functions as the input port, while the second arm acts as the output port. The third arm, which is a length of circular waveguide, is fitted with a short-circuiting plunger and the power flowing through the instrument is indicated by the force on this plunger. The plunger must be carefully designed so that it will move freely along the guide while still having negligible electrical loss. This is done by employing the TE_{01} mode, for which the currents are entirely circumferential, so that there is no flow of current across the gap between the plunger and the waveguide wall. The plunger itself is constructed in the form of concentric wire rings in order to minimise the effects of thermal convection currents in the air. In this way the necessity of performing the measurements in vacuum is avoided. By suitably positioning the plunger it is possible to make the instrument appear as a reflectionless transmission line when seen from the input or output end.

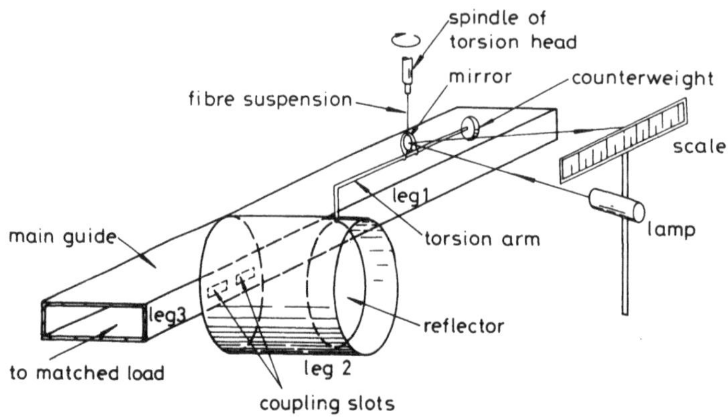

Fig. 9.8 *Radiation pressure wattmeter due to Cullen. The force is exerted on a movable reflector consisting of concentric wire rings*

The force on the plunger is measured by converting it into a torque, which causes a beam of light to be deflected. A movable arm to which the plunger is attached is suspended by a quartz fibre of diameter 10 μm. The read-out is provided by a mirror which is fixed to the quartz fibre and which reflects the light beam onto a scale. The relation between the angular deflection of the arm and the torque can be determined by attaching weights of known moment of inertia and measuring the period of oscillation.

The power absorbed by a matched load at the output port can be expressed in terms of the force on the plunger by determining the values of the components in the equivalent circuit of the junction, or alternatively Cullen's general theory can be applied (see ref. [21] and Section 9.3). The accuracy obtained was very modest, agreement with thermal methods of power measurement being within 17 percent, but the device nevertheless served to demonstrate the correctness of the principles for calculating radiation pressure in waveguides and led to the development of other devices.

9.4.2. Torque vane wattmeter

One way of avoiding thermal convection problems in force-operated instruments is to design the instrument so that the force is produced mainly by electric fields acting on charged conductors rather than magnetic fields acting on short-circuited currents. The currents in the region of the moving element are then smaller, resulting in lower resistive heating. This idea was pursued in the design of the torque vane wattmeter [23], which consists of a movable conducting vane mounted in a section of waveguide in such a way that it is free to rotate (Fig. 9.9). The torque exerted on the vane by the fields may be regarded as being produced chiefly by the attraction between the electric charges on the vane and induced charges on the waveguide walls. The power flowing through the instrument can be expressed in terms of the torque by

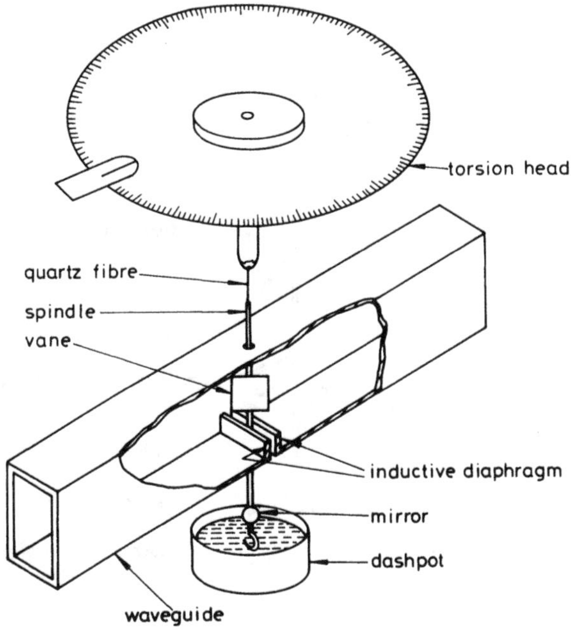

Fig. 9.9 *Principle of torque vane wattmeter. The torque on the vane is due mainly to attraction between opposite charges on the vane and waveguide walls*

means of Cullen's general theory. The vane is suspended by a quartz fibre and the angular displacement detected by a mirror arrangement similar to that used for the radiation pressure wattmeter. A later version of the torque vane wattmeter incorporated a double vane [22] which enabled the error due to load mismatch to be reduced (Fig. 9.10). This version was manufactured commercially.

9.4.3. Instruments using resonant cavities

Instruments using resonant cavities can have higher sensitivities than non-resonant devices. Fig. 9.11 shows a type of wattmeter in which the power is absorbed in the losses of the cavity. Assuming that the Q factor is high, the power dissipated in the cavity is calculated from the force on the vane, the rate of change of resonant frequency with vane position, and the Q factor. The Q factor is determined either by observing the rate of decay of electrical oscillations or by making impedance measurements at the input as a function of frequency. A problem which has been reported is detuning of the cavity as the vane rotates, which can in some circumstances lead to instability in the vane position [24]. One way of further improving the sensitivity of force-operated instruments is to modulate the power at the resonant frequency of the mechanical resonance of the moving element [25,26]. Power levels of

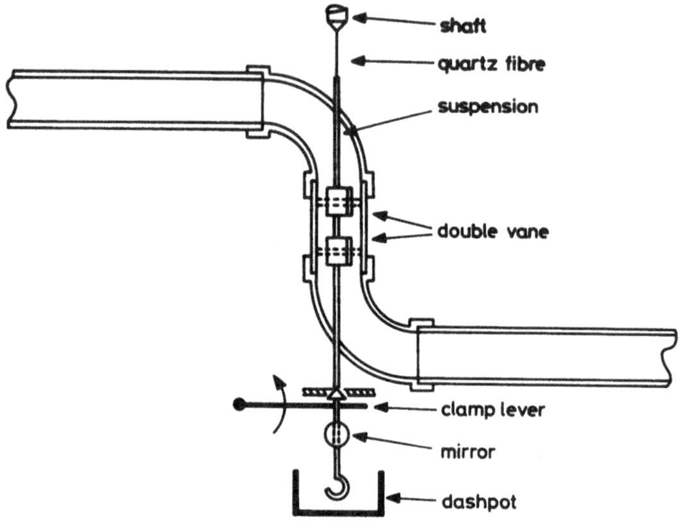

Fig. 9.10 *Double vane wattmeter. The use of two vanes reduces the error due to load mismatch*

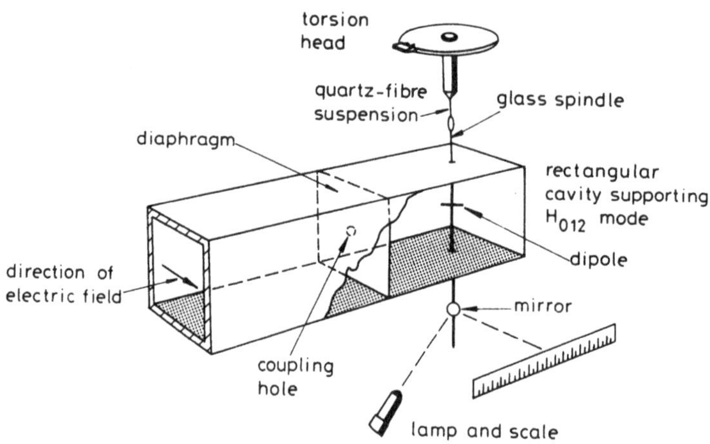

Fig. 9.11 *Resonant cavity torque-operated wattmeter. The power to be measured is absorbed in the losses of the cavity*

the order of a milliwatt can be measured. In this mode of operation the angular deflection θ is independent of the Q factor and is given by:

$$\theta = \sqrt{\frac{P}{2f}\frac{M}{K}} \tag{9.19}$$

where P is the microwave power, f the frequency, K the specific couple of the quartz fibre suspension and M the mechanical magnification factor (defined as the reciprocal of the fractional decrease in the amplitude of successive swings of the moving element in natural oscillation). The instrument may be regarded as a microwave reciprocating engine, converting some of the microwave power into mechanical power.

Barlow and Cross [27] used a resonant cavity with a moving dielectric element in a power meter operating at 35 GHz. The cavity was formed from a length of circular waveguide and it resonated in the TE_{01n} mode. The movable element was a dielectric sleeve placed concentrically with the cavity and positioned towards one end. The force on the dielectric may be explained qualitatively as being due to the gradient in the electrical stored energy. The dielectric is pulled into the region of highest field in a way similar to that shown in Fig. 9.3. By combining two such cavities with two directional couplers a reflectometer type of instrument was constructed which indicated the difference between the forward and reflected power and thus gave a direct measure of the power absorbed in the load.

The use of open resonators of measured Q-factor has also been proposed for the millimetre wave region [28]. The cavity could be formed from two parallel plates, making a Fabry-Perot interferometer. The power would then be calculated either by a direct measurement of the force on one of the plates or by a compensation method involving the application of a dc voltage across the plates.

9.4.4 Side-wall power meter

A simple force-operated power meter is shown in Fig. 9.12. A thin metal diaphragm forming part of a capacitor microphone [29] is incorporated in the narrow wall of a rectangular waveguide in which the TE_{10} mode propagates.

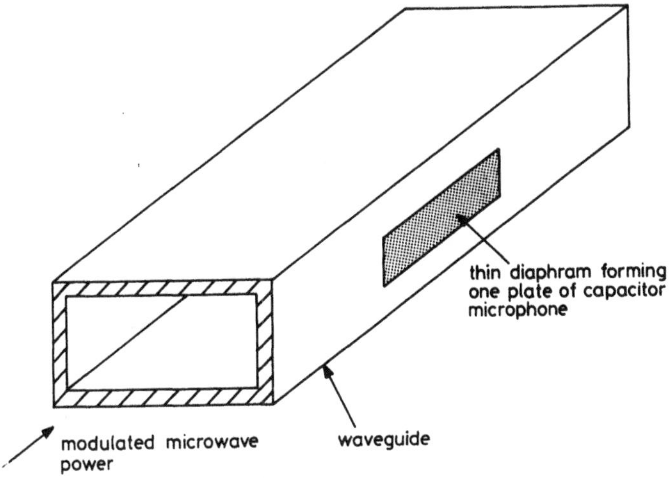

Fig. 9.12 *Sidewall force-operated power meter. The diaphragm forms one plate of a capacitor microphone*

The diaphragm experiences a force due to the magnetic field at its surface and, since there is no counterbalancing electric field, the wall will tend to be pushed outwards. The signal to be measured is modulated, thus producing an audio frequency output. A coaxial instrument of this type was also constructed by mounting the diaphragm in the outer conductor of a coaxial line. In this case current-interrupting gaps must be incorporated, since otherwise the forces due to the electric and magnetic fields would cancel each other. In principle such an instrument can be used as a broad band rf/dc substitution instrument, although there are practical difficulties which limit the accuracy. Ferrite vanes have also been used [30].

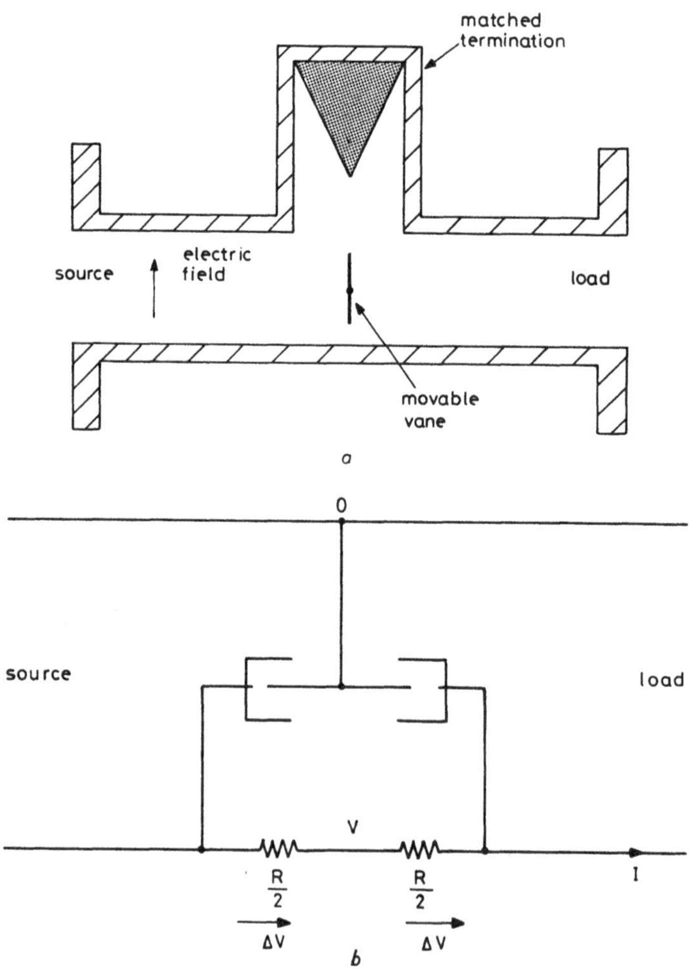

Fig. 9.13 *Microwave equivalent of the electrostatic wattmeter (a) and equivalent circuit (b). The torque exerted on the vane is proportional to the power delivered to the load, irrespective of load reflection coefficient*

9.4.5 Microwave electrostatic wattmeter

Fig. 9.13 shows a waveguide instrument which is the microwave equivalent of the electrostatic wattmeter [31–34]. Like the radiation pressure wattmeter, it is a feed-through power meter based on a waveguide T-junction, but the third arm of the junction is terminated in a matched load rather than a short circuiting plunger. This matched load appears in the equivalent circuit as a centre-tapped series resistance. The torque on the moving element is given by:

$$\begin{aligned} T &= k_1 [(V + \Delta V)^2 - (V - \Delta V)^2] \\ &= 2k_1 \Delta V\, V \\ &= k_2 I\, V \end{aligned} \tag{9.20}$$

where V, I, and ΔV are defined in Fig. 9.13 and k_1 and k_2 are constants. Thus the torque is proportional to the net power, although for the microwave version the theory does not give the constant of proportionality.

9.4.6 Instruments based on angular momentum

One of the first proposals for an angular momentum wattmeter was by Cullen [21], who pointed out that the moving element in the common form of phase shifter [35] shown in Fig. 9.14 will experience a torque. The wave is first converted to circular polarisation, then the sense of the polarisation is reversed

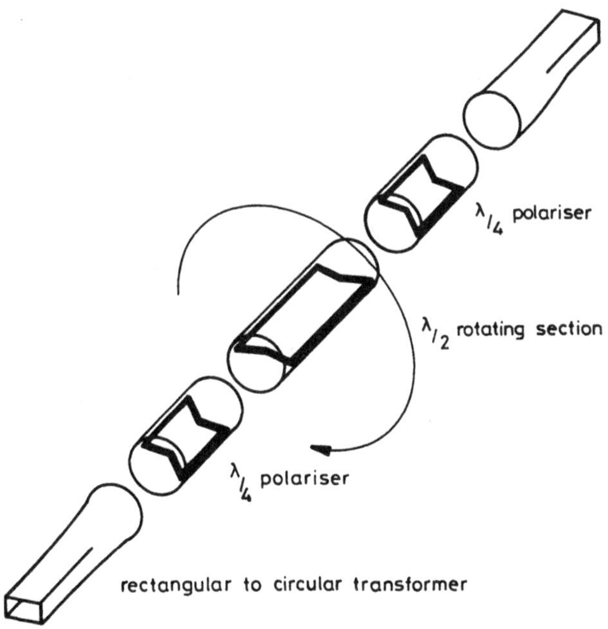

Fig. 9.14 *Wattmeter utilising the torque due to the reversal of the angular momentum of a circularly polarised wave. The arrangement is similar to a common type of variable phase shifter*

by means of a half-wave rotatable dielectric vane, and finally the wave is converted back to a linearly polarised one. The torque arises from the reversal of the angular momentum and is independent of the angular position of the vane. The theoretical formula giving the torque in terms of the power flowing through the device can be derived directly from the change in angular momentum. An instrument based on this principle was later constructed and found to agree with thermal methods of power measurement, although few details of the results are given in the paper [36].

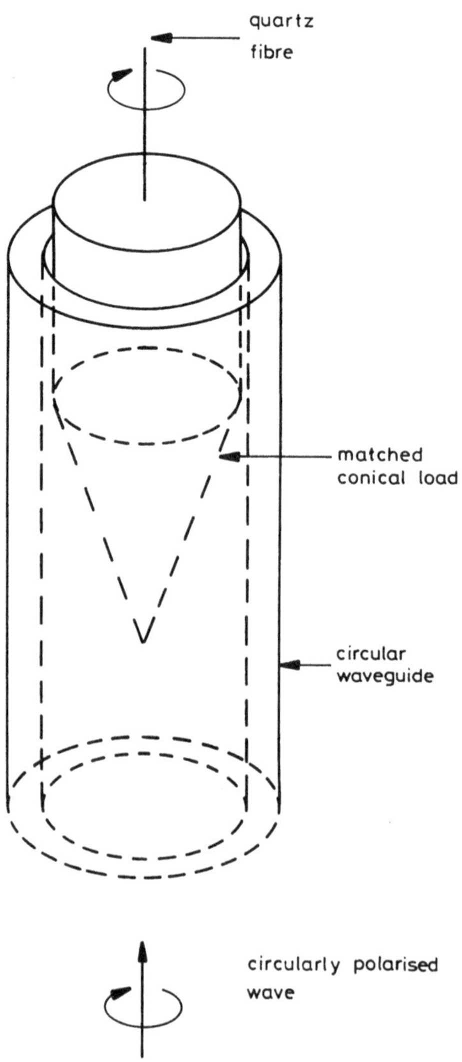

Fig. 9.15 *Power meter based on the torque produced by the absorption of angular momentum by a conical load*

In other instruments which have been built the torque is produced by the absorption of angular momentum in a dissipative load [37]. One would expect that thermal convection currents due to the heat generated by the absorbed power would pose a serious problem in such a device, but the effects are greatly reduced by the circular symmetry of the waveguide and load provided that the instrument is operated with its axis vertical (Fig. 9.15). Measurements on a power meter operating at 35 GHz were claimed to be the first accurate verification of the formula for angular momentum in waveguide [11]. A second instrument was built at 10 MHz, where the angular momentum for a given power is much greater [38] (see equation 9.12).

9.4.7 Other instruments

References [39–49] relate to various types of force-operated power meter.

9.5 Discussion

Of the large number of experimental force-operated power meters which have been constructed, none has been developed to the stage where it could compete either with the best calorimeters in terms of accuracy or with everyday instruments such as thermistor, thermoelectric and diode power meters in terms of convenience. Consequently their practical application has been extremely limited and their main contribution to the measurements field lies in the fact that the comparisons which have been carried out with thermal instruments provide confirmation that the latter are not seriously in error. From a more general point of view such comparisons also provide added confidence that the formulas for calculating forces due to electromagnetic fields are correct, in particular those using the concepts of linear and angular momentum.

9.6 References

1 Arnold, A.H.M., 'Audio frequency power measurements by dynamometer wattmeters', *Proc. IEE*, **102**, Pt B, 2, pp. 192–203, Mar. 1955.
2 Cullen, A.L., 'Absolute power measurement at microwave frequencies', *Nature*, **163**, p. 413, 1949.
3 Cullen, A.L., 'Absolute measurement of microwave power by radiation pressure', *Nature*, **165**, p. 726, 1950.
4 Cullen, A.L., 'Absolute measurement of microwave power in terms of mechanical forces', *Nature*, **167**, p. 812, 1951.
5 Cullen, A.L., 'Absolute power measurement at microwave frequencies', *Proc. IEE*, **99**, Pt. IV, 2, pp. 100–110, April 1952.
6 Maxwell, J.C., 'A treatise on electricity and magnetism', 1st edition, Oxford University Press, 1873, p. 391.
7 Bartoli, A., 'On movement produced by light and heat', Le Mounier, Florence, 1876 (in Italian).

8 Vigoureux, P., 'Radiation pressure', *Proc. IEE*, **125**, 7, pp. 709–713, July 1978.
9 Barlow, H.E.M., 'New proposals for the measurement of microwave power by radiation pressure', *Proc. IEE*, **133**, 7, pp. 1243–1247, July 1966.
10 Barlow, H.M., 'Some new high-frequency wattmeters operated by forces arising from a rate of change of electromagnetic momentum', *IEEE Trans.*, **IM-15**, 4, pp. 333–337, Dec. 1966.
11 Steven, D.H., and Cullen, A.L., 'Angular-momentum wattmeter for the 8 mm waveband', *Proc. IEE*, **110**, 11, pp. 1968–1974, 1963.
12 Lebedew, P., 'Reports presented to the International Congress of Physics', Paris, 1900, 2, p. 133 (in French).
13 Nichols, E.F., and Hull, G.F., 'The pressure due to radiation', *Proc. American Academy of Arts and Sciences*, **38**, p. 559, 1903.
14 Jones, R.V., 'Radiation pressure in a refracting medium', *Nature*, **167**, pp. 439–440, 1951.
15 Cook, J.J., Flowers, W.F., and Arnold, C.B., 'Measurement of radiation pressure at light frequencies', *Proc. IRE*, **50**, 7, p. 1693, July 1962.
16 Husson, S., 'Mechanical action exerted on a conductor by electromagnetic waves', *Comptes Rendus*, **191**, p. 33, 1930 (in French).
17 Fritz, K., 'Measurement of the ponderomotive radiation force on resonators in an electromagnetic field', *Annalen der Physik*, **11**, p. 987, 1931 (in German).
18 Carrara, N., and Lombardini, P., 'Radiation pressure of centimetre waves', *Nature*, **163**, p. 171, 1949.
19 Beth, R.A., 'Mechanical detection and measurement of the angular momentum of light', *Phys. Rev.* **50**, pp. 115–125, 1936.
20 MacLean, W.R., 'The resonator action theorem', *Quarterly Journal of Applied Mathematics*, **2**, p. 329, 1945.
21 Cullen, A.L., 'A general method for the absolute measurement of microwave power', *Proc. IEE*, **99**, Pt IV, 2, pp. 112–120, April 1952.
22 Cullen, A.L., Rogal, B., and Okamura, S., 'A wide-band double-vane torque-operated wattmeter for 3 cm microwaves', *IRE Trans.*, **MTT-6**, 2, pp. 133–136, April 1958.
23 Cullen, A.L., and Stephenson, I.M., 'A torque-operated wattmeter for 3-cm microwaves', *Proc. IEE*, 99, pt. IV, 4, pp. 294–301, Dec. 1952.
24 Bailey, R.A., 'A resonant-cavity torque-operated wattmeter for microwave power', *Proc. IEE*, **103**, Pt.C, 3, pp. 59–63, Mar. 1956.
25 Cullen, A.L., and French, H.A., 'An instrument for the absolute measurement of low-level microwave power in the 3 cm band', *Proc. IEE*, **104C**, 5, pp. 456–464, Mar. 1957.
26 French, H.A., and Cullen, A.L., 'An oscillator torque-operated wattmeter for the 8 mm waveband', *Proc. IEE*, **109B**, 48, pp. 511–514, Nov. 1962.
27 Barlow, H.E.M., and Cross, P.H., 'Radiation pressure instrument for absolute measurement of power at 35 GHz', *Proc. IEE*, **117**, 4, pp. 853–856, April 1970.
28 Beers, Y, 'Measurement of millimetre wave power by radiation pressure', NBS Report 7913, July 1963.
29 Sucher, M., and Fox, J., 'Handbook of microwave measurements', Vol. 1, Polytechnic Press, 1963.
30 Demyankov, I.F., Mathew, M.G. and Koryuishii, T., 'Relative sensitivity improvement in side wall moving vane type microwave power meter', *IEEE Trans.*, **IM-26**, 1, pp. 92–95, Mar. 1977.
31 Barlow, H.E.M., 'A microwave electrostatic wattmeter', *Proc. IEE*, **110**, 1, pp. 85–87, 1963.
32 Wizner, W., 'A new method for absolute measurement of microwave power', Ph. D. Thesis, Univ. of London, 1968.
33 Barlow, H.E.M., and Wizner, W.W., 'Microwave electrostatic wattmeter', *Proc. IEE*, **117**, 1, pp. 249–254, Jan. 1970.
34 Barlow, H.M., and Wizner, W.W., 'A new method for absolute measurements of microwave power', *Prace Przemysl. Inst. Telekomun.*, **20**, 67 pp. 1–12, 1970 (in Polish).

35 Fox, A.G., 'An adjustable waveguide phase changer', *Proc. IRE*, **35**, 12, p. 1489, Dec. 1947.
36 Barlow, H.E.M., and Williams, A.E., 'Power measurements at 10 Gc/s by reversal of the angular momentum of a circularly polarised field', *Proc. IEE*, **113**, 7, pp. 1248–1252, July 1966.
37 Barlow, H.E.M., 'Measurement of power by absorption of the angular momentum of a circularly polarized wave', *Proc. IEE*, **110**, 5, pp. 865–868, 1963.
38 Barlow, H.E.M., and Tay, E.S., 'Power measurement at 10 Mc/s by absorption of the angular momentum of a rotating field', *Proc. IEE*, **113**, 7, pp. 1253–1255, July 1966.
39 Braginsky, V.B., Minakova, I.I., and Stepunin, P.M., 'The absolute measurement of small powers in the millimeter waveband', *Instr. and Exp. Tech.*, **1963**, 5 (translation of Pribory i Tekh. Eksp., 5, p. 130, Sept–Oct. 1963).
40 Orlov, V.G., 'Absolute calibration of ponderomotive wattmeters with losses', *Meas. Tech.*, **64**, 5, pp. 431–433 (translation of Izmeritel'naya Tekhnika, pp. 46–48, May 1964).
41 Valitov, R.A., Zhilkov, V.S., Kukush, V.D., Sen'ko, A.P., and Terekhov, M.V., 'Comparing the readings of high precision calorimetric power meter and a stationary ponderomotive wattmeter', *Meas. Tech.*, **1970**, 4, p. 626, April 1970.
42 Kukush, V.D., 'Errors due to losses during absolute calibration of ponderomotive wattmeters', *Radiotekhnika*, **18**, pp. 80–84, 1971.
43 Wizner, W., 'Microwave instruments applying ponderomotive forces', *Pr. Przem. Inst. Telekomun.*, **22**, 76, pp. 24–33, 1972 (in Polish).
44 Zhilkov, V.S., and Sirotnikov, A.I., 'The error of a 2-plate ponderomotive wattmeter due to higher wave types', *Radiotekhnika*, **22**, pp. 100–103, 1972 (in Russian).
45 Zhilkov, V.S., Shpagin, Yu.V., Serednii, V.P., Bliznyuk, V.G., Naryzhnyi, and V.G., Klimashevskii, V.S., 'Combined ponderomotive SHF power meter', *Radiotekhnika*, **24**, pp. 100–104, 1973 (in Russian).
46 Zhilkov, V.S., 'Considering the losses in an electromagnetic resonator when giving support to the ponderomotive method of SHF power measurement, II', *Radiotekhnika*, **28**, pp. 151–3, 1974 (in Russian).
47 Dem'yankov, I.F., Khazhmuradov, M.A., Stoyan, Yu.G., and Kukush, V.D., 'Error minimisation in ponderomotive microwave wattmeters', *Radiotekhnika*, **32**, pp. 132–139, 1975 (in Russian).
48 Serednii, V.P., and Zhilkov, V.S., 'A highly-accurate ponderomotive meter for moderate and large microwave powers', *Radiotekhnika*, **38**, pp. 120–124, 1976 (in Russian).
49 Dem'yankov, I.F., Berlin, G.S., and Pereyaslavets, V.A., 'Mechanical calibration procedure for a microwave pressure wattmeter with a mechanotron transducer', *Radiotekhnika*, **42**, pp. 129–133, 1977 (in Russian).

Chapter 10

Other types of power meter

Many different physical phenomena have been considered as a possible basis for power measurement. The present chapter collects together various approaches which represent alternatives to the more common power meters described in Chapters 2 to 9. The instruments can be divided into three groups. The devices described in Sections 10.1 to 10.4 are thermal instruments employing temperature sensors other than the commonly used thermocouple, resistance thermometer, and bolometer. Sections 10.5 to 10.7 describe sensors based on non-thermal effects and Sections 10.8 to 10.11 describe instruments intended for use as standards.

10.1 Pyroelectric detectors

Pyroelectric detectors make use of insulating materials possessing a spontaneous temperature-dependent electric polarisation [1–4]. For these materials, surfaces normal to the direction of polarisation exhibit a temperature-dependent surface charge which can be detected in principle by attaching a pair of electrodes connected to a high impedance voltmeter (Fig. 10.1). In practice, only changes in the surface charge can be observed in this manner,

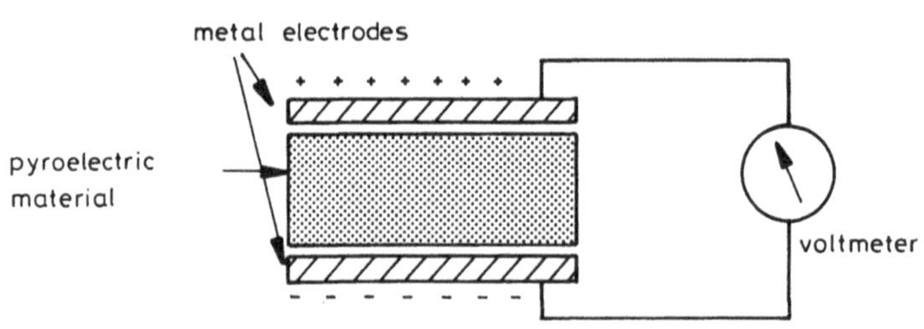

Fig. 10.1 *Principle of the pyroelectric detector*

since the charge corresponding to a constant temperature is eventually neutralised by stray charges. Thermal detectors based on the pyroelectric effect have been used in various parts of the electromagnetic spectrum [5-7], ranging from the higher microwave frequencies to the ultra-violet. They are relatively simple and inexpensive, but require a modulated signal. A popular material for the submillimetre and infra-red regions is TGS (triglycine sulphate), but for microwave applications the polymer polyvinylidene fluoride (PVF_2) is used [8]. This possesses the advantage that it is easily made into a variety of forms including sheets and films, although its pyroelectric coefficient is considerably lower than that of TGS. Films are made by conventional polymer extrusion techniques followed by mechanical drawing and 'poling'. The poling process consists of the application of an electric field of about 8 MV m^{-1} across the film at a temperature of about 100°C.

Fig. 10.2 shows an experimental pyroelectric detector [9] intended for operation at 100 GHz. The pyroelectric element consists of a 9 μm thick PVF_2 film on which are deposited two NiCr electrodes. These electrodes also serve as the absorbing resistors for the microwave power. The element is placed in the centre plane of a waveguide to form a matched load, with a suitable arrangement for making dc connection to the electrodes. The output voltage is amplified by a high impedance FET amplifier. Measurements taken at 100 GHz with 30 Hz square wave modulation gave voltage responsivities (output voltage/input power) of 230 and 300 V/W respectively for two different element widths. The noise equivalent power (NEP) (see Chapter 8) was 8 nW for a 1 Hz bandwidth.

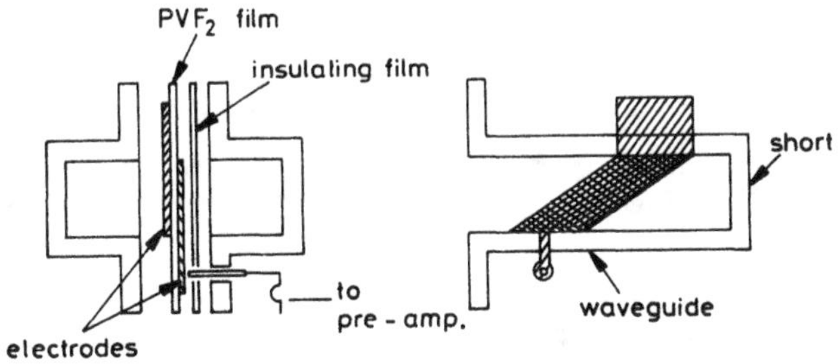

Fig. 10.2 *Construction of a waveguide pyroelectric detector for use at 100 GHz. The microwave power, which is modulated at a frequency of typically 30 Hz, is absorbed in the electrodes. The pyroelectric material is a PVF_2 film*

The NEP of the ideal pyroelectric detector is independent of the modulation frequency of the microwave signal provided that this is higher than the reciprocal of the thermal time constant, as is usually the case [1]. Given a suitable frequency-compensated amplifier, therefore, the pyroelectric detector can follow rapidly modulated signals without the signal-to-noise ratio becoming unacceptable. This property makes it useful for measurement of pulse-modulated power. By contrast, the signal-to-noise ratio of most thermal sensors such as thermistors, barretters, and thermoelectric power meters deteriorates with increasing modulation frequency. This is because, as the modulation frequency increases, the sensor is unable to follow the rapid variations and the output falls, whereas the thermal noise generated by the resistance of the sensor has a flat frequency response. In a non-ideal pyroelectric detector the maximum useful bandwidth is limited by the thermal noise associated with the leakage conductance of the pyroelectric material and by the amplifier noise.

10.2 Differential air thermometer power meter

The differential air thermometer power metre [10] was developed during the 1950s. The principle on which it is based is similar to that of the older differential air milliammeter [11–13]. Two glass cells are connected by a capillary tube containing a liquid pellet (Fig. 10.3). The rf power is arranged to

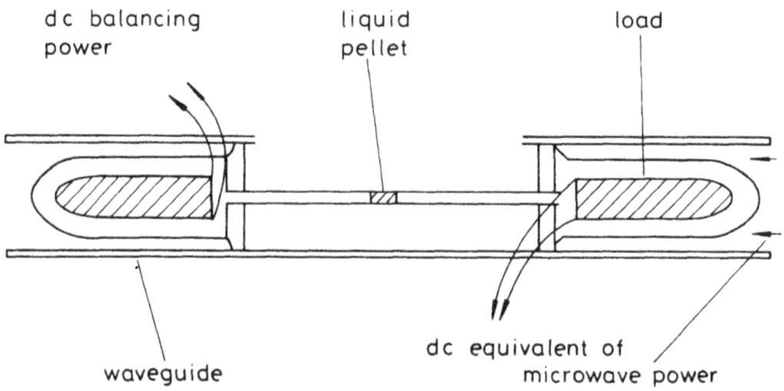

Fig. 10.3 *Differential air thermometer power meter. The absorbed microwave power heats the load cell, causing the air in it to expand and the liquid pellet to move*

heat one cell, causing the air in it to expand and the liquid pellet to move. The other cell is heated by a dc power which is adjusted to restore the pellet to its original position.

10.3 Temperature-limited thermionic diode

In the temperature-limited thermionic diode [14] the rf power to be measured is applied to the heater of the diode. The rise in temperature of the cathode is detected by monitoring the anode current. Richardson's law applies:

$$I = AT^2 \exp\left(-\frac{B}{T}\right) \qquad (10.1)$$

where I is the anode current, T the absolute temperature of the cathode, and A and B are constants which are characteristic of the cathode material. A commercial version incorporating a feedback system had a claimed uncertainty of 2 percent up to 300 MHz when used as an rf/dc substitution instrument [15].

10.4 Photometer

In the photometer, the rf power to be measured is dissipated in the filament of a lamp and the emitted light is sensed by means of a photocell. Photometers were replaced for power measurement by bolometric instruments, although a similar principle was later used in an rf current measuring instrument in which the infra-red radiation from a heater was focussed onto a thermopile [16].

10.5 Gas discharge

A capillary tube containing neon gas will glow if exposed to microwave power at a sufficiently high level [17,18]. This technique can be applied both to continuous wave and to pulse power measurement. The light output may be measured with a photocell. In a dc-biased neon lamp there is an increase in current when the lamp is exposed to microwaves. This phenomenon has been used for the indication and detection of microwave power [19–23]. The detection mechanism has been studied by Kopeika [24,25].

10.6 Hall effect

Some experimental work was carried out in the 1950s on the use of the Hall effect for rf and microwave power measurement. When a magnetic field is applied to a current-carrying conductor, an electric field is developed in a direction perpendicular to both the magnetic field and the current [26,27]. This electric field, which arises from the sideways force on the moving charges, is proportional to the product of the current and the magnetic field. It is given by the vector equation:

$$\boldsymbol{E} = R\,\boldsymbol{J} \wedge \boldsymbol{B} \tag{10.2}$$

where $\boldsymbol{E}$ is the electric field, R the Hall coefficient, $\boldsymbol{J}$ the current density, and $\boldsymbol{B}$ the magnetic flux density. The Hall effect can be used as the basis for an analogue multiplier by arranging for $\boldsymbol{J}$ and $\boldsymbol{B}$ to be proportional to the two quantities to be multiplied together. For power measurement these two quantities are either the instantaneous voltage and current or the instantaneous electric and magnetic fields. Thus it is possible to construct a feedthrough instrument which responds directly to the net power flow. The

Hall coefficient varies inversely with the density of the charge carriers and is therefore much greater for semiconductors than for metals. Its value is positive if the current flow is due to holes and negative if the current flow is due to electrons.

The application of the Hall effect to power measurement has been demonstrated at frequencies ranging from 50 Hz to approximately 10 GHz [28–35]. Fig. 10.4 illustrates the application of the principle to a waveguide power sensor. A small slab of semiconductor is placed at the centre of the guide parallel to the narrow face. The electric field produces a current I in the

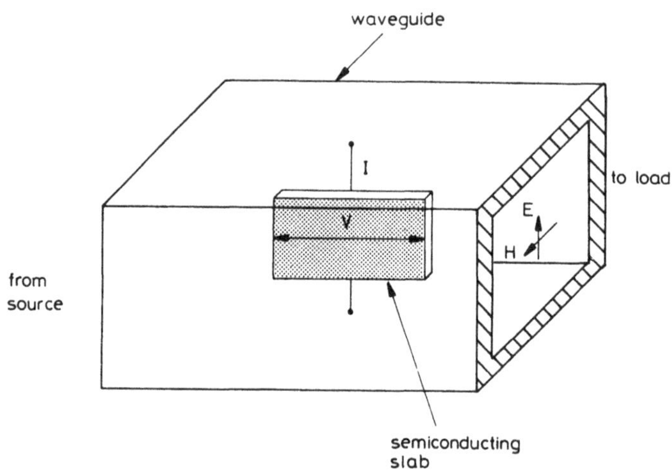

Fig. 10.4 *Principle of the Hall effect power meter. The voltage V is proportional to the net power flow provided that the current I is in phase with the electric field which gives rise to it*

semiconductor and this current interacts with the transverse magnetic field of the wave. The resulting Hall emf V, which acts along the axis of the guide, is proportional to the instantaneous net power flow, irrespective of the load impedance, and is given by:

$$V = \left(\frac{R}{t}\right) I B \tag{10.3}$$

where t is the thickness of the slab. Two leads attached to electrodes on the semiconductor are brought out through the narrow faces of the guide so that the dc component of V, which is proportional to the average power, may be measured. It is important that the current I should be exactly in phase with the electric field and to ensure this a tuning arrangement, not shown in the figure, may be incorporated, comprising a length of coaxial line and an adjustable piston. The piston must be adjusted at each frequency of operation so that a zero reading is obtained with the guide terminated in a short circuit.

Unfortunately the Hall effect power meter suffers from a number of practical difficulties. The sensitivity is very low: two versions described in ref. [31] using germanium gave values of 7.5 μV/W and 5 μV/W at a frequency of 9.360 GHz. Some of the power is absorbed in the semiconductor material, causing heating, which leads to a change in the Hall coefficient. Rectifying contacts give rise to unwanted voltages, as do thermoelectric effects caused by non-uniform heating of the semiconductor at high values of load reflection coefficient. These problems prevented the development of this type of instrument into a practical rf and microwave power meter. Other ways of utilising the Hall effect are given in ref. [36].

10.7 Magnetic effects

The magnetoresistive effect [37] is the increase in the dc resistance of a semiconductor in the presence of a magnetic field. Some work has been carried out on its application to rf and microwave power measurement [38–43]. Other types of sensor using magnetic effects are described in refs. [44–52].

10.8 Electron beam power standard

The principle of the electron beam power standard is that a beam of electrons is velocity-modulated by the electric field of the microwave signal and the microwave power level is determined by measuring the depth of modulation. One of the advantages of the method is that the beam does not significantly disturb the fields. An instrument of this kind was developed by Thomas [53]. Fig. 10.5 shows the basic arrangement. A monochromatic beam of electrons traverses the narrow dimensions of a rectangular waveguide. When the microwave signal is present in the guide, the electric field accelerates or decelerates the electrons according to their time of arrival and the time spent in the field. Maximum modulation occurs when the electrons pass through the field in a time equal to a half period of the microwave signal. The depth of modulation is determined by measuring the stopping potential required to bring the most energetic electrons in the beam to rest, both with and without the presence of microwave power. This is done by adjusting the repeller voltage until the current just ceases to flow. The instrument is not a true power meter, as it responds only to electric field and the power reading is correct only for a matched load. Disadvantages of Thomas's version of the instrument were the high operating power level of several hundred watts and the uncertainty about the field around the apertures in the waveguide.

Many years later Oldfield and Ide at the Royal Signals and Radar Establishment in Britain built a more refined electron beam power standard

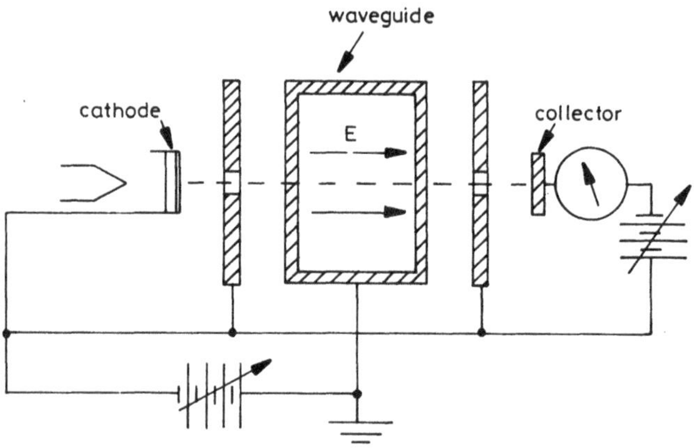

Fig. 10.5 *Thomas's electron beam power standard. The electrons are velocity-modulated by the microwave field and the microwave power is deduced by determining the stopping potential which will just bring the most energetic electrons to rest*

[54,55]. The purpose of this instrument, which operated at a frequency of approximately 9 GHz, was to provide an independent check on calorimetric power standards, which by this time had reached an advanced state of development. By the use of a resonating cavity the field was magnified so that the power level could be reduced to tens of milliwatts. The instrument was circularly symmetrical and resembled a reflex klystron in construction (Fig. 10.6). The main parts were: the electron gun, the magnetic lens, the cavity,

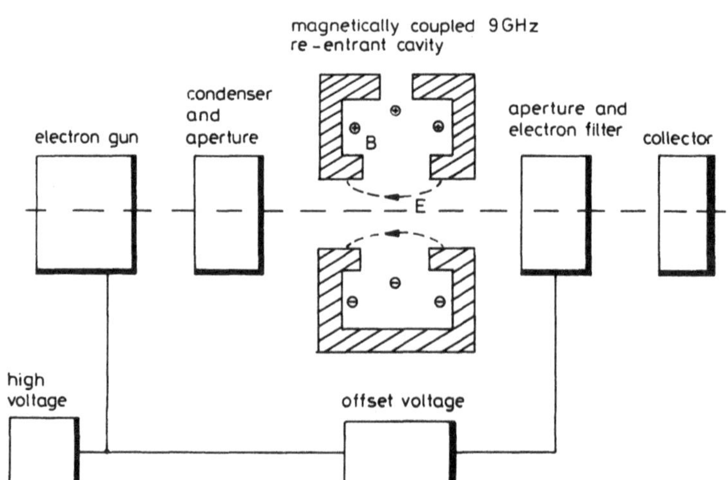

Fig. 10.6 *Oldfield and Ide's electron beam power standard. All components through which the beam passes are of circular cross section*

and the electron filter which enabled the stopping potential to be determined. The electron gun was of the type used in electron microscopes and it produced a fine beam of electrons at a current of 0.1 μA and an energy of about 12 keV. This beam, which had a diameter of 40 μm at the anode, was made parallel by the magnetic lens, the system apertures and the drift spaces. The electric field across the 3 mm wide cavity gap modulated the energy of the electrons. In order to relate this field to the microwave power it was necessary to know the cavity field distribution and the Q factor. The electron filter was of a special design. The electrons passed through an aperture in the barrier electrode and were then re-accelerated to a separate collector at or near earth potential. This scheme had the merit that surface charging of the collector caused by contamination was unimportant, although the need for an aperture in the barrier electrode leads to perturbation of the electron trajectories and uncertainty in the value of the stopping potential. These effects were minimised by choosing a special T-shaped configuration for the barrier electrode. A further contribution to uncertainty arose from the energy spread, typically 1 eV, within the electron beam due to thermal and electrostatic effects. However, using the electron energy spread characteristic determined in the absence of microwave power, and a convolution technique, the stopping potential was resolved to 0.015 eV.

Comparisons between Oldfield and Ide's electron beam standard and a microcalorimeter gave agreement of 0.7 percent or better, which is within the estimated uncertainties.

10.9 Superconductivity

It has been known since 1911 that below a critical temperature, the electrical resistance of certain metals falls to zero. Superconductors also exhibit the Meissner effect, that is the magnetic flux inside the conductor is zero. An equally important discovery was that the magnetic flux linking a superconducting loop is quantised. These properties are responsible for a number of phenomena which were predicted theoretically by Josephson [56,57] in 1962 and are known as the Josephson effects. Techniques based on the Josephson effects have contributed greatly to many areas of measurement science [58,59]. They have made possible highly accurate dc voltage standards (see Chapter 1). sensitive magnetometers, and efficient detectors and mixers for the submillimetre and infra-red regions. Their practical impact in the rf and microwave area, however, has so far been fairly small. The use of SQUIDs [60,61] (superconducting quantum interference devices) in rf metrology was first investigated in the early 1970s and this work can be seen as the first attempt at making rf quantities directly traceable to standards based on quantum units. In theory a SQUID can be used to measure rf current in terms of the magnetic flux quantum [62]. Voltage and power can then be derived with the help of impedance standards. In practice, however, power measurement by this method is very difficult because it requires a precise knowledge of the geometry of the SQUID and the ability to calculate the field distribution accurately. The experimental work which has been reported to date has therefore relied on the rf/dc substitution principle, which requires only that the field distribution be the same for rf and dc inputs. Even so, the uncertainties which were achieved were much worse than those obtainable from calorimetric power standards, and the agreement with these standards was no better than 0.1 dB after correcting for rf losses in the input line [63–65]. The upper frequency was also limited because of the need for an interrogating signal at a considerably higher frequency than that of the signal to be measured. At the time an interrogating signal of 10 GHz was used, and the measured signal had a maximum frequency of about 1 GHz. On the other hand, the use of the SQUID as an attenuation standard below 1 GHz has been reasonably successful and comparisons with more conventional methods, such as the waveguide beyond cut-off attenuator, have resulted in agreement in the region of 0.002 dB, or 0.05 percent in power ratio, at 30 MHz [63,64,66–69].

A brief outline of the operation of the SQUID is as follows. The magnetic flux linking a superconducting loop can take only values which are multiples of the flux quantum $h/2e$, where h is Planck's constant and e is the charge of the electron. In this case application of an external magnetic field results not in a change in the flux linking the loop, but in circulating superconducting currents which oppose the change. A SQUID is a superconducting loop which contains a weak link, known as a Josephson junction. This weak link may be a thin insulating gap, one form of which is shown in Fig. 10.7, or a 'cat's whisker'

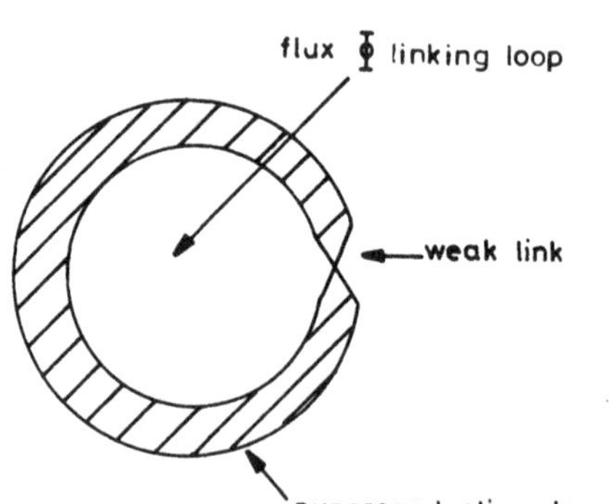

Fig. 10.7 *The basis of the SQUID (Superconducting Quantum Interference Device). The value of the dc current flowing around the superconducting loop is a periodic function of the magnetic flux linking the loop*

point contact. Superconducting currents may flow through the weak link by tunnelling. There is a maximum value I_s which the current cannot exceed without destroying the superconductivity of the weak link. This maximum current, which is a property of the weak link, is known as the critical current. For the device described in refs. [63] and [64] its value is about 16 μA. The presence of the weak link means that the flux linking the loop need no longer be exactly quantised but can be varied by altering the external magnetic field. It is found that the current I in the loop is a periodic function of the magnetic flux Φ linking the loop, and that the period is equal to the flux quantum $h/2e$. The application of a slowly increasing external magnetic field causes both the current I and the inductance L to execute cyclic variations, one complete cycle corresponding to a change of one quantum in the externally applied flux linking the loop. The inductance variations can be sensed by means of a microwave interrogating signal which enables the magnetic flux quanta to be counted.

For larger values of I_s, the impedance presented by the SQUID to the interrogating signal has also a periodically varying dissipative component. This is a hysteresis effect which arises because the curve relating Φ to the external magnetic field is multivalued. Most SQUIDS operate in the dissipative mode and in this case it is the periodic variation of the absorbed power which is sensed, rather than the inductance. This is done by monitoring the reflected power at the interrogating frequency with the aid of a directional coupler and detector. Fig. 10.8 shows one of several possible versions. In this arrangement

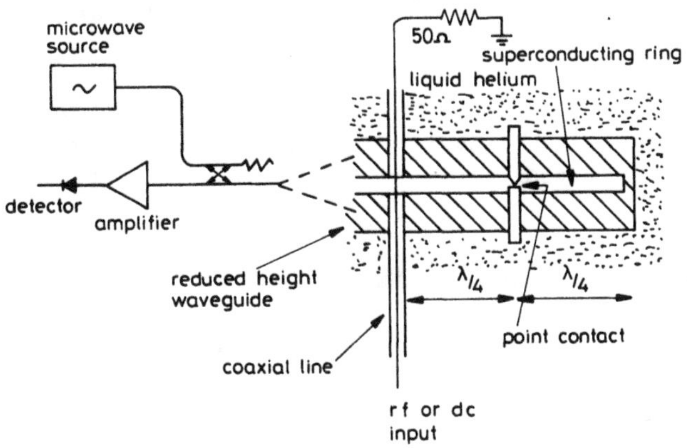

Fig. 10.8 *A practical implementation of the SQUID. The reflection coefficient at the interrogating microwave frequency varies cyclically with the amplitude of the rf or dc current in the coaxial line*

the superconducting loop is formed by connecting a niobium point contact Josephson junction across a short circuited low impedance waveguide made of Babbit (an alloy of tin, lead, antimony and copper). The external magnetic flux is produced by inductive coupling with the rf or dc current flowing in the inner conductor of a coaxial line. For an rf input the quantity of interest is the time average of the reflected power at the interrogating frequency rather than its instantaneous value. It can be shown that, as a function of the rf current amplitude, this average reflected power is represented by a zero order Bessel function plus a constant. In a practical device the constant is eliminated by a technique which uses a 1 kHz modulating signal. The zero crossings of the Bessel function thus correspond to a series of precisely calculable rf current values, which can be found from a table or roots of Bessel functions. These current values are not evenly spaced and the scale which is generated in this way is non-uniform. Power measurements are derived by allowing the current to flow through a known impedance. In a SQUID constructed at the National Bureau of Standards [63,64] the first zero crossing corresponded to an rf power level of 1 nW into a 50 Ω load. The power level for the 100th zero crossing was about 45 dB above this level. A problem from which SQUIDs suffer is that the periodic relationship between the instantaneous values of Φ and I is assumed in the theory to be sinusoidal, whereas a pure sinusoid is rarely obtained in practice and the presence of higher order terms leads to systematic errors.

10.10 Saturation in gases

Some work has been carried out on the use of the saturation of the absorption spectrum of gases in order to provide a known power level [70,71]. In ref. [71] the use of a 7 m long waveguide filled with ammonia is described. The microwave power level at the resonant frequency of the gas molecules (23.87013 GHz) is deduced by noting the decrease in the absorption coefficient due to the power saturation effect. The result is calculated from the molecular constants of the gas. This method is confined to specific frequencies, which depend on the gas used.

10.11 Power measurement using noise standards.

Noise standards make use of the black body radiation from a resistor at a known temperature. The power per unit bandwidth is calculable and therefore by passing the output through a band pass filter with a known bandwidth it is possible to obtain a band-limited noise signal whose total power is calculable. The difficulties are that the power is very small and that the filter characteristics must be known accurately, including any spurious responses. Ref. [72] describes an implementation using a thermal radiator enclosed in an untuned cavity.

10.12 References

1 Putley, E.H., 'The pyroelectric detector in semiconductors and semimetals', ed., Willardson, R.K., and Beer, A.C., Vol. 5, Academic Press, 1970.

2 Lang, S.B., 'Literature guide to pyroelectricity', *Ferroelectrics,* **19**, p. 175, 1978.

3 Liu, S.T., and Long, D., 'Pyroelectric detectors and materials', *Proc. IEEE,* **66**, 1, pp. 14–26, Jan. 1978.

4 Hadni, A., 'Applications of the pyroelectric effect', *J. Phys. E,* **14**, pp. 1233–1240, 1981.

5 Alday, J.R., Everett, G.E., and White, D.J., 'Improved response of pyroelectric millimetre wave detectors', *IEEE Trans.*, **MTT-14**, 2, p. 100, Feb. 1966.

6 Ludlow, J.H., Mitchell, W.H., Putley, E.H., and Shaw, N., 'Infrared radiation detection by the pyroelectric effect', *J. Sci. Instrum.,* **44**, pp. 694–696, 1967.

7 Hadni, A., Thomas, R., Mangin, J., and Bagard, M., 'Pyroelectric detector for submillimeter waves', *Infrared Phys.,* **18**, p. 663, 1978.

8 Gallantree, H.R., 'Review of transducer applications of polyvinylidene fluoride', *Proc. IEE,* **130**, Pt. I, 5, pp. 219–224, Oct. 1983.

9 Iwasaki, T., Inoue, T., and Nemoto, T., 'A matched-load type PVF_2 pyroelectric detector for millimeter waves', *IEEE Trans.,* **IM-28**, 1, pp. 88–89, Mar. 1979.

10 Gordon-Smith, A.C., 'A milliwattmeter for centimeter wavelengths', *Proc. IEE,* **102**, Pt. B, 5, pp. 685–686, Sept. 1955.

11 Schmitz, W., 'Differential air thermometer for ac current measurement', *Z. Hochfrequenz-tech.*, **27**, pp. 18–19, 1926.

12 Strutt, M.J.O., and Knol, K.S., 'Measurement of currents and voltages down to a wavelength of 20 centimeters', *Proc. IRE,* **27**, 12, p. 783, Dec. 1939.

13 Gainsborough, G.F., 'Experiments with thermocouple milliammeters at very high radio frequencies', *J. IEE,* **91**, Pt. 3, 15, pp. 156–161, 1944.

14 Cambell, R.D., 'The diotron, an aid to rms instrumentation', *Electronics*, **23**, 7, 1950.

15 Carlin, H.J., 'Measurement of power', Chap. 3 in 'Handbook of microwave measurements', ed. Sucher, M., and Fox, J., Polytechnic Press, 1963, p. 222.

16 Scott, W.W., 'New coaxial rf-dc ammeter', *IEEE Trans.*, **IM-19**, 4, pp. 318–323, Nov. 1970.

17 Griesheimer, R.N., 'Microwave power measurements', Chapter 3 in 'Technique of microwave measurements', ed. C.G. Montgomery, McGraw-Hill, 1947.

18 Carlin, H.J., 'Measurement of power', Vol. 1, Chapter 3 in 'Handbook of microwave measurements', ed. M. Sucher and J. Fox, Polytechnic Press, 1963.

19 Burroughs, H.A., and Bronwell, A.B., 'An ionized gas energy detector for microwaves', *Proc. NEC,* **7**, pp. 598–600, Chicago, 1951.

20 Severin, P.J.W., and van Nie, A.G., 'A simple and rugged wide band gas discharge detector for millimeter waves', *IEEE Trans.*, **MTT-14**, 9, pp. 431–436, Sept. 1966.

21 Movchan, S.P., Zatenko, N.A., and Kononenko, K.I., 'Waveguide gas discharge indicator of flowing microwave power', *Radio Engineering and Electronic Physics,* **15**, 3, pp. 429–432, 1970.

22 Farhat, N.H., and Kopeika, N.S., 'A low cost millimeter-wave glow-discharge detector', *Proc. IEEE,* **60**, 6, pp. 759–760, Jan. 1972.

23 Farhat, N.H., 'Optimisation of millimeter wave glow-discharge detectors', *Proc. IEEE,* **62**, 2, pp. 279–281, Feb. 1974.

24 Kopeika, N.S., 'On the mechanism of glow discharge detection of microwave and millimeter wave radiation', *Proc. IEEE,* **63**, 6, pp. 981–982, June 1975.

25 Kopeika, N.S., 'Correction to "On the mechanism of glow discharge detection of microwave and millimeter wave radiation"', *Proc. IEEE,* **63**, 12, pp. 1737, Dec. 1975.

26 Barlow, H.E.M., 'Hall effect and its counterpart, radiation pressure, in microwave power measurements', *Proc. IEE,* **104C**, 5, pp. 35–42, Mar. 1957.

27 Chien, C.L., and Westgate, C.R. (eds.), 'The Hall effect and its applications', Plenum, 1980.

28 Barlow, H.E.M., 'The application of the Hall effect in a semi-conductor to the measurement of power in an electromagnetic field', *Proc. IEE,* **102**, Pt. B, 2, pp. 179–185, Mar. 1955.

29 Barlow, H.E.M., 'The design of semi-conductor wattmeters for power frequency and audio-frequency applications', *Proc. IEE,* **102**, Pt. B, 2, pp. 186–199, Mar. 1955.

30 Barlow, H.E.M., and Stephenson, L.M., 'The Hall effect and its application to power measurement at microwave frequencies', *Proc. IEE*, **103**, Pt. B, 7, pp. 110–112, Jan 1956.
31 Barlow, H.E.M., and Kataoka, S., 'The Hall effect and its application to power measurement at 10 Gc/s', *Proc. IEE*, **105B**, 19, pp. 53–60, Jan. 1958.
32 Stephenson, L.M., and Barlow, H.M., 'Power measurement at 4 Gc/s by the application of the Hall effect in a semiconductor', *Proc. IEE*, **106**, Pt. B, 25, pp. 27–30, Jan. 1959.
33 Barlow, H.E.M., 'A new approach to the measurement of HF power by Hall effect in a semiconductor', *Proc. IEE*, **109**, Pt. B, suppl. 23, pp. 746–749, May 1962.
34 Leavenworth, R. A., Maczuk, J., and Schwartz, R.F., 'The application of Hall-effect devices to the measurement of microwave power density', *IRE Trans.*, **I-11**, 3–4, pp. 264–269, Dec. 1962.
35 Gashka, K., and Repshas, K., 'A Hall-effect power meter', *Izv. VUZ Radioelektron.*, **22, 5**, pp. 79–80, May 1979 (in Russian).
36 Barlow, H.M., 'A proposed new method of measuring microwave power and impedance using Hall effect in a semiconductor', *Proc. IEE*, **109B**, 45, pp. 286–289, May 1962.
37 Koike, R., and Barlow, H.E.M., 'Microwave measurements on the magneto-resistive effect in semiconductors', *Proc. IEE*, **109B**, 44, pp. 137–144, Mar. 1962.
38 Kataoka, S., and Kobayashi, S., 'The application of magnetoresistance effect in an intermetallic semiconductor to the measurement of electric power', *Bull. Electrotechnical Lab.*, Tokyo, Japan, pp. 1–8, Feb. 1962.
39 Kataoka, S., 'Application of the magnetoresistance effect in semiconductors to microwave power measurements', *Proc. IEE*, **113**, 6, pp. 948–956, June 1966.
40 Antropov, V.A., Antropova, Kh., Morgacheva, G.A., Kotosonov, N.V., and Shuvaev, A.G., 'On the use of the magneto-resistance effect for measuring UHF transmitting power', *Izv. VUZ Radiofiz.*, **15**, 9, pp. 1393–1397, 1972 (translation in *Radiophys. Quantum Electronics*).
41 Arkhipov, Yu.A., Kats, L.I., Terzhova, V.P., and Shekhter, D.Sh., 'Magnetoresistive feedthrough pulse power meter for the SHF band', *Instrum. and Exp. Tech.*, **18**, 1, Pt. 2, pp. 179–181, Jan./Feb. 1975 (translation of *Prib. Tekh. Eksp.*).
42 Komov, A.N., and Kudryakov, Yu.P., 'Magnetoresistive semiconductor power meter for the 3-cm-wavelength range', *Instrum. and Exp. Tech.*, **21**, 3, Pt. 2, pp. 720–722, May/June, 1978 (translation of *Prib. Tekh. Eksp.*).
43 Komov, A.N., and Kudryakov, Yu.P., 'Semiconductor disks to measure microwave power', *Meas. Tech.*, **22**, 6, pp. 49–50, June 1979.
44 Toda, M., 'Device application of ferromagnetic semiconductor for microwave power and frequency measurements', *Trans. Inst. Electron. Commun. Eng. Japan*, **53**, 7, pp. 16–17, July 1970.
45 Apenina, G.V., Balakov, V.F., Puchkov, V.S., and Volkov, M.A., 'Conversion coefficient of a ferrite sensor-convertor in a wide temperature interval', *Radiotekhnika*, **23**, pp. 57–59, 1972.
46 Bokrinskaya, A.A., Vuntesmeri, Vl.S., Vuntesmeri, Val.S., and Nebosenko, N.L., 'Ferrite power monitor for a one mode waveguide', *Meas. Tech.*, **17**, 9, pp. 1426–1428, Sept. 1974.
47 Lutshkov, V.S., Radchenkov, V.F., and Bilck, G.E., 'Power measurement of oscillations of extremely high frequency by means of ferrite data transducers', *Nachrichtentech. Elektron.*, **26**, 7, pp. 253–254, 1976 (in German).
48 Vuntesmery, V.S., 'Application of galvano-magnetic detectors based on ferromagnetic films to the measurement of microwave power flow in unmatched waveguide', Proc. 8th European Microwave Conf., Paris, Sept. 1978, pp. 513–518.
49 Vuntesmeri, V.S., 'A galvanomagnetic detector using a ferromagnetic film in a rectangular waveguide', *Izv. VUZ Radioelektron.*, **22**, 5, pp. 80–83, May 1979 (in Russian).
50 Vuntesmeri, V.S., 'Galvanomagnetic microwave power flow detectors based on ferromagnetic films', *Izv. VUZ Radioelektron.*, **23**, 3, pp. 28–34, Mar. 1980.

51 Vuntesmeri, V.S., 'Limiting parameters of galvano-gyromagnetic microwave pulse power convertors', *Izv. VUZ Radioelektron.*, **25**, 8, pp. 85–88, 1982 (translation in *Radioelectron and Commun. Syst.*).

52 Moronenko, V.P., 'A microwave power converter based on a thin magnetic film', *Izv. VUZ Radioelektron.*, **25**, 8, pp. 88–89, 1982 (in Russian) (translation in *Radioelectron. and Commun. Syst.*, **25**, 8, pp. 93–94, 1982).

53 Thomas, H.H., 'Microwave power measurements employing electron beam techniques', *Proc. IRE*, **45**, 2, pp. 205–211, Feb. 1957.

54 Oldfield, L.C., and Ide, J.P., 'An electron beam method for absolute measurement of microwave power', IEE Colloquium on the measurement of power at higher microwave frequencies, IEE Digest No. 1979/2, Jan. 1979.

55 Oldfield, L.C., and Ide, J.P., 'A fundamental microwave power standard', *IEEE Trans.*, **IM-32**, 2, pp. 443–449, June 1987.

56 Josephson, B.D., 'Possible new effects in superconducting tunnelling', *Phys. Lett.*, **1**, pp. 251–253, 1962.

57 Shapiro, S., 'Josephson currents in superconducting tunnelling: the effect of microwaves and other observation', *Phys. Rev. Lett.*, **11**, pp. 80–82, 1963.

58 Solymar, L., 'Superconducting tunnelling and applications', Chapman and Hall, 1972.

59 Gallop, J.C., 'The impact of superconducting devices on precision metrology and fundamental constants', *Metrologia*, **18**, 2, pp. 67–92, 1982.

60 Gallop, J.C., and Petley, B.W., 'SQUIDS and their applications', *J. Phys. E*, **9**, pp. 417–429, 1976.

61 Sullivan, D.B., Adair, R.T., and Frederick, N.V., 'RF instrumentation based on superconducting quantum interference', *Proc. IEEE*, **66**, 4, pp. 454–463, April 1978.

62 Kamper, R.A., Simmonds, M.B., Adair, R.T., and Hoer, C.A., 'A new technique for rf measurements using superconductors', *Proc. IEEE*, **61**, 1, pp. 121–122, Jan. 1973.

63 Kamper, R.A., Simmonds, M.B., Hoer, C.A., and Adair, R.T., 'Measurement of rf power and attentuation using quantum interference devices', NBS Tech. Note 643, Aug. 1973.

64 Kamper, R.A., Simmonds, M.B., Adair, R.T., and Hoer, C.A., 'Advances in the measurement of rf power and attenuation using SQUIDS', NBS Tech. Note 661, Sept. 1974.

65 Sullivan, D.B., Frederick, N.V., and Adair, R.T., 'RF power measurements using quantum interference in superconductors', in 'Superconducting Quantum Interference Devices and their Applications', Walter de Gruyter, Berlin, 1977, pp. 355–363.

66 Kamper, R.A., 'RF applications of the Josephson effect', *Microwave J.*, pp. 39–41, April 1976.

67 Petley, B.W., Morris, K., Yell, R.W., and Clarke, R.N., 'The use of a superconducting interference device for the measurement of rf attenuation', NPL Report DES 32, Sept. 1976.

68 Petley, B.W., Morris, K., Yell, R.W., and Clarke, R.N., 'Moulded microwave SQUID for rf attentuation calibration', *Electronics Letters*, **12**, pp. 237–238, 1976.

69 Seppa, H., 'Some new aspects concerning the X band SQUID for the measurement of rf attenuation', *IEEE Trans.*, **IM-32**, 1, pp. 253–259, Mar. 1983.

70 Senitzky, B., and Liebe, M., 'Absolute power measurement at 1.2 mm wavelength', *Rev. Sci. Instrum.*, **37**, pp. 1162–1164, 1966.

71 Hashimoto, S., and Yamaguchi, M., 'Microwave power measurement using absorption spectrum of ammonia gas', *IEEE Trans.*, **IM-25**, 4, pp. 348–352, Dec. 1976.

72 Llewellyn-Jones, D.T., and Gebbie, H.A., 'A new technique for power measurement in the near millimetre wave region using the untuned cavity', Colloquium on the Measurement of Power at Higher Microwave Frequencies, London, Jan. 1979, p. 8/1.

Part C

Calibration and comparison techniques

Chapter 11

Basic techniques for calibrating power meters

The simplest and most obvious method for comparing two power meters, or calibrating one against the other, is to connect each in turn to a stable source (Fig. 11.1). If the power meters have identical reflection coefficients, then each will absorb exactly the same amount of power from the source.

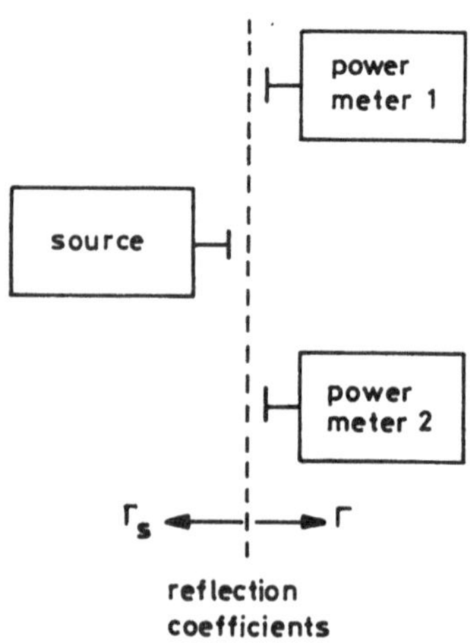

Fig. 11.1 *Comparison of power meters by the method of simple substitution*

Unfortunately such an assumption is justified only for relatively crude measurements, since the actual reflection coefficients of the power meters being compared will usually differ significantly from one another [1]. This fact may be taken into account as follows. Let P_{a1} and P_{a2} be the powers absorbed by the two power meters and let P_{z0} be the Z_0 available source power, defined as the power which the source would deliver to a load whose impedance is equal to the characteristic impedance Z_0 of the system. It is assumed that Z_0 is real. Then the power which would be delivered to an arbitrary load of reflection coefficient Γ is given by:

$$P_{a1} = \frac{(1 - |\Gamma|^2)}{|1 - \Gamma_s \Gamma|^2} P_{z0} \tag{11.1}$$

where Γ_s is the source reflection coefficient. The ratio of the absorbed powers P_{a1} and P_{a2} is therefore:

$$\frac{P_{a2}}{P_{a1}} = \frac{1 - |\Gamma_2|^2}{1 - |\Gamma_1|^2} \frac{|1 - \Gamma_s \Gamma_1|^2}{|1 - \Gamma_s \Gamma_2|^2} \tag{11.2}$$

where Γ_1 and Γ_2 are the power meter reflection coefficients. Greatest accuracy is obtained when Γ_1, Γ_2, and Γ_s are small, as this is the condition under which uncertainty in the measured reflection coefficients has least effect on the absorbed powers. This is a consequence of the fact that equation 11.2 contains only squares and products of reflection coefficients. If Γ_s is zero then only the magnitudes of Γ_1 and Γ_2 are required. Otherwise the magnitudes and phases of all three reflection coefficients are necessary. In cases where the phase angles of the reflection coefficients are not known equation 11.2 may be replaced by:

$$\frac{P_{a2}}{P_{a1}} = \frac{1 - |\Gamma_2|^2}{1 - |\Gamma_1|^2} \frac{(1 \pm |\Gamma_s||\Gamma_1|)^2}{(1 \mp |\Gamma_s||\Gamma_2|)^2} \tag{11.3}$$

from which the maximum and minimum possible values can be found, which define the uncertainty interval.

For small values of Γ_s, Γ_1, and Γ_2 the following approximation can be used for equation 11.1:

$$P_{a1} \simeq (1 - |\Gamma|^2)[1 + 2|\Gamma_s||\Gamma| \cos(\angle\Gamma_s + \angle\Gamma)]P_{z0} \tag{11.4}$$

where $\angle\Gamma_s$ and $\angle\Gamma$ are the phase angles of Γ_s and Γ.

The effects of the source mismatch can be made to cancel approximately if the measurement is repeated with a lossless quarter wavelength line interposed between the source and load. This introduces an additional 180 degree phase shift in the reflection coefficient seen by the source, which reverses the sign of the cosine term in equation 11.4. The mean of the two results is then taken.

11.1 Levelling and monitoring

The need to measure the source reflection coefficient creates a number of problems. The reflection coefficient of a generator is more difficult to measure than that of a passive load and it is also more likely to vary with time, necessitating frequent remeasurement. Moreover, many signal generators and amplifiers have a non-linear output impedance, which makes equation 11.2

invalid because this is derived using linear network theory. Such difficulties are avoided by the use of a levelling circuit [2]. The generator output is sampled by means of a three-port directional coupler (or power splitter) and a diode detector, as shown in Fig. 11.2. A feedback circuit maintains the output from the detector at a constant predetermined level. The effective source now consists of the signal generator, the directional coupler, the detector, and the feedback circuitry. The output reflection coefficient of this composite source is

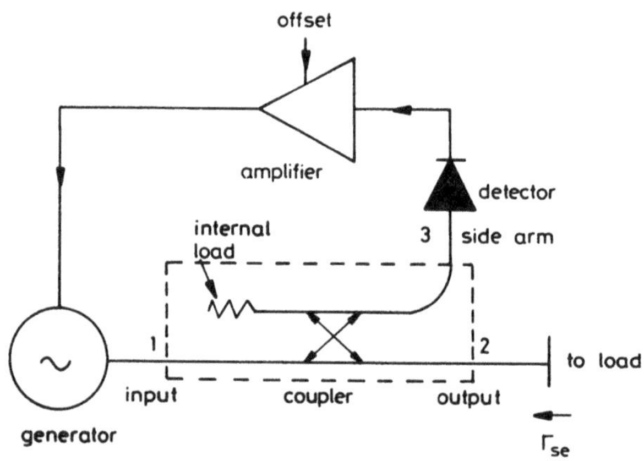

Fig. 11.2 *A levelled source. The equivalent source reflection coefficient Γ_{se} depends only on the properties of the directional coupler*

known as the equivalent source reflection coefficient Γ_{se} and it is this reflection coefficient which appears at the output terminals of the coupler when the levelling system is in operation. The main reason for preferring a levelled source is that, for high loop gain, Γ_{se} depends only on the coupler, which can be a relatively stable, linear device. The generator characteristics are eliminated from the measurements, assuming no spurious effects due to harmonics or frequency pulling. The value of Γ_{se} may be expressed in terms of the scattering parameters of the coupler as follows (see Appendix 4):

$$\Gamma_{se} = s_{22} - \frac{s_{21}s_{32}}{s_{31}} \tag{11.5}$$

where ports 1,2,3, are the input, output and side arm ports of the coupler respectively. For a perfect coupler Γ_{se} is zero for infinite feedback gain. Equation 11.5 applies not only to directional couplers but to any linear three-port network used in a similar way as part of a levelling system.

Levelling can often produce a substantial reduction in amplitude noise as well as correcting for slow drifts. The improvement is particularly noticeable in systems incorporating broad band travelling wave tube amplifiers, which can contribute appreciably to the noise of the system. On the other hand a noisy detector diode in the levelling loop may result in a worsening of the signal-to-noise ratio, since it will cause amplitude modulation of the generator. It is particularly important in levelling circuits to include low-pass filters to suppress harmonics arising from the generator and the detector diode respectively. These harmonics can be particularly troublesome if they lie outside the specified frequency range of the couplers, since for the out-of-band frequencies the attenuation between main arm and side arm may be quite low. It is also important to maintain the frequency of the signal constant, preferably by means of a phase-locked loop or by the use of a synthesiser as a signal generator; otherwise changes in frequency will be converted into changes in power level by the frequency-dependence of the characteristics of the couplers and other components.

Substantially the same effect as levelling may be achieved by continuously monitoring the power level during the course of the measurements and subsequently correcting the results for any changes which may have occurred. The monitoring is usually done by an assembly consisting of a coupler (or power splitter) and a power meter. A simple method of applying the correction for changes in level is to use normalised powers, which are obtained by dividing the readings of the power meters being compared by the corresponding readings of the monitoring power meter. These normalised powers may be thought of as the readings which would be obtained if the source were to be levelled so as to maintain the power indicated by the monitoring power meter constant and equal to unity. In this way it can be seen that monitoring and levelling are equivalent and that, provided normalised powers are substituted for actual powers, equations 11.2 and 11.3 can still be used. Monitoring is useful when the changes to be corrected for are small, because the use of normalised powers entails the assumption that the power meters are linear. A disadvantage is that it does not eliminate the effects of rapid changes in signal level or noise, because the monitoring power meter will not necessarily respond in the same way as the power meter under test to a sudden increase or decrease in power. One can of course use a combination of levelling and monitoring. The operation of the levelling circuit need not then be perfect, since any residual changes in level will be corrected for.

11.2 Properties of directional couplers

Directional couplers [3–11] feature prominently in power meter calibration methods and it is worth while devoting some space to their properties. An ideal coupler is a lossless matched reciprocal four-port network, the main use of which is for monitoring the forward and reflected signals in a transmission line. In the most common practical form one of the four ports is permanently terminated internally in a matched load, the resulting device being known as a three-port coupler. Power entering at the input port (see Fig. 11.2) divides between the output and the side arm. Power entering at the output port divides between the input port and the internal load, producing ideally no output signal at the side arm port, although in practice a small leakage signal is present. The coupling factor is defined as the ratio of the incident power at the input port to the power emerging at the side arm port when both this and the output port are terminated in matched loads. Typical coupling factors are in the range 3 dB to 30 dB, the most common values being 10 and 20 dB. Unwanted coupling between the output port and the side arm is characterised by the directivity of the coupler, which may be defined as the ratio of those two signals which, if applied to the input and output ports respectively, would produce the same response at the side arm port.

Couplers are divided into quadrature types, in which the coupling mechanism introduces a 90 degree phase shift, and in-phase types. Coaxial quadrature couplers rely on cancellation between signals coupled via the electric and magnetic fields. An example is shown in Fig. 11.3. Electric coupling excites in the side arm two waves travelling in opposite directions

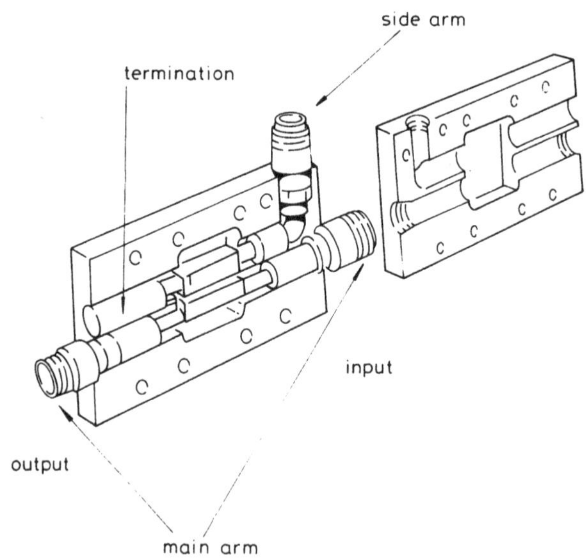

Fig. 11.3 *Coaxial quadrature type coupler*

with electric fields in phase with one another. Magnetic coupling excites two waves whose electric fields are in anti-phase. By adjustment of the relative values of the two types of coupling, cancellation can be achieved between the waves travelling in one direction, leaving a resultant wave travelling in the other direction. This resultant wave travels in the opposite direction to the wave in the main arm which gives rise to it. The coupling is usually distributed along the line [12] in order to achieve a coupling factor which is approximately constant with frequency. Improved frequency coverage can be obtained by exponentially tapering the coupling with distance. With distributed designs bandwidths of up to four octaves can be obtained with a directivity better than 26 dB and a coupling factor which varies by no more than ± 1 dB over the band [10,13–15]. This technique is normally limited to frequencies above 100 MHz by the maximum practical length of the coupler, but it can be extended down to 1 MHz by coiling a twisted pair of insulated wires into a helix [16]. The highest directivity obtainable with broad band coaxial couplers is around 35 dB. The directivity is relatively insensitive to frequency and consequently coaxial couplers can often be used at frequencies below their intended operating band provided that the departure of the coupling factor from its specified value can be tolerated. Stripline couplers are based on similar principles to coaxial ones. Fig. 11.4 shows an example.

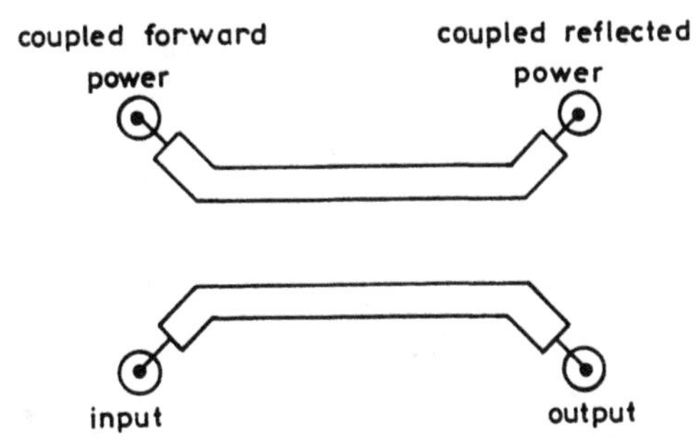

Fig. 11.4 *Microstrip quadrature type coupler. Only the top layer is shown. A dielectric layer and ground plane are assumed*

Waveguide quadrature couplers are usually of the multi-hole variety [17–21] as shown in Fig. 11.5. This type derives its directional properties by destructive and constructive interference between signals coupled via the different holes.

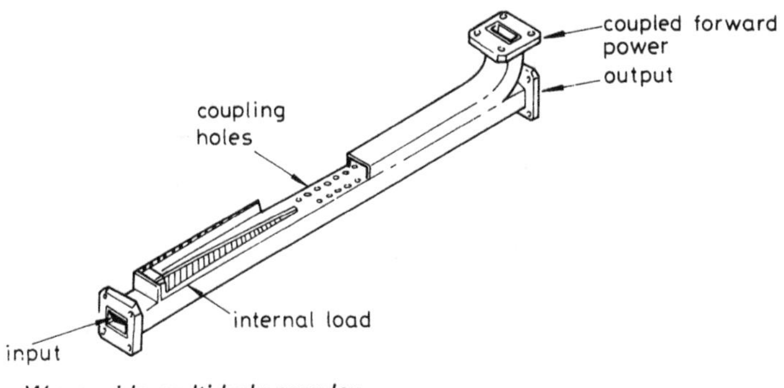

Fig. 11.5 *Waveguide multi-hole coupler*

The wave induced in the side arm propagates in the same direction as the wave in the main arm which gives rise to it. Directivities of more than 40 dB can be achieved relatively easily over a waveguide band, while for the best couplers the figure is in the region of 55 dB.

In-phase lumped element couplers can be constructed for frequencies below 1 GHz using ferrite cored transformers, as shown in Fig. 11.6. The operation relies on cancellation of the two signals arriving by different paths and the circuit is inherently broad-band. The practical bandwidth is limited by the performance of the transformers. Operating frequency ranges of 10 MHz to 1 GHz with a reflection coefficient of 0.13 maximum are achievable. The directivity for such broad bandwidths is usually not very high and may be no better than 20 dB.

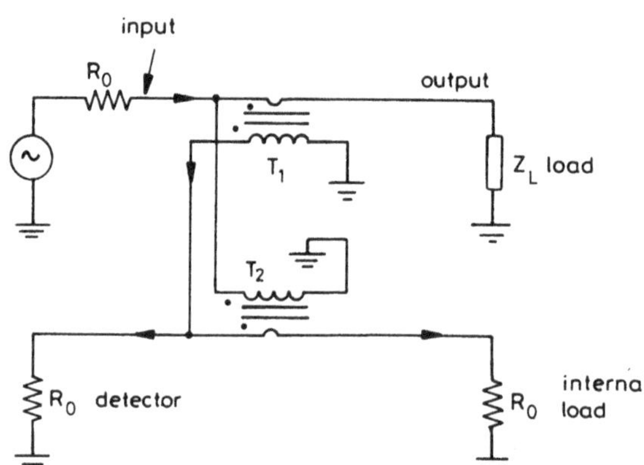

Fig. 11.6 *In-phase directional coupler using transformers*

The coupling factor of a microwave directional coupler does not necessarily vary monotonically with frequency but may undulate. This makes interpolation between measurement frequencies uncertain. Adjustable slot length couplers have been built to overcome this problem [22]. The properties of microwave directional couplers, especially waveguide ones, tend to be relatively stable with time because they are determined mainly by the geometry of the coupler. This argument does not, however, apply to the transformer type shown in Fig. 11.6.

The equivalent source reflection coefficient Γ_{se} of a coupler when used in a levelling system can be found from equation 11.5 if the complete scattering parameters are known. Unfortunately the parameters which are usually quoted in the specifications of directional couplers do not contain phase information. Without the phase information Γ_{se} cannot be calculated exactly but it is possible to derive an upper limit for its magnitude by assuming that the phases combine in the least favourable way. This upper limit is given by:

$$|\Gamma_{se}| \leq |s_{22}| + \left|\frac{s_{21}s_{32}}{s_{31}}\right| \tag{11.6}$$

11.3 Coupler-power-meter assemblies

A coupler-power-meter assembly is a feed-through instrument consisting of a three-port directional coupler and a power meter [23–25]. The power meter is permanently attached to the side arm of the coupler, as shown in Fig. 11.7, and tuners may be attached to improve the match. Such an assembly can be used as part of a levelling system or as a monitor. The power meter may be of the bolometric, thermoelectric or diode type. Bolometric instruments, of which the most widely used is the thermistor, are particularly suited to this

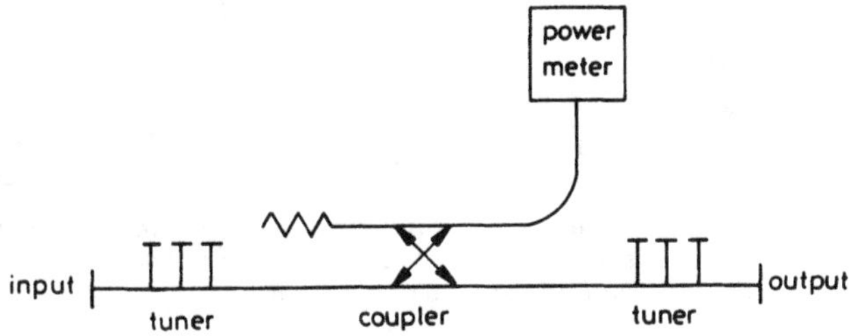

Fig. 11.7 *A tuned coupler-power-meter assembly*

application for two reasons. Firstly they are rf/dc substitution devices and consequently have relatively stable characteristics. Secondly the substituted dc power is applied without the need to disconnect the power meter from the coupler. Couplers with coupling factors of 3, 10, 20 or 30 dB are typical, the choice depending on the relative power levels required at the output and the side arm. When a thermistor power meter is employed, the level at the side arm is usually in the range 1 to 10 mW. There is no need for coupling factors of less than 3 dB, since the same result can be achieved simply by interchanging the roles of the side arm and output port. Some commercial instruments using 3 dB couplers fitted with tuners are described in ref. [26].

The power P_a absorbed by a load of reflection coefficient Γ_l connected to the output of a coupler-power-meter assembly is related to the reading P of the sidearm power meter by

$$P_a = KP\frac{1 - |\Gamma_l|^2}{|1 - \Gamma_{se}\Gamma_l|^2} \tag{11.7}$$

where the constant K is known as the calibration factor or calibration constant and Γ_{se} is the equivalent source reflection coefficient. The value of Γ_{se} can be found in principle by measuring the scattering parameters of the coupler and substituting these in equation 11.5. This entails disconnecting the power meter from the coupler, however, which may cause a change in the value of K. An alternative method which avoids the removal of the power meter is illustrated in Fig. 11.8. The procedure is as follows. First disconnect the generator and load from the coupler-power-meter assembly. Then connect a variable impedance (for example, a tuner terminated in a matched load) to the input port and adjust this so that when a signal is applied at the output port no

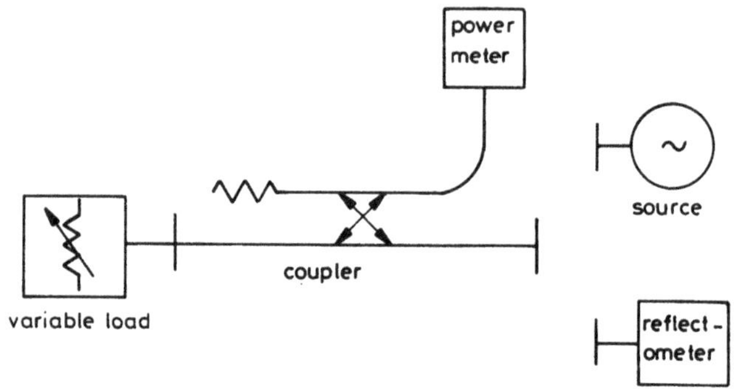

Fig. 11.8 *A method of measuring the equivalent source reflection coefficient of a coupler-power-meter assembly. The variable load is adjusted until the power meter reads zero with the source connected*

response is observed at the sidearm of the coupler. This adjustment is carried out by using the permanently attached power meter as a detector. After completion of the adjustment the reflection coefficient at the output port is equal to Γ_{se} and it can be measured by any convenient means, for example a reflectometer. Once the value of Γ_{se} is known, K can be found from equation 11.7 by connecting a standard power meter of known reflection coefficient to the output of the coupler. The assembly is then completely calibrated and may be used to calibrate an unknown power meter. A further method of calibrating coupler-power-meter assemblies using a six-port reflectometer is given in ref. [27]. In cases where the magnitude but not the phase of Γ_{se} is known equation 11.7 may be used to calculate the corresponding mismatch uncertainty by considering the phase angles which give the extreme values.

11.4 Use of resistive power splitters

Resistive power splitters can be used in place of directional couplers for levelling or monitoring. They have the advantage for this purpose of a broad frequency coverage extending down to dc and often a better match than coaxial couplers, but their properties may be temperature-dependent and the power handling capacity is limited by the fact that they have a loss of 6 dB and dissipate a proportion of the power. The type used for power meter comparison contains two resistors, each equal to the characteristic impedance Z_0 of the system and connected in series with the loads at the two output ports as shown in Fig. 11.9. When used in a levelling system, by connecting output 2 to a detector, the common point of the two resistors acts as a virtual earth and the equivalent source impedance at output 1 is therefore Z_0. This type should not be confused with the three-resistor power splitter, which has a resistor equal to $Z_0/3$ in series with each of the three ports.

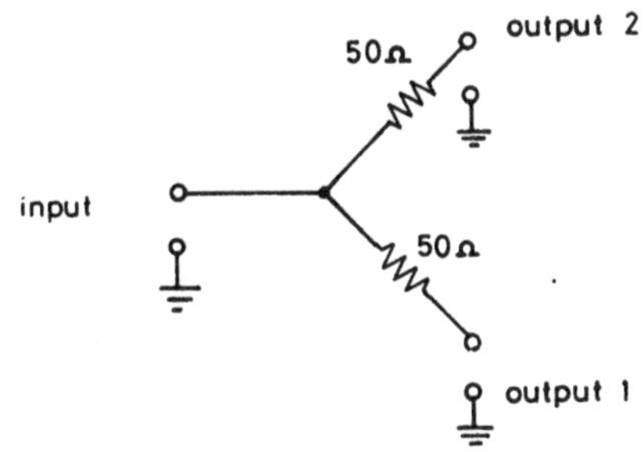

Fig. 11.9 *A resistive power splitter*

In the past a change from lumped resistive circuits to distributed directional couplers was usually made in the range 100 to 500 MHz. However, improvements in the performance of resistive components has extended their use into the microwave region and power meter calibrations are now regularly carried out using power splitters at frequencies up to 18 GHz [28–30]. The mismatches can, however, cause errors of half a percent or more at the higher frequencies if not corrected for. A presently available power splitter [31] has a maximum power of 1 W and an equivalent source VSWR of 1.15 up to 18 GHz with reduced performance up to 26 GHz.

One way of using a resistive power splitter is the simultaneous comparison method. The two power meters to be compared are connected to the two output arms of the power splitter. Theoretically the powers incident on the two instruments are then equal. Correction for mismatch when using the simultaneous comparison method is more complicated than for the direct substitution method in which one port is used to monitor the signal level and the power meters to be compared are connected in turn to the other output port. This is because the mismatches at both output ports of the power splitter then need to be taken into account. Unless perfect symmetry can be assumed it is necessary to determine for each output port the value of Γ_{se} which would be obtained if the other port were to be used for levelling. For this reason the simultaneous comparison method is not often used for measurements where the highest precision is required. The method is, however, quite convenient if mismatch errors are not to be corrected for, provided that errors due to asymmetry are eliminated by interchanging the two power meters to be compared.

If the scattering parameters of a power splitter are known, the equivalent source reflection coefficient of either output port can be found from equation 11.5. The method described for coupler-power-meter assemblies in Section 11.3 cannot be used without modification, as the reflection coefficient which the tuner would be required to produce is not a passive one. An alternative procedure, illustrated in Fig. 11.10, makes use of a sliding detector with a known high reflection coefficient. The modulus of the equivalent source

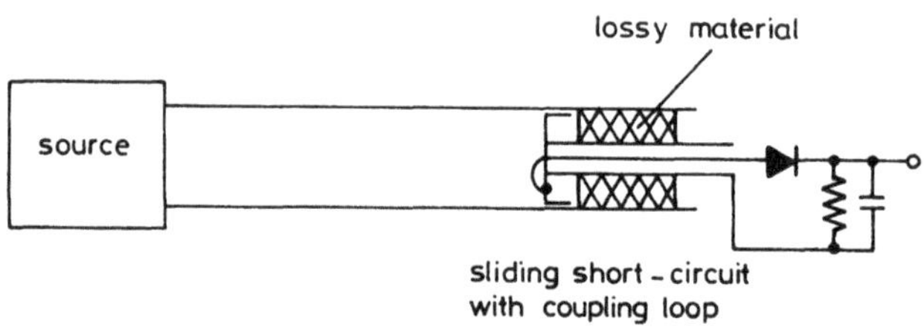

Fig. 11.10 *Alternative method of measuring equivalent source reflection coefficient. The variation in the detector output is noted as the short circuit is moved*

reflection coefficient is found, using equation 11.1, from the ratio of maximum to minimum reading as the position of the detector is varied. The phase angle is found from the position corresponding to the minimum. Another implementation of this technique uses a slotted line with a sliding short connected to its output. The probe and a sliding short are mechanically coupled so that the separation between them remains constant as the probe is moved.

11.5 Calibration at high and low power levels

The calibration of power meters at levels of more than a few watts or less than a milliwatt may involve the comparison of two power meters whose power ranges do not overlap. This problem can be tackled in several ways.

11.5.1 Calibrated attenuator method

A calibrated attenuator is connected in front of the more sensitive power meter. A disadvantage of this method is the extra mismatch error introduced by the attenuator. Other difficulties are the effect on the attenuator characteristics of the heat generated within it and the possibility that this heat may reach the power meter under test, causing zero drifts or changes in sensitivity.

11.5.2 Calibrated coupler method

The two power meters to be compared are connected to the output port and the side arm port respectively of a calibrated three-port directional coupler. The coupler characteristics are presumed to be independent of power level and the coupler can therefore be calibrated at any convenient power level [32] provided that the necessary standards of attenuation and reflection coefficient are available. If tuners are fitted to the coupler these are adjusted as described in reference [33].

11.5.3 Power ratio method

A coupler-power-meter assembly is calibrated at a single power level and its calibration factor is then assumed to be independent of level. Other power levels can then be established. The advantage of this method is that no additional mismatch errors are introduced beyond those which would occur for a comparison at the same power level, but the dynamic range is rather limited. For a coupler-bolometer assembly it is about 20 dB. The accuracy depends on the linearity of the power meter used in the coupler-power-meter assembly. The linearity of dual element bolometric power meters, which are common in coaxial systems, cannot be assumed because of the dual element substitution error (Chapter 6), although measurements made on commercial thermistor power meters suggest that this error is small in practice [34]. It is possible to extend the range of the power ratio method by means of the cascaded coupler technique [35] (Fig. 11.11). Suppose that a 10 mW power standard is available and that it is desired to calibrate an unknown power meter at a level of 100 W, that is, 40 dB above the standard. Two couplers are required together with two additional power meters (3 and 4), each with a dynamic range of 20 dB. The

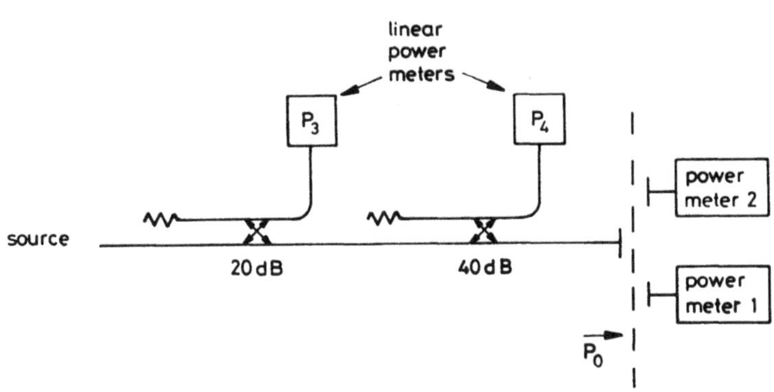

Fig. 11.11 *The cascaded coupler method of comparing two power meters operating at widely differing power levels*

coupling factor of coupler 1 is chosen so that power meter 3 is at the lower end of its range when the output power P_o is 10 mW. The coupling factor of coupler 2 is chosen so that power meter 4 is at the lower end of its range when power meter 3 reaches its maximum. If, for example, power meters 3 and 4 each have an upper limit of 10 mW, then we require for the first coupler a coupling factor of 20 dB and for the second coupler a coupling factor of 40 dB. The first step in the procedure is to connect the standard power meter to the output of the second coupler, adjust the level for an output P_o of 10 mW, and note the ratio P_3/P_o. The standard is then replaced by the unknown power meter, the signal level increased by 20 dB, and the ratio P_4/P_3 noted. One can then calculate the ratio P_4/P_o, given by

$$\frac{P_4}{P_o} = \left(\frac{P_4}{P_3}\right)\left(\frac{P_3}{P_o}\right) \tag{11.8}$$

The value of P_4/P_o is independent of the first coupler and power meter 3, and these can therefore be removed to prevent a possible overload. Increasing the level by a further 20 dB gives the required 100 W level for the output power P_o, for which the precise value can be calculated from P_4 and P_4/P_o.

11.6 References

1 Beatty, R.W., and Macpherson, A.C., 'Mismatch errors in microwave power measurements', *Proc IRE*, 41, 9, pp. 1112–1119, Sept. 1953.

2 Engen, G.F., 'Amplitude stabilization of a microwave signal source', *IRE Trans.*, **MTT-6**, 2, pp. 202–206, April 1958.

3 Mumford, W.W., 'Directional couplers', *Proc. IRE*, **35**, 2, pp. 160–165, 1947.

4 Kyhl, R.L., 'Directional couplers', Chapter 14 in 'Technique of Microwave Measurements', ed. C.G. Montgomery, McGraw-Hill, New York 1947.

5 Riblet, H.J., 'A mathematical theory of directional couplers', *Proc. IRE*, **35**, 11, pp. 1307–1313, Nov. 1947.

6 Schwartz, R.F., 'Bibliography on directional couplers', *IRE Trans.*, **MTT-2**, 2, pp. 58–63, July 1954.

7 Lombardini, P.P., Schwartz, R.F., and Kelly, P.J., 'Criteria for the design of loop type directional couplers for the L band', *IRE Trans.*, **MTT-4**, 4, pp. 234–239, 1956.

8 Koike, R., 'Recent advances in the use of coupled transmission lines as directional couplers', *Proc. IEE*, **108B**, 37, p. 120, Jan. 1961.

9 Kahn, W.K., 'Directional Couplers', Chapter 12 in 'Handbook of microwave measurements', Vol. 2, Sucher, M., and Fox, J. (eds), Polytechnic Press, 1963.

10 Prickett, R., 'New coaxial couplers for reflectometers detection and monitoring', *Hewlett-Packard J.*, **16**, 6, Feb. 1965.

11 Caswell, W.E., 'The directional coupler 1966', *IEEE Trans.*, **MTT-15**, 2, pp. 120–122, Feb. 1967.

12 Monteath, G.D., 'Coupled transmission lines as symmetrical directional couplers', *Proc. IEE*, **102**, Pt. B, 3, pp. 383–392, May 1955.

13 Hunton, J.K., Poulter, H.C., and Reis, C.S., 'High directivity coaxial directional couplers and reflectometers', *Hewlett-Packard J.*, **7**, 2, Oct. 1955.

14 Hudson, P.A., 'A high directivity, broadband coaxial coupler', *IEEE Trans.*, **MTT-14**, 6, pp. 293–294, June 1966.

15 Anderson, R.W., and Dennison, O.T., 'An advanced new network analyser for sweep measuring amplitude and phase from .1 to 12.4 GHz', *Hewlett-Packard J.*, **18**, pp. 2–10, 1967.

16 Somlo, P.I., 'Precision impedance measurement and Smith-chart display with easily fabricated directional couplers from 1 MHz upwards', *Proc. IREE* (Australia), **35**, pp. 341–345, 1974.

17 Early, N.C., 'A wide band directional coupler for waveguide', *Proc. IRE*, **34**, 11, pp. 883–886, 1946.

18 Miller, S.E., and Mumford, W.W., 'Multielement directional couplers', *Proc. IRE*, **40**, 9, p. 1071, 1952.

19 Barnett, E.F., and Hunton, J.K., 'A precision directional coupler using multihole coupling', *Hewlett-Packard J.*, **3**, 7/8, Mar./April 1952.

20 Barnett, E.F., 'More about HP precision directional couplers', *Hewlett-Packard J.*, **4**, 5/6, Jan./Feb. 1953.

21 Hensperger, E.S., 'The design of multihole coupling arrays', *Microwave J.*, **2**, 8, p. 38, 1959.

22 Hudson, P.A., and Salisbury, L.F., 'An adjustable slot length UHF coaxial coupler with decade bandwidth', *IEEE Trans.*, **MTT-19**, 9, pp. 781–783, Sept. 1971.

23 Kothari, P.C., Bhatnagar, H.M., Agrawal, V.K., and Chandra, K., 'A feed-through microwave power measuring transfer standard at 10 GHz', *J. Inst. Electron. and Telecomm. Eng. (India)*, **26**, 11, pp. 579–583, Nov. 1980.

24 Sergeev, I.A., Mikhailova, L.S., and Naumov, B.I., 'Coaxial incident power converters', *Izmer. Tekh.*, **24**, 9, pp. 49–52, Sept. 1981 (in Russian), (translation in *Meas. Tech.*, 24, 9, pp. 773–776, Sept. 1981).

25 Ide, J.P., 'A broadband waveguide transfer standard for the dissemination of UK national microwave power standards', RSRE Memo 3390, Ministry of Defence, RSRE, Malvern, Worcestershire, Jan. 1982.

26 Sorger, G.U., Weinschel, B.O., and Raff, S.J., 'System for transfer of calibration factor for coaxial bolometer mounts with one percent transfer inaccuracy', *IEEE Trans.*, **IM-15**, 4, pp. 343–358, Dec. 1966.

27 Weidman, M.P., 'A semi-automated six-port for measuring millimeter-wave power and complex reflection coefficient', *IEEE Trans.*, **MTT-25**, 12, pp. 1083–1085, Dec. 1977.

28 Powell, R.C., Banning, H.W., and Byloff, J.R., 'An automated power meter calibration system', IEEE MTT-S International Microwave Symposium Digest, Dallas, June 1982, pp. 357–359.

29 Belski, A., 'Power meter accuracy scrutinized by traceable calibration method', *Microwave Syst. News*, **13**, 10, pp. 146–163, Oct. 1983.

30 Weinschel, B.O., Hancock, C.W., and Powell, R.C., 'Techniques explained for CW power measurement', *Microwave Syst. News & Commun. Technol.*, **16**, 5, pp. 78–90, 93–96, 99–103, May 1986.

31 Catalogue, Weinschel Engineering 1986.

32 Clark, R.F., Jurkus, A., and McLaren, G.D., 'Calibration and use of a reference standard directional coupler for measurement of large coupling factors', *IEEE Trans.*, **IM-32**, 4, pp. 501–506, Dec. 1983.

33 Engen, G.F., 'A method of improving isolation in multichannel waveguide systems', *IRE Trans.*, **MTT-8**, pp. 460–461, July 1960.

34 Orford, G.R., and Abbott, N.P., 'Some recent measurements of linearity of thermistor power meters', IEE Colloquium on Microwave Measurement, London, May 1981, p. 5/1–4.

35 Bramall, K.E., 'Accurate microwave high-power measurements using cascaded coupler method', *J. Res. NBS*, **75C**, 3/4, pp. 185–192, July–Dec. 1971.

Chapter 12

Power meter calibration using reflectometers

Reflectometers enable measurements of power and reflection coefficient to be carried out simultaneously. Scalar reflectometers give only amplitude information. Vector reflectometers, on the other hand, measure complex quantities and display the reflection coefficient either in terms of modulus and phase or in terms of real and imaginary parts. Six-port reflectometers also measure complex quantities, but in these the phase information is deduced from ratios of amplitude readings rather than by the use of a phase meter or phase-sensitive detector.

12.1 Untuned scalar reflectometers

A typical untuned scalar reflectometer consists of two directional couplers and two power meters or detectors, as shown in Fig. 12.1. It is classed as a four-port instrument, the four ports being the input port (port 1), the measurement port (port 2), and the side arm ports of the two couplers (ports 3 and 4). The power

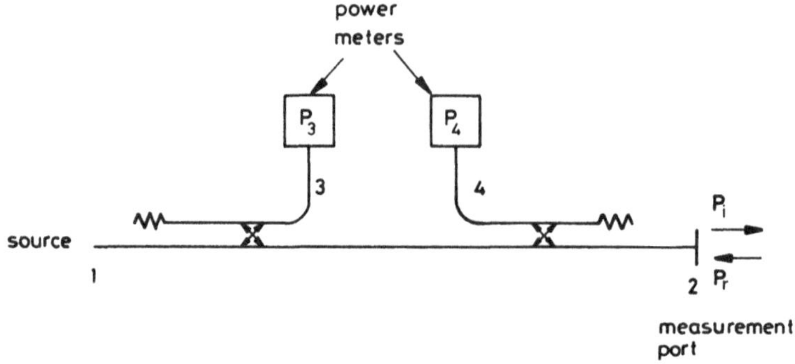

Fig. 12.1 *Untuned reflectometer based on two directional couplers*

meters must be linear, but need not be calibrated. For perfect couplers the power meter readings P_3 and P_4 are proportional to the incident and reflected powers P_i, P_r, respectively, at the measurement port. Thus

$$\begin{aligned} P_i &= K_1 P_3 \\ P_r &= K_2 P_4 \\ |\Gamma_l|^2 &= \frac{K_2}{K_1} \frac{P_4}{P_3} \end{aligned} \tag{12.1}$$

where K_1 and K_2 are constants and Γ_l is the reflection coefficient of the load at port 2. The net power P_{ab} absorbed by the load is therefore

$$P_{ab} = K_1 P_3 - K_2 P_4 \tag{12.2}$$

The constants K_1 and K_2 are found by substituting in equation 12.2 two sets of readings of P_{ab}, P_3, and P_4 to yield two simultaneous equations. These readings are obtained by connecting in turn to the measurement port a lossless termination (for which P_{ab} is zero) and a standard power meter. In practice equation 12.2 is only an approximation because of the mismatches and imperfect directivities of the couplers. One consequence is that the values obtained for K_1 and K_2 are dependent on the phase angle of the reflection coefficient of the lossless termination used in the calibration. Correction has been made for such effects by the use of a 'spacer' technique, which involves the connection between the measurement port and the test object of one or more fixed lengths of precision line [1,2].

An alternative scalar reflectometer circuit is shown in Fig. 12.2. It consists of a resistive power splitter followed by a resistive bridge, often referred to as a reflection bridge or rho-bridge [3–5]. The directivities of such bridges can be in the range 40–60 dB. The same equations 12.1 and 12.2 apply, although unlike the ideal two-coupler circuit of Fig. 12.1 ports 2 and 3 are not decoupled from one another. Other resistive circuits are given in refs. [6–9].

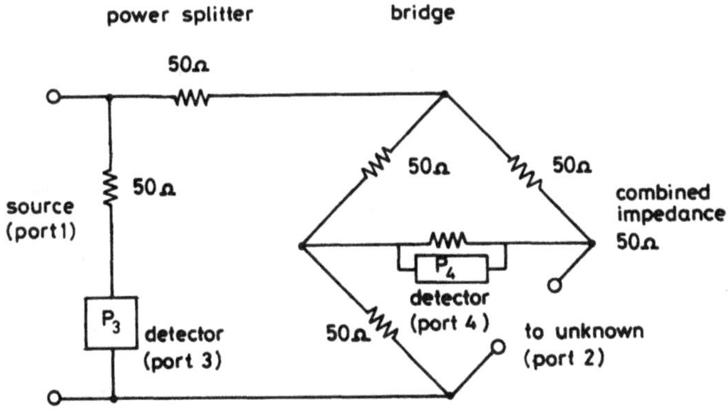

Fig. 12.2 *Untuned reflectometer comprising a resistive power splitter and a reflection bridge*

12.2 Tuned scalar reflectometers

Fig. 12.3 shows a tuned scalar reflectometer [10–12]. It is used in a similar way to the untuned reflectometer with the exception that, before the commencement of the measurement, the tuners are adjusted to cancel out the

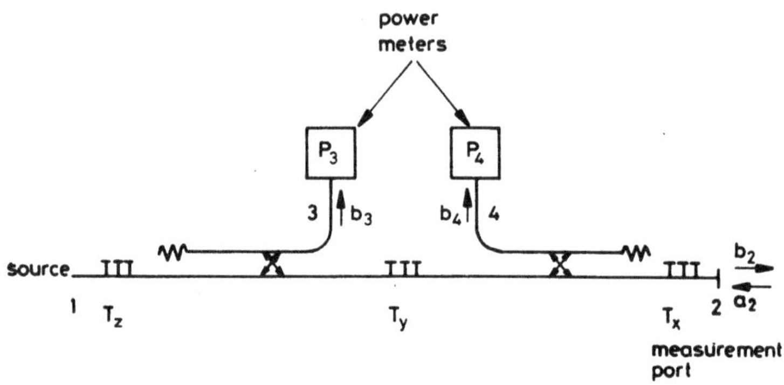

Fig. 12.3 *Tuned reflectometer*

coupler imperfections. The theory behind the adjustment procedure is as follows. Each of the emergent voltage waves b_3 and b_4 at ports 3 and 4 can be expressed as a linear combination of the forward and reflected waves b_2, a_2 at port 2. Thus

$$\begin{aligned} b_3 &= Aa_2 + Bb_2 \\ b_4 &= Ca_2 + Db_2 \end{aligned} \tag{12.3}$$

where A, B, C, D are constants. The ratio b_4/b_3 can therefore be expressed in the form

$$\frac{b_4}{b_3} = \frac{C\Gamma_l + D}{A\Gamma_l + B} \tag{12.4}$$

where Γ_l is the reflection coefficient of the load at port 2. If the couplers were perfect, A and D would be zero. The reflectometer must be tuned so as to make these two constants as small as possible.

Two methods of adjusting the tuners will be described. The first is helpful in clarifying the principles, but the second is the one normally used in practice. In the first method tuner T_x is adjusted for a null at port 4 when port 2 is terminated in a perfectly matched load. This sets D to zero. The generator is then connected to port 2, a load (not necessarily matched) is connected to port 1 and T_z is adjusted to produce a null at port 3. T_y is now adjusted so that the reflection coefficient observed at port 2 vanishes. It can be shown that this sets A to zero and that the required conditions have been achieved [11]. The order in which the adjustments are carried out is important, since the adjustment of

T_x is independent of the adjustment of T_y but the converse is not true. Tuner T_z is not essential for the operation of the reflectometer and is required only to enable the second tuning operation to be carried out. It may be readjusted later, if desired, without affecting the result obtained by adjustment of T_y.

The second method requires a sliding short circuit in addition to a matched load. Tuner T_x is adjusted as before to produce a null at port 4 with a matched load at port 2. The sliding short circuit is then connected to port 2 and tuner T_y adjusted until the ratio of the two power meter readings is independent of the position of the sliding short circuit. It follows from equation 12.4 that, since D is zero as a result of the first adjustment, A must also be zero and the required condition is again achieved. This second method is more convenient than the first, as it is not necessary to disconnect the generator from the reflectometer. It also has the advantage that if it is not intended to use the reflectometer for reflection coefficient measurements, but only for absorbed power measurement, adjustment of T_x is not necessary and the matched load can then be dispensed with. The absorbed power P_{ab} will still be given correctly by an expression of the form of equation 12.2 but K_1 and K_2 will no longer have the same physical meaning and it will be impossible to determine the incident and reflected powers separately. Thus the effective efficiency of a bolometer could be measured accurately in this way but not its calibration factor. An alternative form of the tuned reflectometer which is suitable for lower power levels and which enables approximately equal powers to be obtained at ports 2 and 3 is shown in Fig. 12.4. In this arrangement the second coupler is fed via its side

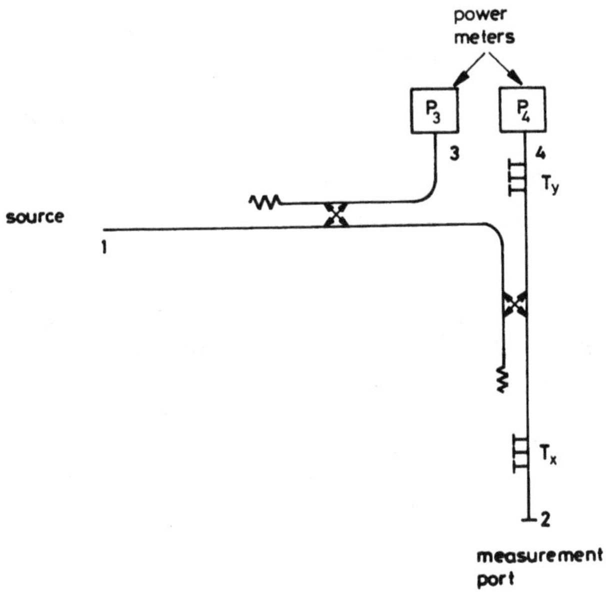

Fig. 12.4 *Version of the tuned reflectometer suitable for low power levels. Approximately equal powers are obtained at ports 2 and 3*

arm and the power meter P_4 monitoring the reflected power is connected to what in conventional coupler terminology would be called the output port. The theory of scalar reflectometers is equally valid for negative values of absorbed power P_{ab}. This fact may be used to make a variable impedance power meter by replacing the generator by an adjustable load [13]. The instrument reads absorbed power, irrespective of its impedance setting.

In all tuned reflectometers the adjustment procedure is very tedious and, although such instruments have been widely used in top echelon standards laboratories, they are not very convenient for calibration work of a routine nature.

12.3 Vector reflectometers

RF and microwave instruments capable of measuring both amplitude and phase [14] entered routine use in the late 1960s. Probably the best known of these instruments is the heterodyne network analyser [15,16], which measures complex scattering parameters (Fig. 12.5). It is essentially a vector reflectometer with additional switches and loads to enable both transmission and reflection coefficient measurements to be performed. The signals are heterodyned down to a convenient frequency where the amplitude and phase measurements can be carried out more easily. The inclusion of phase information enables the coupler characteristics to be corrected for by calibration [17,18], resulting in an instrument which is more convenient and more easily automated than the tuned reflectometer. For a modern computer controlled instrument [19] the uncertainty in reflection coefficient after calibration is specified as 0.006 in magnitude and 1.7° in phase at $\Gamma = 0.2$. For $\Gamma = 1$ the corresponding uncertainties are 0.023 and 1.2° respectively. Fig. 12.6 shows an alternative vector reflectometer based on a resistive bridge and suitable for frequencies up to 500 MHz. It requires a balanced rf supply, which is provided with the aid of a transformer. The circuitry needed for computer controlled measurements [20–25] is omitted from Figs 12.5 and 12.6. Two penalties which are paid for the greater convenience of the vector reflectometer are additional sources of error, due to the more complicated electronic circuitry, and in earlier instruments much poorer long-term stability. These problems stimulated research into alternative forms of network analyser, particularly into those based on six-port reflectometers (see Section 12.4).

To enable a vector network analyser to be used for power meter calibrations some modification may be necessary, since it is normally intended to measure ratios rather than absolute levels. Fig. 12.7 shows one possible arrangement

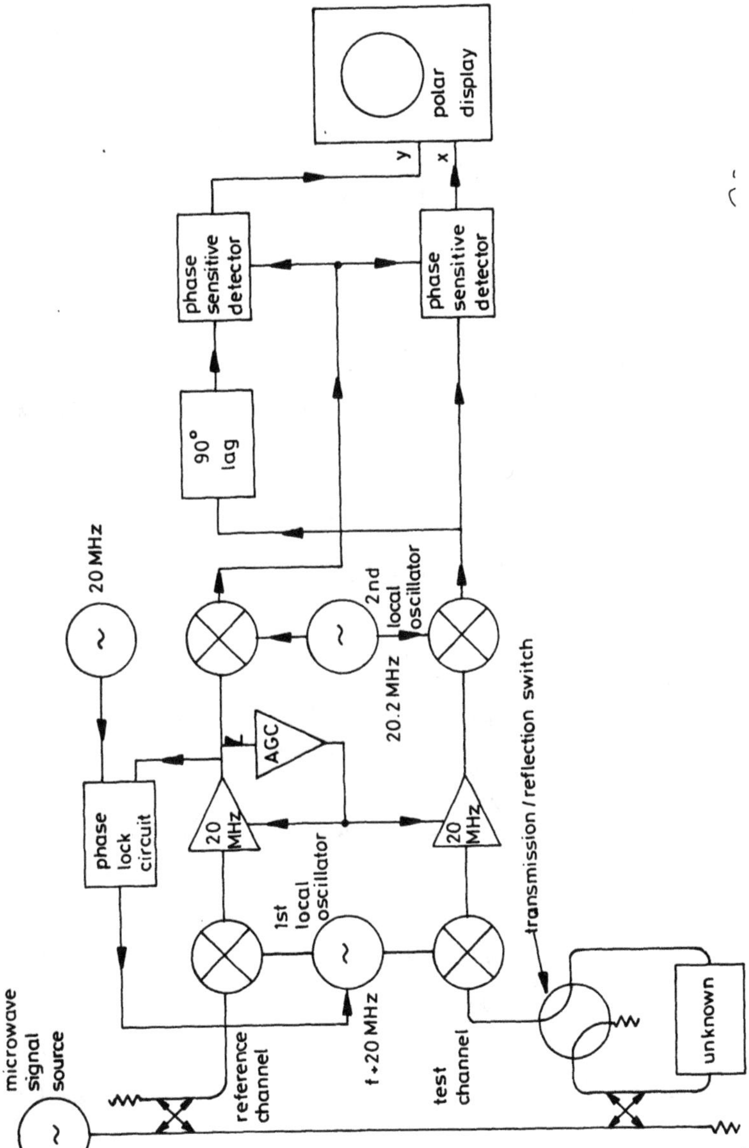

Fig. 12.5 *Twin channel super-heterodyne vector network analyser (simplified). The test and reference signals are down-converted first to 20 MHz and then to 200 kHz before being fed to phase-sensitive detectors to obtain a polar display*

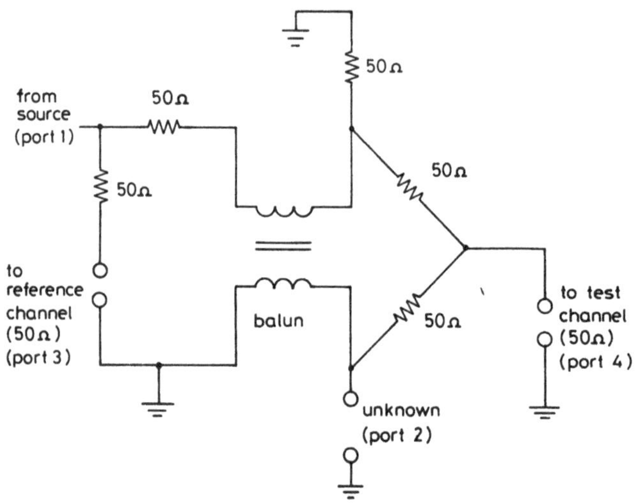

Fig. 12.6 *Alternative vector reflectometer based on a reflection bridge. This circuit is suitable for frequencies up to 500 MHz*

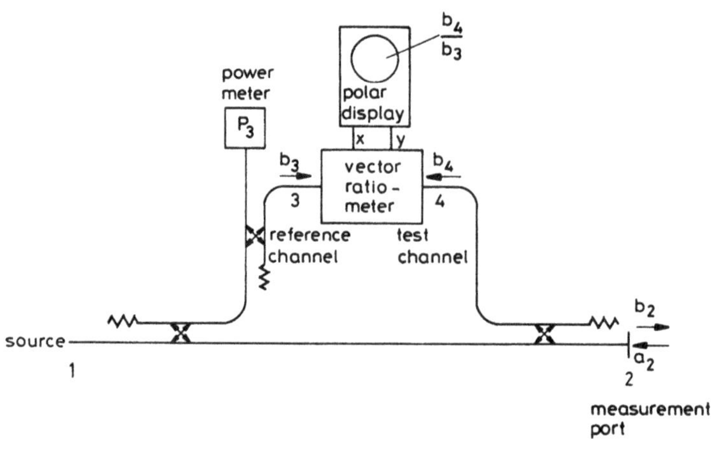

Fig. 12.7 *Modification of network analyser for power meter calibration. The vector ratiometer measures the complex ratio between b_3 and b_4 while the power meter establishes absolute levels*

[26]. The signals a_2 and b_2 at the measurement port can be expressed in terms of b_3 and b_4 by:

$$
\begin{aligned}
a_2 &= A'b_3 + B'b_4 \\
b_2 &= C'b_3 + D'b_4
\end{aligned}
\qquad (12.5)
$$

where A', B', C', D' are complex constants. Assuming the use of normalised waves (see Appendix 1), the net power P_{ab} absorbed by the load connected to the measurement port is given by:

$$\begin{aligned} P_{ab} &= |b_2|^2 - |a_2|^2 \\ &= |C'b_3 + D'b_4|^2 - |A'b_3 + B'b_4|^2 \end{aligned} \qquad (12.6)$$

and the reflection coefficient Γ_l of this load is given by:

$$\Gamma_l = \frac{a_2}{b_2} = \frac{A'b_3 + B'b_4}{C'b_3 + D'b_4} \qquad (12.7)$$

For perfect couplers, A' and D' would be zero. The values of b_3 and b_4 are obtained from the readings of the power meter and the vector voltmeter. The magnitude of b_3 is given by:

$$|b_3|^2 = k\,P_3 \qquad (12.8)$$

where k is a constant. The value of k may be put equal to unity without loss of generality, in which case k is absorbed into the constants A', B', C', D'. Similarly it is unimportant whether the reading of the vector ratiometer is equal to the ratio of b_4 to b_3 or merely proportional to this ratio. There must, however, be no cross-coupling between the test and reference inputs.

Calibration of the reflectometer consists of determining A', B', C', and D'. This can be done by using three standard loads of known reflection coefficient and one standard power meter. By substituting in equation 12.7 the corresponding values of Γ_l and b_4/b_3 obtained when the three standard loads are connected in turn to the measurement port, three equations are obtained which enable the values of three of the unknown complex constants to be expressed in terms of the fourth. The magnitude of the fourth constant is found by connecting the standard power meter to the measurement port and substituting in equation 12.6. Its phase is unimportant.

The impedance standards used in the calibration may be a fixed short circuit and two offset short circuits of known and different phase angles. Alternatively, a fixed short, an offset short and a matched load may be used. Greater accuracy is obtained by replacing the matched load by a sliding termination.

A simpler partial calibration can be carried out if the reflectometer is to be used only for power measurements. This approach, which is an example of the power equation method [27–33], allows the net power absorbed by the load to be determined but not the incident and reflected powers, nor the reflection coefficient [34]. The theory is as follows. Equation 12.6 may be rewritten in a form which contains only one complex constant plus two real constants. The result is

$$\begin{aligned} P_{ab} &= |b_2|^2 - |a_2|^2 \\ &= K|b_3|^2 \left[1 - \left| \frac{\frac{b_4}{b_3} - R_c}{R} \right|^2 \right] \end{aligned} \qquad (12.9)$$

where P_{ab} is the absorbed power, R_c is the complex constant, and R and K are the two real constants. The values of these constants may be found by using a sliding short circuit and a standard power meter whose reflection coefficient need not be known. First the sliding short is connected to the measurement port and the locus of b_4/b_3 is determined as the short circuit is moved. From equation 12.9, this locus is given by:

$$\left| \frac{\frac{b_4}{b_3} - R_c}{R} \right|^2 = 1 \qquad (12.10)$$

which is the equation of a circle. The values of R_c and R are given by the centre of the circle and its radius, as illustrated in Fig. 12.8. When R_c and R have been

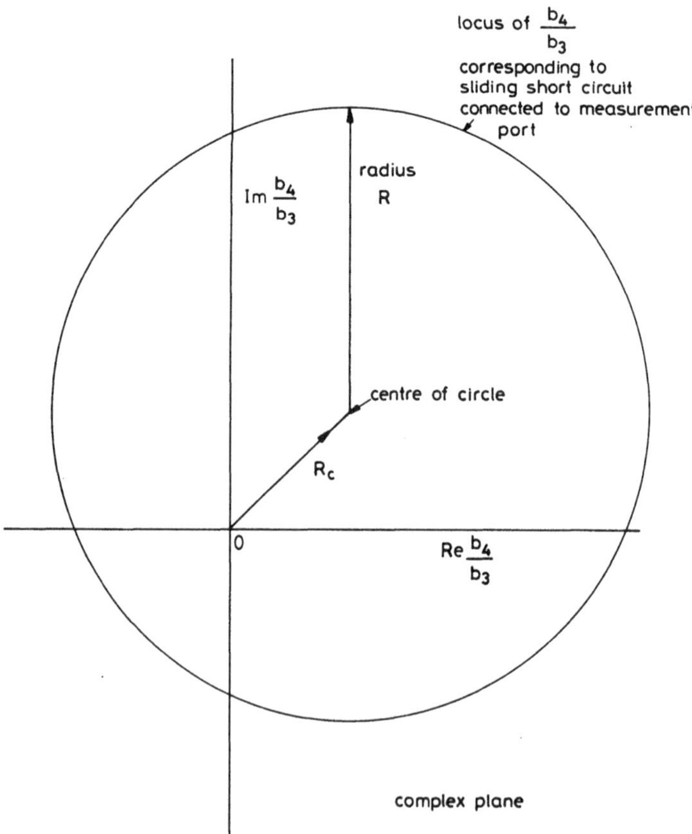

Fig. 12.8 *Circular locus of b_4/b_3 in the complex plane for a sliding short circuit connected to the measurement port. The constants R and R_c are obtained from the radius and the position of the centre of the circle respectively*

found, K can be determined by connecting the standard power meter to the measurement port and substituting in equation 12.9. The constants R_c, R, K are unchanged if a lossless network is interposed between the measurement port and the load. Therefore lossless discontinuities introduced by connectors have no effect on the calibration procedure or the measured power, although lossy contacts will cause errors. The sliding short is used as a standard of zero power in this application rather than as an impedance standard. It may be replaced by three fixed lossless terminations of different, but unknown, phase angles (for example approximately 0°, 90° and 180°). These enable three points on the circle to be obtained, which are sufficient to determine its centre R_c and radius R. A limitation of the method is that it does not take account of the effect of losses in the ideally lossless terminations. These losses become serious below 1 GHz because of the increased line lengths. A short circuit, open circuit, and a coaxial capacitor must then be used.

The alternative arrangement of Fig. 12.9 may be used to reduce the effect of network analyser non-linearity on the absorbed power measurement. It may be regarded as an untuned reflectometer with the addition of a vector ratiometer. One way of using the extra power meter reading would be to derive the magnitudes of b_3 and b_4 from P_3 and P_4 respectively and the phase

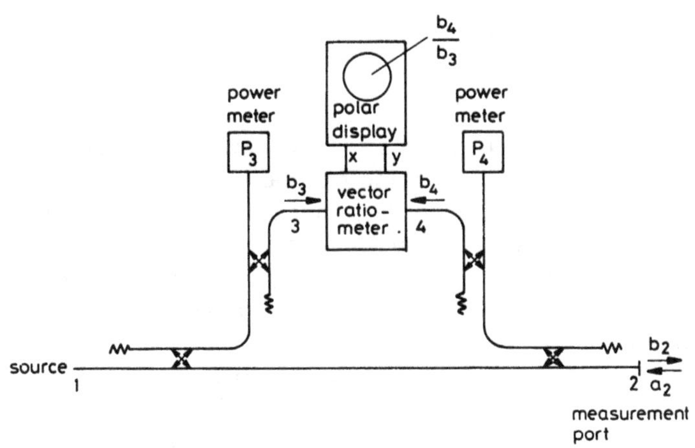

Fig. 12.9 *Network analyser incorporating two power meters*

difference between them from the reading of the ratiometer. However, although this may be satisfactory for large load reflection coefficients it is very insensitive for reflection coefficients close to zero because P_4/P_3 is proportional to the square of $|(b_4/b_3)|$ and a very large dynamic range is required of the power meters. A better scheme makes use of the X and Y output voltages of the ratiometer, which for the present purpose may be taken as being equal to the real and imaginary parts of b_4/b_3. The power absorbed by the load may be expressed as

$$P_{ab} = P_3\,(A'' + B''X + C''Y) + D''P_4 \qquad (12.11)$$

in which A'', B'', C'', D'' are real constants. These constants are found by connecting in turn the three lossless terminations (for which $P_{ab} = 0$) and the standard power meter and substituting in equation 12.11 for P_{ab}, P_3, X, Y, P_4 to obtain four simultaneous equations. For perfect couplers $B'' = C'' = 0$ and the ratiometer plays no part in determining the absorbed power. As the couplers depart from ideality B'' and C'' increase and the ratiometer outputs X and Y become increasingly important. Thus the ratiometer does no more than is necessary to correct for the imperfections of the couplers, and the effects of the extra sources of uncertainty which it introduces are therefore minimised. The method is similar to the six-port approach, with P_3, P_3X, P_3Y, P_4 as the four power meter readings (Section 12.4). However, the equivalence is not perfect, since P_3X and P_3Y may be either positive or negative, unlike actual power meter readings which are always positive. The method is insensitive to errors caused by a difference in gain in the X and Y channel outputs of the ratiometer or a phase error in the internal 90 deg phase shifter, both of which turn circles in the X, Y display into ellipses. Such errors are absorbed into the constants A'', B'', C'', D''.

12.4 Six-port reflectometers

The practical use of six-ports for rf and microwave measurements dates from the early 1970s [35–37], although the basic ideas are much older than this. Interest in the subject grew rapidly during the late 1970s, spurred on by the fact that relatively inexpensive desk-top computers had recently become available [38–48]. These enabled computer-corrected measurements, which are an essential feature of the six-port approach, to be carried out in a very convenient way. The main area of application of six-ports is in standards laboratories, as there has been relatively little commercial exploitation.

The principle behind the six-port is that of deriving phase information by measuring amplitudes. An example of this principle is the well-known three-voltmeter or three-ammeter method [49,50], in which the phase difference between two ac voltages or currents is obtained by comparing the magnitude of their vector sum with their individual magnitudes. The method is based on the cosine rule, as illustrated in Fig. 12.10. The solution obtained in this way is ambiguous because the technique cannot distinguish between positive and negative phase angles. A six-port reflectometer consists of an arbitrary six-port network with power meters permanently connected to four of its ports, as shown in Fig. 12.11. The remaining two ports are the measurement port and the input port. The properties of the network are

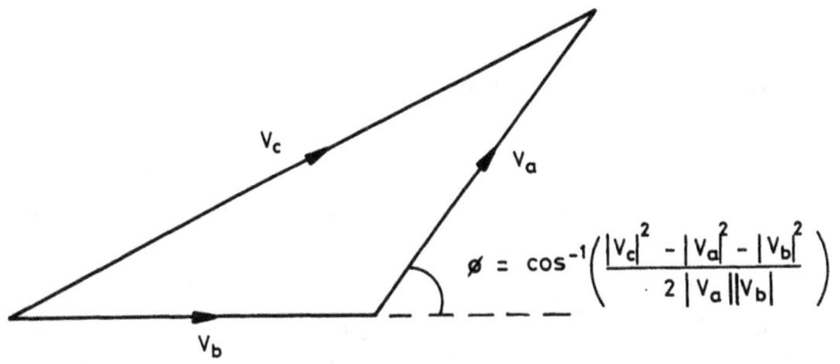

Fig. 12.10 *Principle of the three-voltmeter method. The phase angle between voltages V_a and V_b is deduced using the cosine rule*

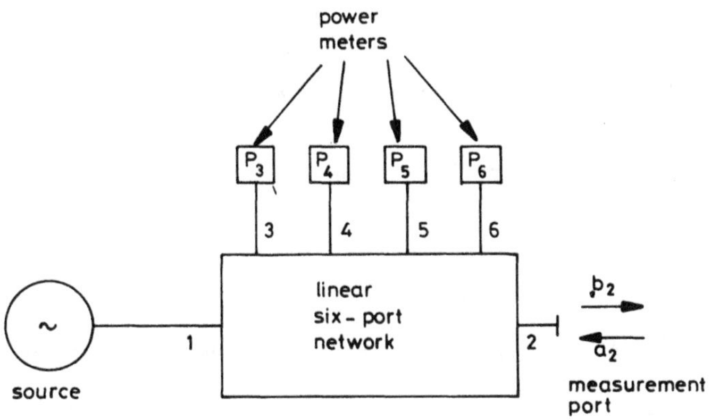

Fig. 12.11 *General six-port reflectometer*

determined by calibration. The use of four power meters rather than the minimum requirement of three removes ambiguities, but it also introduces redundancy, because four power readings together contain slightly more information than is strictly necessary to calculate a unique result.

12.4.1 Theory of six-ports

From simple linear network theory, each of the power meter readings P_3, P_4, P_5, P_6 in Fig. 12.11 can be expressed in terms of the normalised forward and reflected waves b_2, a_2 at the measurement port:

$$\begin{aligned} P_i &= |A_i a_2 + B_i b_2|^2 \\ &= |A_i|^2 |a_2|^2 + |B_i|^2 |b_2|^2 + A_i B_i^* a_2 b_2^* + A_i^* B_i^* a_2^* b_2 \end{aligned} \qquad i = 3-6 \quad (12.12)$$

where A_i and B_i are complex constants. Twelve real constants are sufficient to characterise the complete reflectometer. These are the four moduli $|A_i|$, the four moduli $|B_i|$, and the four phase differences ($\angle A_i - \angle B_i$). The individual phases of A_i and B_i are not important because the power meters respond only to amplitude. The twelve real constants take into account not only the six-port network itself but also the input impedances and sensitivities of the power meters. Equations 12.12 may be regarded as a system of four linear equations relating the variables $|a_2|^2$, $|b_2|^2$, $a_2b_2^*$, $a_2^*b_2$ to the four power meter readings. The fact that a_2b_2 and $a_2^*b_2$ are complex conjugates is ignored by treating these two quantities as if they were independent variables. Inverting equations 12.12 gives

$$\begin{aligned} |a_2|^2 &= \Sigma\alpha_i P_i \qquad i = 3-6 \\ |b_2|^2 &= \Sigma\beta_i P_i \\ a_2b_2^* &= \Sigma(C_i + jS_i)P_i \\ a_2^*b_2 &= \Sigma(C_i - jS_i)P_i \end{aligned} \tag{12.13}$$

where α_i, β_i, C_i and S_i are real constants. Subtracting the first of equations 12.13 from the second gives the power P_{ab} absorbed by the load at the measurement port:

$$\begin{aligned} P_{ab} &= |b_2|^2 - |a_2|^2 \\ &= \gamma_3P_3 + \gamma_4P_4 + \gamma_5P_5 + \gamma_6P_6 \end{aligned} \tag{12.14}$$

where

$$\gamma_i = \beta_i - \alpha_i \qquad i = 3-6 \tag{12.15}$$

Dividing the third of equations 12.13 by the second gives the reflection coefficient of the load:

$$\Gamma_l = \frac{a_2}{b_2} = \frac{a_2b_2^*}{|b_2|^2} = \frac{\Sigma(C_i + jS_i)P_i}{\Sigma\beta_i P_i} \qquad i = 3-6 \tag{12.16}$$

The various constants in equations 12.12 to 12.16 must be determined by calibration.

The idea of using an arbitrary six-port for microwave measurements was proposed by Engen and Hoer [36]. The original suggestion related to power measurement, but it was realised that the same device would also measure reflection coefficients, voltages and currents [37]. For the measurement of absorbed power it is sufficient to know the four constants γ_i, which can be determined by substituting in equation 12.14 four sets of corresponding values of P_{ab}, P_3, P_4, P_5, P_6. These values can be obtained by connecting to the measurement port in turn three different lossless terminations (for which P_{ab} is zero) and a standard power meter. Komarek [51] applied this method to the calibration of bolometer mounts in the frequency range 1-18 GHz, using the circuit shown in Fig. 12.12. The four power meters P_3, P_4, P_5, P_6 give readings which, for ideal components, would be proportional to $|b_2|^2$, $|a_2|^2$, $|(a_2+b_2)^2|$

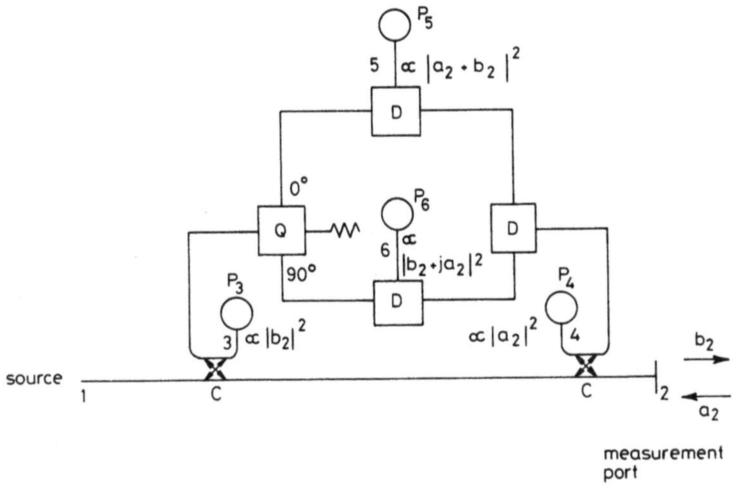

Fig. 12.12 *One possible configuration for a six-port*

and $|(b_2+ja_2)|^2$ respectively, neglecting for simplicity the phase shifts introduced by the lengths of line. The precise configuration of the circuit and the properties of the components are relatively unimportant, because the theory makes no assumptions about the six-port beyond linearity, but certain conditions must be avoided. For example, although the phase shift in the quadrature hybrid need not be exactly 90°, values in the region of 0° and 180° are to be avoided, because small errors in the measurements then lead to large errors in the results. In most applications the six-port will be used for both power and reflection coefficient measurements. This requires a more complete calibration than simply determining the four constants in equation 12.14. Many calibration methods have been proposed and some of these are outlined later in this section.

A graphical method of analysis has been developed which allows the operation of a six-port to be visualised more easily [40]. Equations 12.12 are first rewritten in terms of the forward wave b_2 at the measurement port and the reflection coefficient Γ_l of the load:

$$P_3 = |B_3|^2 |b_2|^2 |1 - \Gamma_s\Gamma_l|^2 \qquad (12.17a)$$
$$P_4 = |A_4|^2 |b_2|^2 |\Gamma_l - q_4|^2 \qquad (12.17b)$$
$$P_5 = |A_5|^2 |b_2|^2 |\Gamma_l - q_5|^2 \qquad (12.17c)$$
$$P_6 = |A_6|^2 |b_2|^2 |\Gamma_l - q_6|^2 \qquad (12.17d)$$

where

$$\Gamma_s = -\frac{A_3}{B_3} \qquad (12.18)$$

$$q_i = -\frac{B_i}{A_i} \qquad i = 4, 5, 6$$

All of the constants in equations 12.17 have simple physical interpretations. Constants $|B_3|^2$, $|A_4|^2$, $|A_5|^2$, $|A_6|^2$ represent scaling factors associated with each of the four power meters. In a well-designed six-port these are chosen so that the expected readings lie comfortably within the power meter ranges. The complex constant q_4 is the value of load reflection coefficient Γ_l for which P_4 is zero, and similarly for q_5 and q_6. The complex constant Γ_s is the equivalent source reflection coefficient which would be seen at the measurement port if P_3 were held constant by a levelling loop. It is usual to design the circuit so that Γ_s is small. Power meter 3 is known as the reference power meter. If each of equations 12.17*b*, 12.17*c*, 12.17*d* is divided by equation 12.17*a*, then b_2 is eliminated and three new equations are obtained relating Γ_l to the power ratios P_4/P_3, P_5/P_3, P_6/P_3.

$$\frac{P_i}{P_3} = \left|\frac{A_i}{B_3}\right|^2 \left|\frac{\Gamma_l - q_i}{1 - \Gamma_s \Gamma_l}\right|^2 \qquad i = 4, 5, 6 \quad (12.19)$$

Assuming that the constants B_3, Γ_s, A_i, q_i have been determined by calibration, then for given measured values of P_3, P_4, P_5, P_6, each of equations 12.19 defines a circle for Γ_l in the complex reflection plane. For Γ_s equal to zero, the centres of the three circles are q_4, q_5, q_6, and for this reason q_4, q_5, q_6 are known as the q-centres. Fig. 12.13 shows the positions of the q-centres for the circuit of Fig. 12.12. When Γ_s is not exactly zero but is small, the centres of the three circles lie close to the q-centres but do not coincide exactly with them.

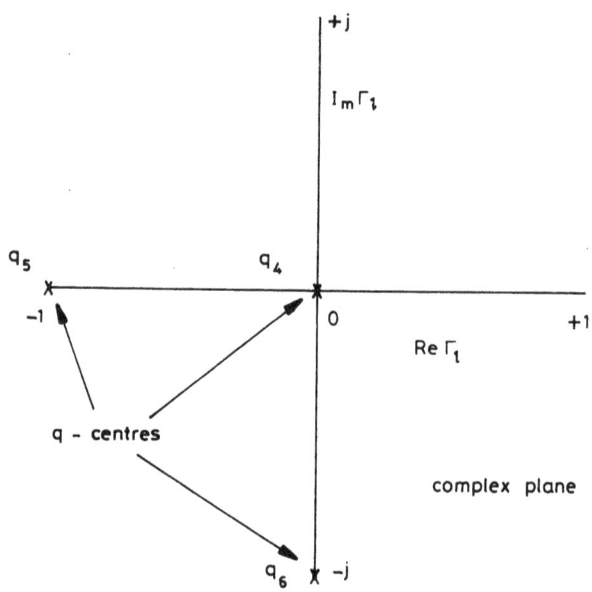

Fig. 12.13 *Position of q-circles for the circuit of Fig. 12.12, assuming perfect components*

In theory the three circles intersect in a common point, which corresponds to the load reflection coefficient Γ_l which is being measured, but in a practical situation this will not be the case. Instead, three closely spaced points of intersection will be obtained, as illustrated in Fig. 12.14, and it will not be obvious what value should be taken for Γ_l. This problem is a consequence of the redundancy in the system and it arises because of inconsistencies in the power meter readings. One method of obtaining a unique answer is to take the circles in pairs and to draw the three common chords, as shown in Fig. 12.14. It is a geometrical property of three intersecting circles that their three common chords meet at a single point, known as the radical centre, and it can be shown that this point corresponds to the value of Γ_l given by equation 12.16. Other methods will result in slightly different values for Γ_l. Engen [52] proposed a least squares method for determining the most likely value, taking into account the noise in the power meters. When Γ_l has been found, the forward power $|b_2|^2$ can be determined from any one of equations 12.17. Other quantities such as reflected power, impedance, voltage and current can then be calculated from the forward power and the reflection coefficient.

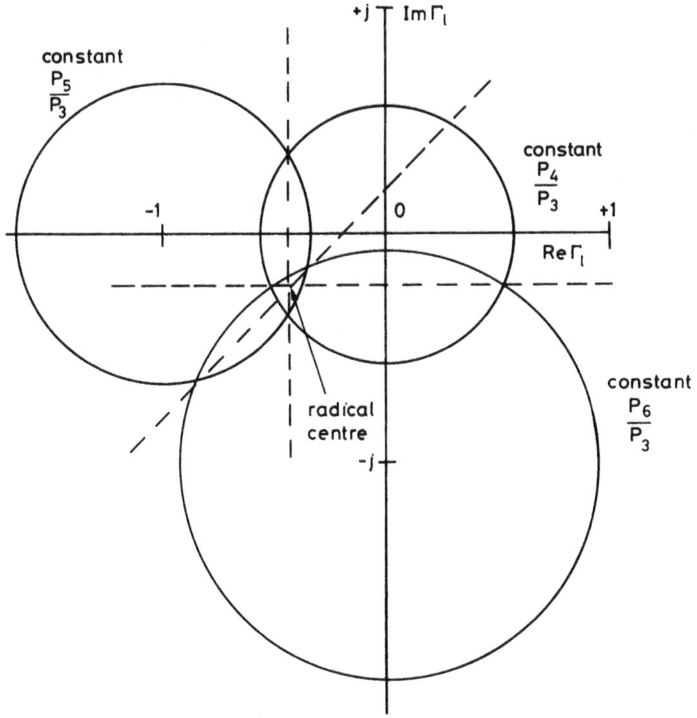

Fig. 12.14 *Effect of inconsistencies in the power meter readings in a six-port. The three circles do not intersect in a common point. The three common chords, however, do meet in a point, known as the radical centre*

The positions of the q-centres determine the accuracy of the reflectometer. They must be well separated from one another, since otherwise the circles will intersect at glancing angles and the exact points of intersection will be difficult to determine. The arrangement of q-centres shown in Fig. 12.13, located at $0, -1, -j$, is not the best one for reflection coefficient measurements because a q-centre at the origin leads to poor resolution for small values of Γ_l [40]. Some alternative schemes which avoid this problem place all three centres on the unit circle, for example at j, -1, and $-j$. Sometimes an extra power meter is added, making seven ports in all with four q-centres. Fig. 12.15a shows such a circuit, which is built up from quadrature hybrids and in-phase power dividers [41,42]. The q-centres of ports 4-7 are located at $1, j, -1, -j$, with port 3 as the reference port. The extra power meter enables an approximate real-time display of reflection coefficient to be obtained in a very simple way. This is done by keeping P_3 constant by means of a levelling circuit and deriving signals proportional to $(P_4 - P_5)$ and $(P_6 - P_7)$ from the analogue outputs of the other four power meters. These signals are fed to the X and Y inputs of an oscilloscope or polar display unit as shown in Fig. 12.15*b*. The instrument can at the same time be used as a conventional six-port by disregarding one of the power readings P_4, P_5, P_6, P_7. Other schemes use q-centres which lie outside the unit circle [53,54].

The majority of six-ports are built up from hybrids and three- or four-port directional couplers. Examples in addition to those already mentioned are given in refs. [55–60]. However, other methods of construction are used. Some or all of the couplers may be replaced by voltage probes [61]. Special couplers have also been developed for six-port applications [62–66]. Microstrip circuits [67–69] and, in the millimetre wave region, dielectric waveguide circuits [70] have been used. A difficulty in designing broadband six-ports is that, as the frequency changes, the positions of the q-centres move, with the result that the circuit may become ill-conditioned at certain frequencies. Avoiding this problem requires control over the phase shifts of the components.

Six-ports can also be constructed using resistive elements provided that there is at least one reactive component to provide the necessary phase shift. A disadvantage is that the need for the resistance values to be positive limits the possible positions of the q-centres. In addition, since the bandwidth of most six-ports is limited by the phase-shifting element, one of the chief attractions of resistive networks – that of wide frequency coverage – is lost. In some circumstances however the six-port approach might usefully be applied to resistive networks, especially where the reactive component of the load is known to be small. For example, in comparing a 50 Ω power meter with a 75 Ω power meter at frequencies below 100 MHz the reactances of the power meters may in some cases be small enough to be included in the comparison uncertainties, whereas the large difference in resistance must be corrected for.

A variation on the six-port reflectometer is the multistate reflectometer, which incorporates a provision for changing its parameters either by switching,

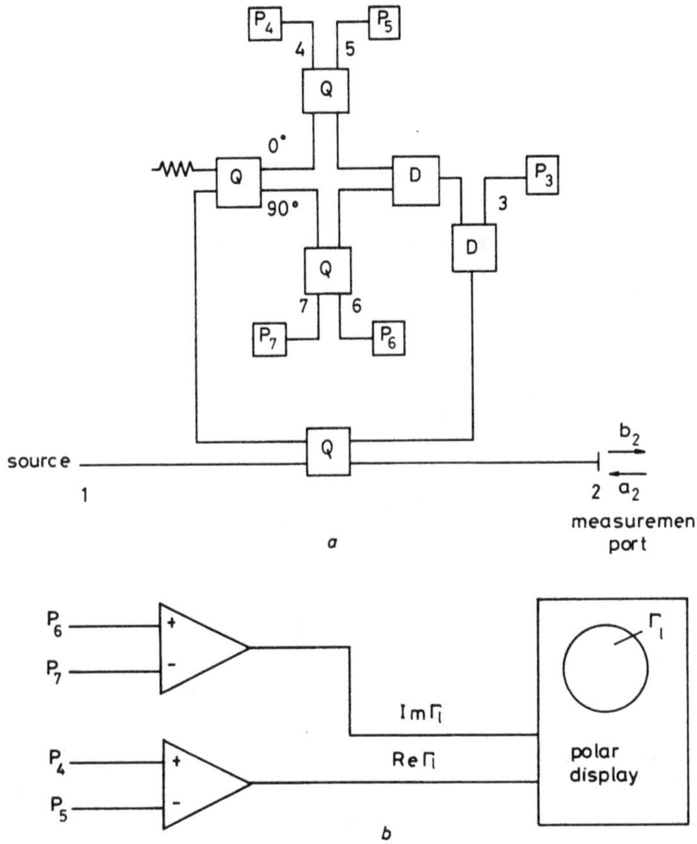

Fig. 12.15 *Seven-port network with q-centres at 0, j, −1, −j. Q = quadrature hybrid, D = in-phase power divider*

in the case of coaxial circuits, or by means of a resettable sliding short circuit for waveguide instruments [71,72]. Only two power meters are used, the necessary additional power readings being obtained by making measurements with the reflectometer in three or more different states. In general the behaviour of a multistate reflectometer is more complicated than that of a true six-port because the equations contain more unknown constants. The reason is that the value of Γ_s in each of the three equations 12.17*b, c, d* will be different. There are, however, certain configurations for which the multistate reflectometer simplifies to become exactly equivalent to a conventional six-port. Fig. 12.16 shows one such circuit that has been implemented in rectangular waveguide [73]. The equivalence is due to the fact that in this circuit the equivalent source reflection coefficient which would be obtained by using the reference port for levelling depends only on the properties of the output coupler and cannot therefore be affected by the position of the short circuit.

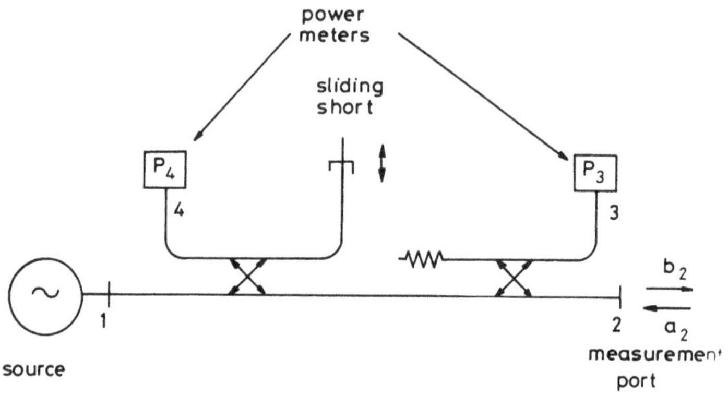

Fig. 12.16 *Multistate reflectometer. Readings are taken with the sliding short in at least three different positions*

The main use of the multistate reflectometer is for applications where speed is not essential and economy is sought.

12.4.2 Six-port calibration methods

For a complete calibration of a six-port reflectometer the usual approach is first to calibrate the instrument so that it will correctly measure reflection coefficients and then to establish absolute signal levels with the aid of a standard power meter. One of the simplest methods for the reflection coefficient part of the calibration is to connect a number of standard loads of known reflection coefficient Γ_l to the measurement port and to substitute corresponding values of P_i and Γ_l in equation 12.16. Each standard load produces two real equations by equating real and imaginary parts:

$$\begin{aligned} \mathrm{Re}\,\Gamma_l &= \frac{\Sigma C_i P_i}{\Sigma \beta_i P_i} \qquad i = 3-6 \\ \mathrm{Im}\,\Gamma_l &= \frac{\Sigma S_i P_i}{\Sigma \beta_i P_i} \end{aligned} \tag{12.20}$$

There are effectively eleven constants in equations 12.20, rather than the apparent twelve, since it is only the ratios between C_i, S_i, β_i which are of significance, not their absolute values. It might be supposed that in order to determine the eleven effective constants five and a half standards would be sufficient (a half standard being defined as one for which either the real or the imaginary part is known), since each standard gives two real equations. Woods [74] has shown that this approach is too simple, as singularities are encountered in solving the equations. These singularities can be avoided by regarding equations 12.20 as two independent equations. The fact that the denominators of the two equations are identical is ignored, and the values of β_i are determined twice over, thus increasing the number of effective constants to

14 (7 per equation). The number of standards necessary in order to be able to solve for the fourteen effective constants is thus seven. This scheme is relatively straightforward but it requires more standard loads than are necessary. Five standard loads are sufficient if one substitutes in equations 12.19 instead of in 12.20, since three equations are then obtained for each standard. One of the five standards need be known only approximately, since it is required merely to resolve the ambiguity arising from the non-linear nature of the equations. In reference [75] the use of four offset short circuits for the four accurately known standards is suggested (see also [76]), but there are also advantages in making one of the standards a matched load [77,78]. Another scheme requiring five standards is that due to Woods [79]. The reflectometer is treated as though it were composed of three independent four-port scalar reflectometers. Yet another calibration scheme makes use of the redundancy of the system to derive relations which result in a reduction in the number of independent constants. The reflectometer can then be calibrated as if it were a four-port vector reflectometer, for which three standard loads are sufficient. This scheme is known as six-port to four-port reduction [80,81] and it requires, in addition to the three known loads, a number of different but unknown loads. If two six-ports are available, self-calibration techniques such as the TRL technique (through-reflect-line) can be used [82,83]. This requires only one impedance standard, in the form of a section of line of known characteristic impedance but unknown length [84].

12.4.3 Calibration of coupler-power-meter assemblies by six-port techniques
Coupler-power-meter assemblies can be calibrated by incorporating them as the output stage of a six-port. Both the calibration constant and the equivalent source reflection coefficient can be derived in terms of the constants characterising the six-port. Reference [54] gives details.

12.5 Self-calibration methods

Much of the theory of reflectometers concerns the determination of the corrections for such coupler imperfections as mismatch, poor directivity and unknown coupling factor. Usually it is the values of constants appearing in the equations which must be determined rather than the coupler parameters themselves, but the result is the same. There are many methods of making such corrections, most of them involving the connection to the measurement port of a number of standard impedances, for example open circuits, short circuits and matched loads. Different methods require different numbers of impedance standards and it is relevant to ask what is the absolute minimum number for a specific measurement application. This problem is discussed in ref. [85] in connection with the direct evaluation of a coupler. It is clear that open and short circuits are not essential, as one can simulate these by connecting two

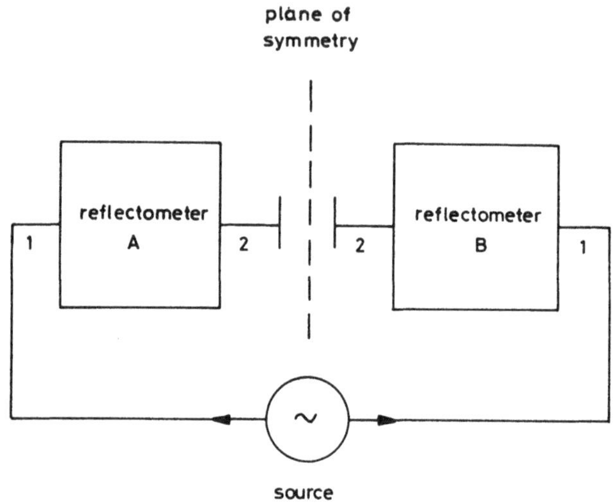

Fig. 12.17 *Simulation of open and short circuit by self-calibration methods*

identical measurement systems face to face as shown in Fig. 12.17. No standards are required to recognise when two systems are identical. If the generator outputs in Fig. 12.17 are in phase then by symmetry there is no current at the interface and an open circuit is simulated. If the systems are in antiphase a short-circuit is simulated. Such techniques are referred to as self-calibration techniques. Considerations such as these lead to the following conclusions. Couplers can be evaluated using only one standard impedance. They can then be assembled into reflectometers and used to measure a variety of quantities. For some applications the value of the one standard impedance cancels, so that it is possible, but not necessarily convenient, to carry out the measurements without the need for any form of impedance standard whatsoever. This applies to the measurement of voltage ratios, current ratios, impedance ratios and ratios of volt-amps (product of voltage and current, whether complex or moduli only). Thus, for example, two voltmeters could be compared without access to known impedances. However, for power meter comparisons this is not true. This fact can be seen by expressing the absorbed power as the product of volt-amps (moduli only) and power factor. In order to measure power factor a half-standard is required in the form of an impedance with a known ratio of reactance to resistance (that is a known phase angle). This could be a pure unknown resistance or a pure unknown reactance. Errors in the phase angle of the impedance half-standard do not cancel for power meter comparisons because the relation between the phase angle and the power factor is nonlinear, being a cosine law. Of course, if the incident and reflected powers are also to be determined, rather than just the absorbed power, then a whole impedance standard is required.

The above considerations are often helpful in understanding the various types of reflectometer. They also raise some interesting questions on the best way to disseminate standards. For example, if secondary standards were maintained for volt-amps and impedance, rather than power and impedance, then each could be disseminated independently of the other. In practice of course this advantage might be offset by other considerations, such as relative stability of the various parameters for which it would be possible to keep standards.

12.6 References

1 Harris, I.A., 'An accurate method for the comparison of two RF power meters which have similar or differing power ranges', Progress in Radio Science 1966–69, Vol. 2, ed. J.A. Lane, J.W., Findlay, and C.E. White, pp. 249–254, URSI, Brussels.

2 Oldfield, L.C., and Ide, J.P., 'Measurement of complex reflection coefficients in W-band using a 4-port reflectometer and precision waveguide spacers', IEE Colloquium on Advances in S-Parameter Measurement at Millimetre Wavelengths, London, May 1983, p. 8/1–6.

3 Rytland, W.A., and Geldea, D.R., 'Network analysis in the range 100 kHz to 110 MHz', *Hewlett-Packard J.*, **21**, 4, pp. 2–11, Dec. 1969.

4 Wiltron Co., 'Broadband VSWR measurements with simple bridges', *Microwave J.*, **13**, 8, p. 30, Aug. 1970.

5 Hewlett-Packard, 'Programmable scalar network analyzer and high directivity bridge systems', *Microwave J.*, **26**, 3, p. 118, Mar. 1983.

6 Pession, G., and Gorio, T., 'Measurement of power and efficiency of radio transmitting apparatus', *Proc. IRE*, 19, 3, pp. 377–400, 1931.

7 Schrack, R.A., 'Radio-frequency power measurements', NBS Circ. 536, Mar. 1953.

8 Hoer, C.A., and Agy, D., 'A broad band resistive divider type directional coupler', *IEEE Trans.*, **IM-19**, 4, pp. 336–343, Nov. 1970.

9 Jenkins, A, and Cullen, A.L., 'Resistive 4-port directional coupler', *Proc. IEE*, **129**, Pt. H, 3, pp. 94–98, June 1982.

10 Engen, G.F., and Beatty, R.W., 'Microwave reflectometer techniques', *IRE Trans.*, **MTT-7**, 3, pp. 351–355, July 1959.

11 Engen, G.F., 'A transfer instrument for the intercomparison of microwave power meters', *IRE Trans.*, **I-9**, 1, pp. 202–208, Sept. 1960.

12 Anson, W.J., 'A guide to the use of the modified reflectometer technique of VSWR measurement', *J. Res. NBS*, **65C**, pp. 217–223, Oct.–Dec. 1961.

13 Engen, G.F., 'A variable impedance power meter and adjustable reflection coefficient standard', *J. Res. NBS*, **68C**, pp. 7–24, Jan.–Mar. 1964.

14 Beatty, R.W., 'Automatic measurement of network parameters – a survey', NBS Monograph No. 151, 1976.

15 Anderson, R.W., and Dennison, D.T., 'An advanced new network analyser for sweep measuring amplitude and phase from .1 to 12.4 GHz', *Hewlett-Packard J.*, **18**, 6, pp. 2–10, Feb. 1967.

16 Adam, S.F., 'Automatic microwave network measurements', *Proc. IEEE*, **66**, 4, pp. 384–391, April 1978.

17 Fitzpatrick, J., 'Error models for systems measurements', *Microwave J.*, **21**, 5, pp. 63–66, May 1978.

18 Weinschel, B.O., and Weinert, F.K., 'I.F. vector substitution increases accuracy', *Microwave Syst. News*, **10**, pp. 94–106, 1980.

19 Donecker, B., 'Accuracy predictions for a new generation network analyzer', *Microwave J.*, **27**, 6, pp. 127–141, June 1984.
20 Hackborn, R.A., 'An automatic network analyzer system', *Microwave J.*, **11**, 5, pp. 45–52, May 1968.
21 Adam, S.F., 'A new precision automatic microwave measurement system', *IEEE Trans.*, **IM-17**, 4, pp. 308–313, Dec. 1968.
22 Rytting, D.K., and Sanders, S.N., 'A system for automatic network analysis', *Hewlett-Packard J.*, **21**, 6, pp. 2–10, Feb. 1970.
23 Fitzpatrick, J.K., 'A new direction for automatic network analyzer software', *Microwave J.*, **16**, 8, p. 65, Aug. 1973.
24 Maury, M.A. Jr., 'Automated network analyzer microwave measurements past, present and future', *Microwave J.*, **25**, 4, p. 18, April 1982.
25 Fitzpatrick, J., 'A history of automatic microwave network analyzers', *Microwave J.*, **25**, 4, p. 43, April 1982.
26 Hume, F.R., Koide, F.K., and Dederich, D.J., 'Practical and precise means of microwave power meter calibration transfer', *IEEE Trans.*, **IM-21**, 4, pp. 457–466, Nov. 1972.
27 Engen, G.F., 'An introduction to the description and evaluation of microwave systems using terminal invariant parameters', NBS Monograph 112, Oct. 1969.
28 Engen, G.F., 'Power equations: a new concept in the description and evaluation of microwave systems', *IEEE Trans.*, **IM-20**, 1, pp. 49–57, Feb. 1971.
29 Engen, G.F., 'An improved method for microwave power calibration with application to the evaluation of connectors', *J. Res. NBS*, **75C**, 2, pp. 89–93, April June 1971.
30 Engen, G.F., 'An experimental demonstration of the insensitivity of 'power equation' methods to connector imperfections', Proceedings of the 1971 European Microwave Conference, Stockholm, Aug. 1971 p. bl/5.
31 Moyer, R.D., 'Terminal invariant parameter determinations from complex signal ratio measurements', *IEEE Trans.*, **IM-21**, 4, pp. 532–537, Nov. 1972.
32 Engen, G.F., 'Theory of UHF and microwave measurements using the power equation concept', Rep. No. NBS-TN-637, Nat. Bur. Stand., April 1973.
33 Engen, G.F., 'Comment on "Practical analysis of reflectometers and power equation concepts"', *IEEE Trans.*, **IM-23**, 1, pp. 104–105, Mar. 1974.
34 Komarek, E.L., and Tryon, P.V., 'An application of the power equation concept and automation techniques to precision bolometer unit calibration', *IEEE Trans.*, **MTT-22**, 12, pp. 1260–1267, Dec. 1974.
35 Hoer, C.A., 'The six-port coupler: a new approach to measuring voltage, current, power, impedance and phase', *IEEE Trans.*, **IM-21**, 4, pp. 466–470, Nov. 1972.
36 Engen, G.F., and Hoer, C.A., 'Application of an arbitrary 6-port junction to power measurement problems', *IEEE Trans.*, **IM-21**, 4, pp. 470–474, Nov. 1972.
37 Hoer, C.A., and Engen, G.F., 'Analysis of a six port junction for measuring v, i, a, b, z, Γ and phase', Proc. IMEKO Symp. Acquisition and Processing of Measurement Data for Automation, Dresden, June 1973.
38 Hoer, C.A., and Roe, K.C., 'Using an arbitrary six-port junction to measure complex voltage ratios', *IEEE Trans.*, **MTT-23**, 12, pp. 978–984, Dec. 1975.
39 Engen, G.F., 'Determination of microwave phase and amplitude from power measurements', *IEEE Trans.*, **IM-25**, 4, pp. 414–418, Dec. 1976.
40 Engen, G.F., 'The six-port reflectometer: an alternative network analyzer', *IEEE Trans.*, **MTT-25**, 12, pp. 1075–1080, Dec. 1977.
41 Hoer, C.A., 'A network analyzer incorporating two six-port reflectometers', *IEEE Trans.*, **MTT-25**, 12, pp. 1070–1074, Dec. 1977.
42 Cronson, H.M., and Susman, L., 'A six-port automatic network analyzer', *IEEE Trans.*, **MTT-25**, 12, pp. 1086–1091, Dec. 1977.
43 Engen, G.F., 'The sixport measurement technique, a status report', *Microwave J.*, **21**, 5, pp. 18, 21–22, 24, 84, 87, 89, May 1978.

44 Engen, G.F., 'An overview of the six-port measurement technique', *Proc. IEEE MTT-S Int. Microwave Symp.*, Ottawa, June 1978, pp. 174–175.
45 Hoer, C.A., 'Performance of a dual six port automatic network analyzer', *IEEE Trans.*, **MTT-27**, 12, pp. 993–998, Dec. 1979.
46 Braun, A.E., 'Six port instrumentation faces controversy', *Microwave Syst. News*, **9**, pp. 79–81, 1979.
47 Simmons, J., 'Sixport redux', *Microwave Syst. News*, **10**, p. 16, 1980.
48 Weinschel, B.O., 'Six-port sceptic', *Microwaves*, **19**, pp. 11–12, 1980.
49 Hollis, J.L., 'Measuring r-f power with three ammeters', *Electronics,* **18**, pp. 142–143, June 1945.
50 Marzetta, L.A., 'An evaluation of the three voltmeter method for AC power measurement', *IEEE Trans.*, **IM-21**, 4, pp. 353–357, Nov. 1972.
51 Komarek, E.L., 'Performance characteristics of an automated broadband bolometer unit calibration system', *IEEE Trans.*, **MTT-25**, 12, pp. 1122–1127, Dec. 1977.
52 Engen, G.F., 'A least squares solution for use in the six-port measurement technique', *IEEE Trans.*, **MTT-28**, 12, pp. 1473–1477, Dec. 1980.
53 Engen, G.F., 'An improved circuit for implementing the six-port technique of microwave measurements', *IEEE Trans.*, **MTT-25**, 12, pp. 1080–1083, Dec. 1977.
54 Weidman, M.P., 'A semi-automated six-port for measuring millimeter-wave power and complex reflection coefficient', *IEEE Trans.*, **MTT-25**, 12, pp. 1083–1085, Dec. 1977.
55 Somlo, P.I., and Hunter, J.D., 'A six-port reflectometer and its complete characterisation by convenient calibration procedures', *IEEE Trans.*, **MTT-30**, 2, pp. 186–192, Feb. 1982.
56 Griffin, E.J., 'Six-port reflectometer circuit comprising three directional couplers', *Electronics Letters,* **18**, 12, pp. 491–493, June 1982.
57 Cronson, H.M., and Fong-Tom, Tr.A., 'A 94 GHz diode-based single sixport reflectometer', *IEEE Trans.*, **MTT-30**, 8, pp. 1260–1264, Aug. 1982.
58 Chung, N.S., Kim, J.H., and Shin, J., 'A dual six-port automatic network analyzer and its performance', *IEEE Trans.*, **MTT-32**, 12, pp. 1683–1686, Dec. 1984.
59 Somlo, P.I., Hunter, J.D., and Arthur, D.C., 'Accurate six-port operation with uncalibrated nonlinear diodes', *IEEE Trans.*, **MTT-33**, 3, pp. 281–282, Mar. 1985.
60 Hill, L.D., and Griffin, E.J., 'An automatic stepped frequency 6-port reflectometer for WG22', *Proc. IEE*, **132**, Pt H, 2, pp. 77–81, April 1985.
61 Groll, H.P., and Kohl, W., 'Six port consisting of two directional couplers and two voltage probes for impedance measurement in the millimeter wave range', *IEEE Trans.*, **IM-29**, 4, pp. 386–390, Dec. 1980.
62 Malkomes, M., Kadisch, G., and Schmitt, H.J., 'Optimized microstrip ring-star 5-ports for broadband 6-port measurement applications', Digest, IEEE MTT-S International Microwave Symposium, San Francisco, May–June 1984, pp. 472–474.
63 Cullen, A.L., Judah, S.K., and Nikravesh, F., 'Impedance measurements using a 6-port directional coupler', *Proc. IEE*, **127H**, 2, pp. 92–98, April 1980.
64 Belfort, A.J., and Cullen, A.L., 'First order theory of the five-port symmetrical star junction', *Electronics Letters*, **18**, 19, pp. 841–842, 16 Sept. 1982.
65 Stumper, U., 'New nondirectional waveguide multicoupler as part of a simple microwave six-port reflectometer', *Electronics Letters*, **18**, 18, pp. 757–758, 2nd Sept. 1982.
66 Hanson, E.R.B., and Riblet, G.P., 'An ideal sixport network consisting of a matched reciprocal lossless fiveport and a perfect directional coupler', *IEEE Trans.*, **MTT-31**, 3, pp. 284–288, Mar. 1983.
67 Collier, R.J., and El-Deeb, N.A., 'On the use of a microstrip three-line system as a six-port reflectometer', *IEEE Trans.*, **MTT-27**, 10, pp. 847–853, Oct. 1979.
68 Collier, R.J., and El-Deeb, N.A., 'Microstrip coupler suitable for use as a 6-port reflectometer', *Proc. IEE*, **127H**, 2, pp. 87–91, April 1980.
69 El-Deeb, N.A., 'The calibration and performance of a microstrip six port reflectometer', *IEEE Trans.*, **MTT-31**, 7, pp. 509–514, July 1983.

70 Paul, J.A., and Yen, P.C.H., 'Millimetre wave passive components and six-port network analyzer in dielectric waveguide', *IEEE Trans.*, **MTT-29**, 9, pp. 948–953, Sept. 1981.

71 Griffin, E.J., 'The multistate reflectometer', RSRE Memo No. 3625, Nov. 1983.

72 Oldfield, L.C., Ide, J.P., and Griffin, E.J., 'A multistate reflectometer', *IEEE Trans.*, **IM-34**, 2, pp. 198–201, June 1985.

73 Ide, J.P., Hodgetts, T.E., Potter, S.M., and Griffin, E.J., 'A multistate equivalent of a 6-port reflectometer', IEE Colloquium Digest No 1986/121, 'The technology of measuring microwave and millimetre components', pp. 4/1–4/6, 1986.

74 Woods, D., 'Analysis and calibration theory of the general 6-port reflectometer employing four amplitude detectors', *Proc. IEE*, **126**, 2, pp. 221–228, Feb. 1979.

75 Li, S., and Bosisio, R.G., 'Calibration of multiport reflectometers by means of four open/short circuits', *IEEE Trans.*, **MTT-30**, 7, pp. 1085–1090, July 1982.

76 Hunter, J.D., and Somlo, P.I., 'An explicit six-port calibration method using five standards', *IEEE Trans.*, **MTT-33**, 1, pp. 69–72, Jan. 1985.

77 Riblet, G.P., and Hanson, E.R.B., 'Aspects of the calibration of a single six-port using a load and offset reflection standards', *IEEE Trans.*, **MTT-30**, 12, pp. 2120–2125, Dec. 1982.

78 Somlo, P.I., 'The case for using a matched load standard for six-port calibration', *Electron. Lett.*, **19**, pp. 979–980, 1983.

79 Woods, D., 'Simplified calibration technique for general six port reflectometer requiring only two coaxial airline standards', *Proc. IEE*, **130**, Pt. A, 5, pp. 250–253, July 1983.

80 Engen, G.F., 'Calibration of an arbitrary six port junction for measurement of active and passive circuit parameters', *IEEE Trans.*, **IM-22**, 4, pp. 295–299, Dec. 1973.

81 Engen, G.F., 'Calibrating the six-port reflectometer by means of sliding terminations', *IEEE Trans.*, **MTT-26**, 12, pp. 951–957, Dec. 1978.

82 Hoer, C.A., 'Calibrating two 6 port reflectometers with only one impedance standard', NBS Tech. Note TN1004, June 1978.

83 Engen, G.F., and Hoer, C.A., 'Thru-Reflect-Line: an improved technique for calibrating the dual six-port automatic network analyzer', *IEEE Trans.*, **MTT-27**, 12, pp. 987–993, Dec. 1979.

84 Hoer, C.A., 'Calibrating two sixport reflectometers with an unknown length of precision transmission line', *Proc. IEEE MTT-S Int. Microwave Symp.*, Ottawa, June 1978, pp. 176–178.

85 Allred, C.M., and Manney, C.H., 'The calibration and use of directional couplers without standards', *IEEE Trans.*, **IM-25**, 1, pp. 84–89, Mar. 1976.

Chapter 13

Connectors and adapters

13.1 Connectors

The repeatability of coaxial connectors, and to a somewhat lesser degree of waveguide couplings, imposes a limitation on the accuracy with which power meters and other instruments can be compared. With the very best coaxial connectors presently available the standard deviation of the insertion loss will normally be at least 0.001 dB at a frequency of 8 GHz, corresponding to 0.025 percent in power. The equivalent figure for the very best waveguide couplings is approximately 0.00025 dB at 7.3 GHz [1,2], which assumes the use of hand-lapped flanges with crushable shims (or machine lapping and polishing) and precision dowels. Some reduction in uncertainty is possible by averaging the results of several measurements, in which case the standard deviation will be reduced by a factor of $\sqrt{N}$, where N is the number of measurements taken. However, such a procedure can be self-defeating if carried too far because it increases wear on the connector and results in a worsening of the performance. In practice ten connections and disconnections is probably a reasonable maximum for each set of measurements, corresponding approximately to a threefold improvement. In view of this one must conclude that the connector is to be regarded as a critical component when accurate measurements are to be made.

The connectors used in standards laboratories – and elsewhere when the highest precision is required – are of the precision type [3–5] defined in the IEEE Standard No. 287 [6]. Fig. 13.1 illustrates the main features. The IEEE standard resulted from work which began in the 1950s and which had as its aim the narrowing of the large gap in performance between the existing coaxial connectors and waveguide couplings. Weinschel reviewed the progress made up to 1967 [7] and other work is described in refs [8] and [9]. A very thorough account was later written by Powell [10]. The IEEE standard concentrated on specifying the performance expected of the connectors and avoided giving detailed design drawings, which would inhibit manufacturers from introducing later improvements in the mechanical design.

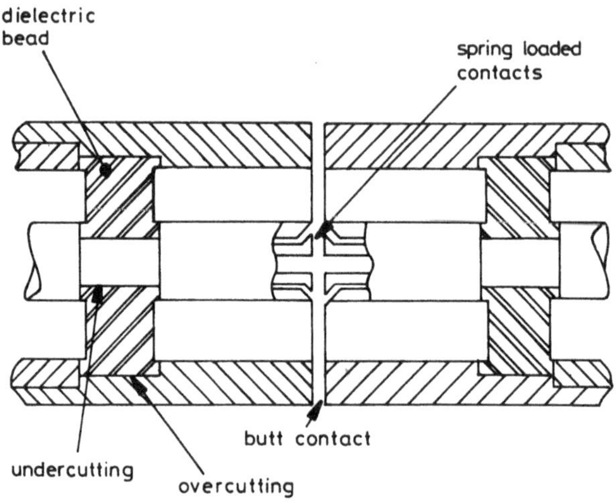

Fig. 13.1 *Main features of a precision type coaxial connector*

Important requirements for a precision connector are:

(*a*) A clearly defined mating plane which is the same for both inner and outer conductors and which coincides with the physical planes of connection and disconnection. This is necessary for the definition and accurate measurement of scattering parameters.

(*b*) Sexless mating. This is necessary to eliminate the mating problems that occur when a two-port component under test has identically sexed connectors at each end. The contacts of precision connectors are of the butt kind, with spring loaded contact surfaces for the inner conductor.

(*c*) Low reflection coefficient together with low and stable contact resistance

(*d*) Good repeatability of both reflection coefficient and insertion loss

(*e*) Low leakage.

Other requirements are that the connector should be mechanically strong, that no moments or rotational forces should be transmitted across the junction, and that when the connector is not in use the mating surfaces should not protrude in such a way that they could be accidentally scratched. The relative importance of the different requirements depends on the measurement application. For example, sexless mating greatly facilitates the evaluation of adapters. Low leakage is important in measuring high values of attenuation.

Great care is necessary in the design of the dielectric bead which supports the inner conductor of the connector, because it is usually the onset of resonances [11] in this bead which determines the upper frequency limit. The problem of bead resonance is made more acute by the need to undercut the inner and overcut the outer conductor so as to compensate for the effect of the dielectric on the characteristic impedance of the line. This undercutting and overcutting lowers the frequency at which the first resonance appears. It also introduces step capacitances, which are compensated for by removing some of the dielectric material. The bead itself is sited at some distance from the mating interface in order to ensure that it is well clear of any evanescent modes which may be generated by imperfect contacts and also to avoid the possibility of interaction through evanescent modes between the two beads of a mated pair. For some applications – for example impedance standards – the imperfections introduced by the bead cannot be tolerated and connectors with unsupported inners are required. These rely on mating with supported types to hold the inner conductor in place. There is some evidence to suggest that unsupported types are less repeatable than the equivalent supported connectors.

A common way of referring to coaxial connectors is to quote the nominal inner diameter of the outer conductor together with the characteristic impedance. The 14 mm 50 Ω precision connector (actually 14.2875 mm) is used up to 8.5 GHz. It was originally manufactured under the commercial designation GR 900 [12] and a 75 Ω version was also produced for use up to 2 GHz. The 7 mm 50 Ω precision connector has an upper frequency of 18 GHz and this has been produced commercially under several designations, for example APC 7, Prezifix AA and WPC7. Because of the smaller conductor size its repeatability is worse than that of 14 mm connectors. Improvements in its design are described by Donecker [13].

Connectors made to rather less exacting requirements than the precision types are known as semi-precision connectors. Examples are GR 890 50 Ω (dc to 8 GHz), Dezifix B 50 Ω, 60 Ω, or 75 Ω (dc to 5 GHz), Prezifix A 50 Ω (dc to 18 GHz) and APC 3.5 50 Ω (dc to 34 GHz). The last named, which was announced in 1976 [14,15], has mechanical interface tolerances similar to precision sexless connectors but is nevertheless of the sexed type. It was designed to be compatible with the SMA connector [16], a relatively inexpensive, widely used solid dielectric non-precision type designed in the late 1950s and limited to 18 GHz. However, when an SMA and APC 3.5 are mated together there is a mismatch due to the air-dielectric interface.

Other non-precision connectors are N 50 Ω (dc to 18 GHz), N 75 Ω (dc to 2 GHz), and BNC 50 or 75 Ω (dc to 2 GHz), all sexed. Fig. 13.2 shows an N-type connector, of which there were numerous, often mutually incompatible, versions prior to the adoption of the MIL-C-39012 standard. The inner and outer mating planes do not coincide and consequently there is no single reference plane. The use of coaxial connectors extends also to frequencies above 40 GHz. For example, operation at frequencies approaching 46 GHz is obtained from 3 mm connectors (actually 2.92 mm) such as the K connector, MPC 3, KMC and WMP 4 [17] and operation up to 50 GHz is obtained from 2.4 mm connectors [18].

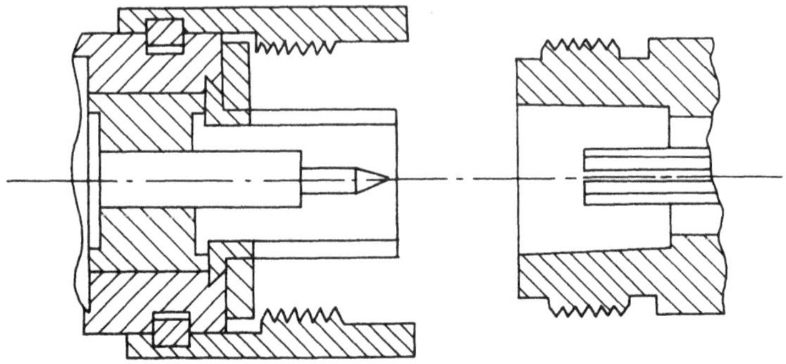

Fig. 13.2 *Details of an N-type connector*

All connectors need to be treated with care if reliable results are to be obtained. They should be regularly cleaned and inspected, and the surfaces protected from damage. It is also important that connectors do not damage the test equipment interfaces to which they are offered for calibration and in order to avoid this possibility gauge kits are available which enable the critical dimensions to be checked. Equipment should be well supported in order to avoid stresses on the connectors and for heavy equipment a system of counterweights or equivalent arrangement may be necessary to allow the connectors to take up their natural position. The use of a torque wrench is recommended for tightening connectors, since overtightening can damage the outer conductor. Whether a torque wrench also improves repeatability is debatable, as manual tightening allows one to feel whether the connector is mating properly.

Table 13.1 *Ranges of insertion loss repeatability (dB) to be expected of coaxial connectors (standard deviation)*

Precision types	1 GHz	8 GHz	18 GHz
14 mm (GR900)	<0.001	0.001–0.002	
7 mm (APC7)	<0.001	0.0015–0.005	0.005–0.020
Non-precision types (best quality)			
N and SMA	<0.001	0.002–0.006	0.005–0.020

The literature on the repeatability of connectors is rather inconclusive and further work is needed to ascertain more reliable performance values. In the past, measurements of repeatability tended to be very time-consuming but improvements in the speed and performance of network analysers have made this less so. Table 13.1 has been compiled from available information [19–25]. It summarises the standard deviations to be expected for the insertion loss for some precision and non-precision types, but should be considered as a guide only, because much depends on the condition of the connectors and the way in which they are used. Worn connectors, for example, may be three times as bad as new ones. It will be seen that, although the precision types tend to have better insertion loss repeatability than non-precision types, the improvement does not appear to be very great. This may be due to the fact that butt contacts are not self-cleaning. One may conclude that the major benefits of precision connectors come from the lower reflection coefficients and the fact that they are sexless, rather than improved insertion loss repeatability. Table 13.1 does not however tell the whole story, because it says nothing about the rate at which connectors wear or how well they stand up to rough handling and misuse. SMA connectors are particularly noted for their tendency to wear quickly and they are also easily damaged by overtightening.

13.2 Adapters

13.2.1 The need for adapters

There are several ways in which power meters fitted with different connectors can be intercompared. In a small number of cases the connectors may be mechanically compatible; for example, 50 Ω and 75 Ω precision 14 mm connectors can be mated together. A 75 Ω power meter can therefore be connected directly to a 50 Ω measurement system and calibrated as if it were a mismatched nominal 50 Ω power meter. The comparison must be carried out in terms of absorbed power rather than incident power; otherwise errors are caused by the capacitance of the 50 – 75 Ω step, which modifies the impedance [26,27]. Other mechanically compatible coaxial connectors are 50 and 75 Ω BNC types. A similar situation applies to some millimetre-wave waveguide couplings where the same flange design is used for more than one waveguide size. An alternative approach would be to modify the connectors in order to achieve mechanical compatibility. In most cases, however, adapters are used. The term 'adapter' will be interpreted as covering a range of items, as shown in Fig. 13.3, including waveguide-to-coaxial transformers, waveguide tapers, stepped lines and resistive matching pads. The loss of the adapter must usually be measured and the additional uncertainty associated with it evaluated. As an example, the loss of a typical 14 mm to 7 mm precision adapter at 8 GHz might be 0.7 percent of the input power and this could be measured with an uncertainty of 0.2 – 0.3 percent.

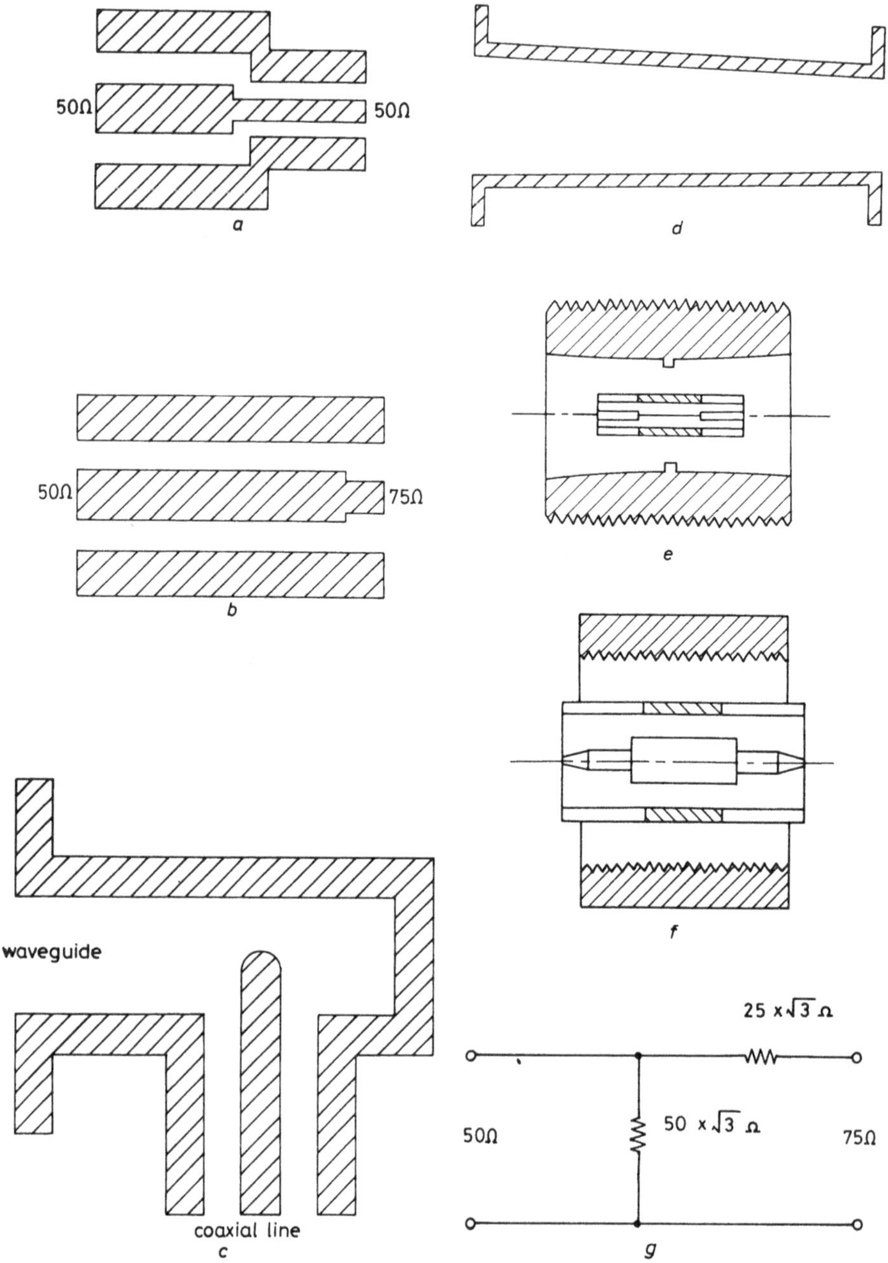

Fig. 13.3 *Examples of adapters. (a) Coaxial 50 Ω/50 Ω adapter, (b) Unmatched 50 Ω/75 Ω adapter consisting of a stepped line, (c) waveguide-to-coaxial transformer, (d) waveguide taper, (e) coaxial adapter enabling two N-type male connectors to be connected together, (f) coaxial adapter enabling two N-type female connectors to be connected together, (g) resistive matching pad*

The parameter which characterises the power loss of an adapter is its efficiency η, defined as:

$$\eta = \frac{\text{power delivered to load}}{\text{power absorbed at input to adapter}} \tag{13.1}$$

Efficiency depends on the reflection coefficient of the load connected to the output port of the adapter and in general it changes when the direction of power flow is reversed. There are two main methods for determining efficiency, referred to as the three-adapter method and the sliding short method, each with its variations. For low-loss adapters simplifications are possible by introducing approximations. The uncertainties due to these approximations are evaluated with the help of the realisability conditions, that is the relations which express the fact that an adapter is known to be a passive, reciprocal, linear device, whose parameters are therefore subject to certain constraints.

For the complete characterisation of an adapter two sets of impedance standards are necessary, one for each connector type. For the measurement of efficiency, on the other hand, the minimum requirement for each connector type is an impedance with a known phase angle (that is a known ratio of reactance to resistance and therefore a known power factor). For further information on the use of such half-standards of impedance see Chapter 12 under 'Self-calibration methods' (section 12.5).

13.2.2 The three-adapter method and its variations

The general three-adapter method is illustrated in Fig. 13.4. Consider first a relatively simple case in which all components are matched. If the characteristic impedance at both ports of the adapter is equal to Z_0, then the efficiency η when terminated in a matched load is given by:

$$\eta_0 = |s_{21}|^2 = |s_{12}|^2 \tag{13.2}$$

In this instance it does not matter which port is considered the output port because the efficiency is the same for both directions of power flow. The value of s_{21} cannot be determined by a simple insertion loss measurement because of the difference in the connectors at the two ports. The three-adapter method overcomes this difficulty provided that the connectors are sexless. The three adapters must be of the same type, but not necessarily identical. The technique consists of connecting the adapters back-to-back in pairs and determining the transmission coefficients s_{21AB}, s_{21BC}, s_{21CA} of these pairs, where A, B, C refer to the three individual adapters. Then

$$\begin{aligned} |s_{21AB}| &= |s_{21A}|\,|s_{21B}| \\ |s_{21BC}| &= |s_{21B}|\,|s_{21C}| \\ |s_{21CA}| &= |s_{21C}|\,|s_{21A}| \end{aligned} \tag{13.3}$$

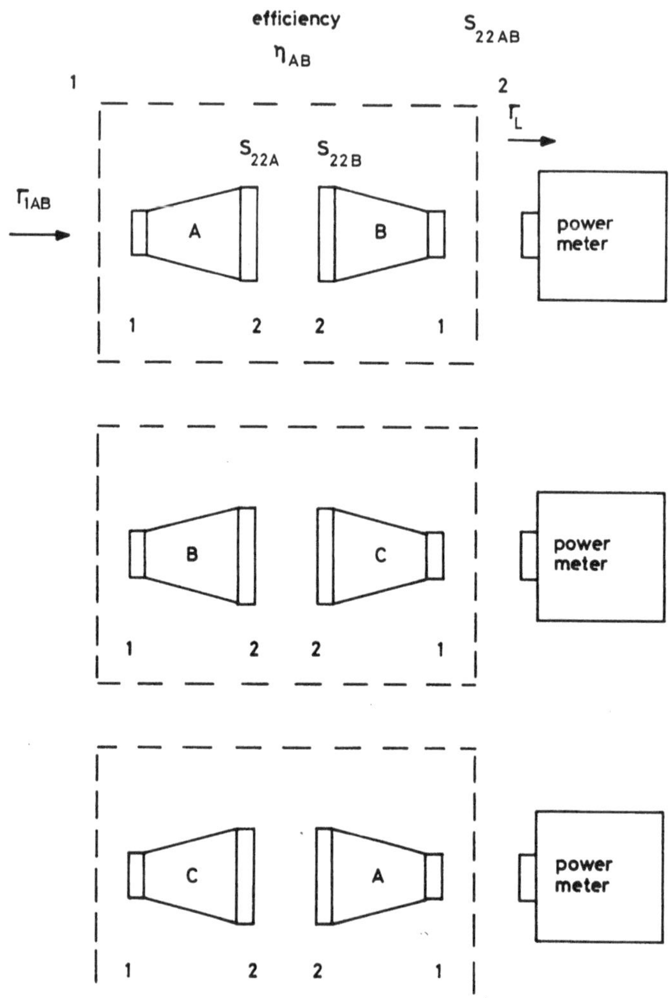

Fig. 13.4 *The three-adapter method. The efficiencies of the individual adapters are deduced from the measured efficiencies of pairs of adapters*

from which η_{0A}, η_{0B}, η_{0C} can be found. Once η_0 is known for a particular adapter, its efficiency η for an arbitrary load can be found. It is given by:

$$\eta = \frac{\eta_0 (1 - |\Gamma_L|^2)}{(1 - |\Gamma_1|^2)} \tag{13.4}$$

$$= \frac{\eta_0 (1 - |\Gamma_L|^2)}{(1 - |s_{21}{}^2 \Gamma_L|^2)}$$

where Γ_L is the load reflection coefficient and Γ_1 the input reflection coefficient when a load of reflection coefficient Γ_L is connected to the output.

A rigorous treatment of the three-adapter method takes into account not only the mismatches at the interfaces but also the fact that the characteristic impedances at the two ports may be different [28]. It will be assumed that the adapter whose efficiency is required is adapter A and that it will be used with port 1 as the input and port 2 as the output. Equations 13.3 now become:

$$
\begin{aligned}
|s_{21AB}| &= \left(\frac{Z_{01}}{Z_{02}}\right)\frac{|s_{21A}||s_{21B}}{|1 - s_{22A}s_{22B}|} \\
|s_{21BC}| &= \left(\frac{Z_{01}}{Z_{02}}\right)\frac{|s_{21B}||s_{21C}|}{|1 - s_{22B}s_{22C}|} \\
|s_{21CA}| &= \left(\frac{Z_{01}}{Z_{02}}\right)\frac{|s_{21C}||s_{21A}|}{|1 - s_{22C}s_{22A}|}
\end{aligned}
\tag{13.5}
$$

where Z_{01}, Z_{02} are the characteristic impedances for ports 1 and 2 respectively. Equations 13.5 can be solved for $|s_{21A}|$, $|s_{21B}|$, and $|s_{21C}|$ if the parameters s_{22A}, s_{22B}, s_{22C} and $|s_{21\ AB}|$, $|s_{21BC}|$, $|s_{21CA}|$ can be determined. The values of s_{22A}, s_{22B}, s_{22C} are measured using a vector reflectometer (for example a network analyser). The values of $|s_{21AB}|$, $|s_{21BC}|$, $|s_{21CA}|$ are deduced from efficiency measurements made on the adapter pairs, using a power meter as a load. The equation giving $|s_{21AB}|$ in terms of the efficiency η_{AB} of the adapter pair AB when terminated in a load of reflection coefficient Γ_L is [28]:

$$
|s_{21AB}|^2 = \eta_{AB}\frac{1 - |\Gamma_{1AB}|^2}{1 - |\Gamma_L|^2}|- s_{22AB}\Gamma_L|^2 \tag{13.6}
$$

where Γ_{1AB} is the input reflection coefficient of the adapter pair AB when terminated in Γ_L and s_{22AB} refers to the load end of the adapter pair. Similar equations apply for adapter pairs BC and CA. Once s_{11A}, s_{22A}, and $|s_{21A}|$ have been determined, the efficiency of adapter A for a matched load of impedance Z_{02} connected to port 2 may be calculated from the formula:

$$
\eta_{0A} = \frac{Z_{01}}{Z_{02}}\frac{|s_{21A}|^2}{1 - |s_{11}|^2} \tag{13.7}
$$

Finally, the efficiency of adapter A for an arbitrary load of reflection coefficient Γ_L is given by:

$$
\eta_A = \frac{Z_{01}}{Z_{02}}\frac{|s_{21A}|^2(1 - |\Gamma_L|^2)}{(1 - |\Gamma_1|^2)|1 - s_{22A}\Gamma_L|^2} \tag{13.8}
$$

where Γ_1 is the reflection coefficient at port 1 when port 2 is terminated in Γ_L. Although $|\Gamma_1|$ could be calculated from the scattering parameters of the adapter and Γ_L, it is usually more convenient to measure it directly, in which case one need not measure s_{11A}.

Several variations of the method are possible. The adapter measurement may be combined with the power meter comparison procedure for which its efficiency is required. As an example, suppose that a waveguide and a coaxial power meter are to be compared using a coaxial test bench. Then only two adapters A and B are required. The procedure, which is illustrated in Fig. 13.5, is as follows:

1. Connect the coaxial power meter (power meter 1) to the source
2. Connect adapter pair AB between source and coaxial power meter.
3. Connect the waveguide power meter (power meter 2) to the source via adapter A.
4. Connect the waveguide power meter to the source via adapter B.

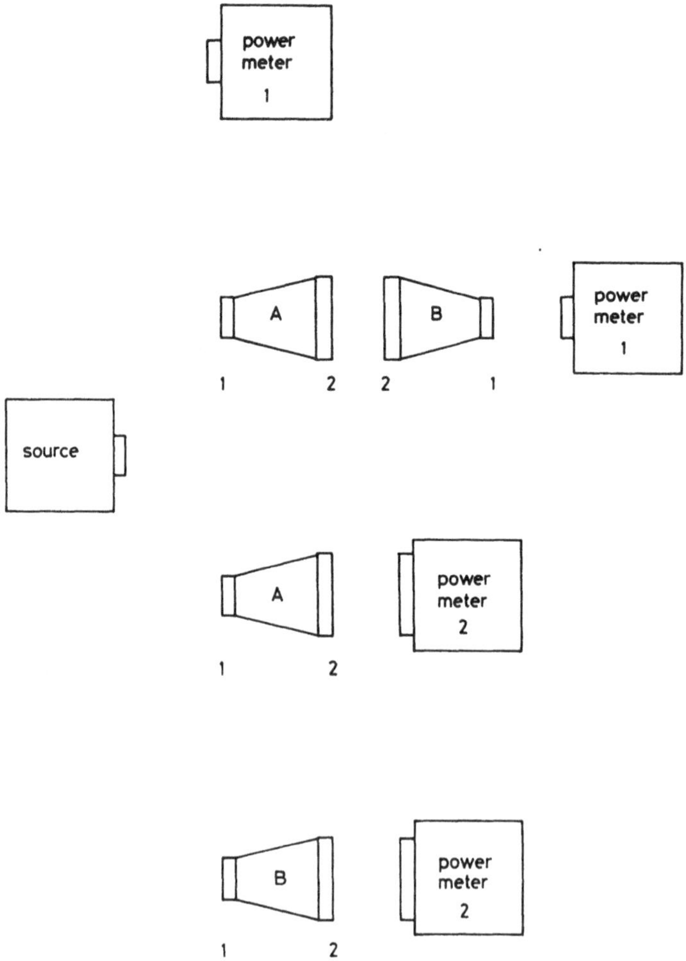

Fig. 13.5 *The two-adapter method. The adapter efficiency is determined as part of the power meter comparison procedure*

The method gives both the result of the intercomparison and the efficiencies of A and B. In the simple case where all components are matched, steps 1 and 2 give the product of the efficiencies of A and B and steps 3 and 4 their ratio, from which the individual efficiencies are easily found. A further possibility is shown in Fig. 13.6. This uses only one adapter but requires both a waveguide

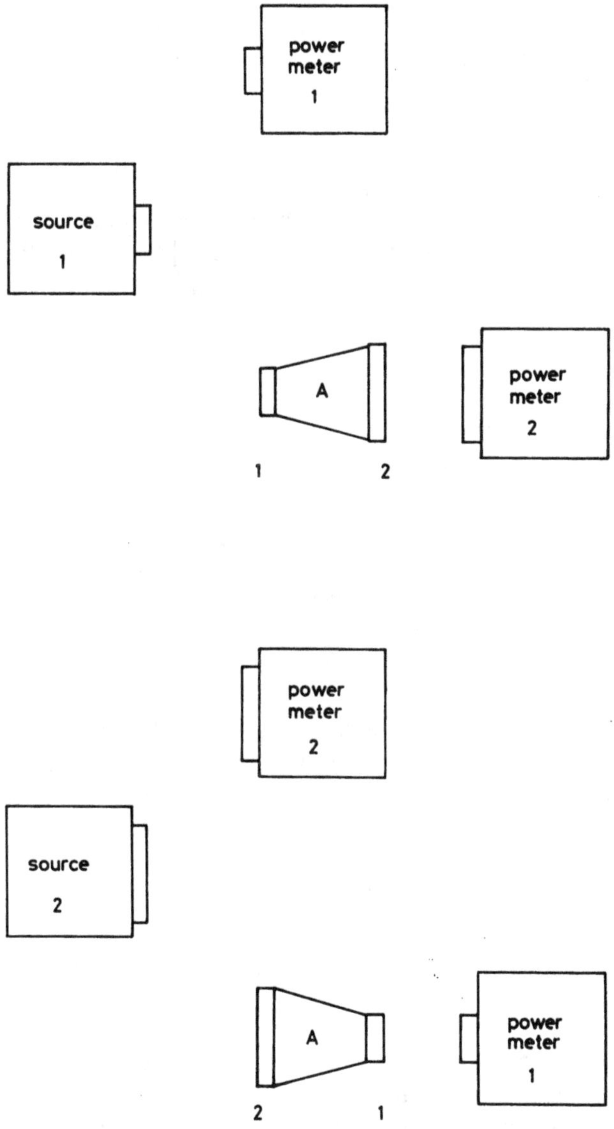

Fig. 13.6 *The singe-adapter method. Two sources are required. For example, source 1 and power meter 1 could be in coaxial line and source 2 and power meter 2 in waveguide. As in the two-adapter method, the evaluation of the adapters is carried out as part of the power meter comparison procedure*

and a coaxial source. The first step in the measurement is to connect the adapter to the waveguide power meter (power meter 2) and carry out a comparison between this adapter/power meter combination and the coaxial power meter using the coaxial source (source 1). In the second step the adapter is connected to the coaxial power meter and a comparison is carried out between this adapter/power meter combination and the waveguide power meter, using the waveguide source (source 2). Ref. [29] gives a simplified treatment which assumes that the efficiency of the adapter is the same for the two parts of the measurement.

13.2.3 The sliding short method

The sliding short method requires measurements of reflection coefficients but no transmission measurements. A linear reciprocal two-port can be represented by the network shown in Fig. 13.7 [30]. The efficiency of this network when terminated at port 2 in an arbitrary load of reflection coefficient

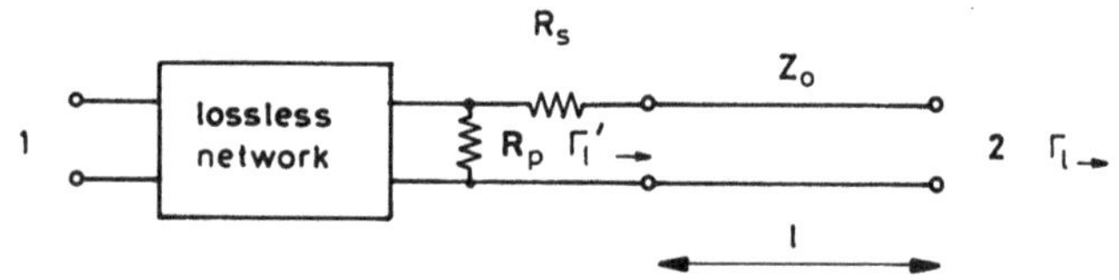

Fig. 13.7 *One possible equivalent circuit for an adapter. The efficiency when terminated in reflection coefficient Γ_l depends only on the parameters R_s, R_p, l, and Γ_l*

Γ_L depends on the value of Γ_L, the resistances R_p and R_s, and the length l of the transmission line. Thus three known parameters suffice. The values of the three parameters are found by connecting to port 1 a sliding short circuit and measuring the locus of the reflection coefficient observed at port 2 [31–34]. As the phase of the sliding short is varied, the reflection coefficient Γ at port 2 will trace out a circle (Fig. 13.8) of radius R given by:

$$R = \frac{R_p Z_0}{(R_s + R_p + Z_0)(R_s + Z_0)} \tag{13.9}$$

and one can show that this radius is equal to the efficiency η_0 when port 2 is terminated in a load of impedance Z_0. A method for calculating the efficiency for an arbitrary load of reflection coefficient Γ_L is given in reference [33]. This involves calculating a new reflection coefficient Γ_N using the formula:

$$\Gamma_N = \frac{\Gamma - \Gamma_L}{1 - \Gamma\Gamma_L} \tag{13.10}$$

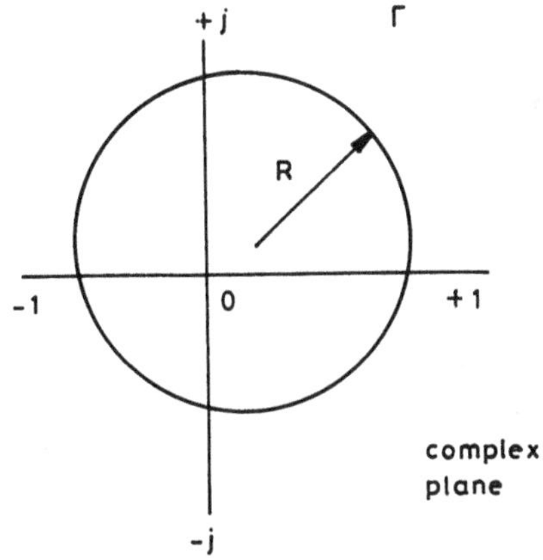

Fig. 13.8 *Locus of the reflection coefficient at port 2 of the circuit shown in Fig. 13.7 for a sliding short connected to port 1*

where Γ is the measured reflection coefficient. The locus of Γ_N, like that of Γ, is a circle. The efficiency for a load Γ_L is given by

$$\eta = R_n \frac{1 - |\Gamma_L|^2}{|1 - \Gamma_L{}^2|} \tag{13.11}$$

where R_n is the radius of the new circle. The sliding short may be replaced by three fixed lossless terminations, since three points are sufficient to define a circle. If the adapter is used with the direction of power flow reversed, that is with the load connected to port 1 and the source connected to port 2, then the same equivalent circuit of Fig. 13.7 can be used but equations 13.9, 13.10, and 13.11 must be modified. In this case Γ_L is replaced by minus the input reflection coefficient at port 2 and η is replaced by 1/η. Alternatively the equivalent circuit can be reversed and new values determined for R_p, R_s, and l. The choice will depend on whether one wishes to express the efficiency in terms of the load reflection coefficient or the input reflection coefficient of the adapter.

High accuracy is required for the reflection coefficient measurements [34]. The loss of the sliding short, or the fixed lossless terminations, must be taken into account and procedures for doing this are discussed in references [34] and [35].

13.2.4 Approximation for low-loss adapters

For low-loss adapters the efficiency is often assumed to be independent of the load for reasonably small values of load reflection coefficient. The use of this approximation may be illustrated by the following example.

Assume that the efficiency η_0 of an adapter has been accurately determined for a load of impedance Z_0. Suppose now that this adapter is used with a load with reflection coefficient Γ_L under the assumption that its efficiency η is still equal to η_0. The difference between η and η_0 thus represents the error in the efficiency. The limits for this error can be found by calculating the maximum and minimum values of η which are consistent both with the realisability conditions and with the known value of η_0. The realisability conditions are a consequence of the fact that the adapter is a passive reciprocal two-port device. They may be expressed in various ways depending on the parameters chosen to represent the adapter [36,37], but in terms of the equivalent circuit shown in Fig. 13.7 they have a particularly simple form:

$$\begin{aligned} R_p &> 0 \\ R_s &> 0 \end{aligned} \tag{13.12}$$

Let Γ_L' be the reflection coefficient of the load, modified so as to include the phase shift of the transmission line l in Fig. 13.7. For a given value of Γ_L maximum efficiency can be shown to occur when R_p is equal to infinity and the phase of Γ_L' is zero:

$$\eta_{max} = \frac{1}{\left(\dfrac{1}{\eta_0} - 1\right)\left(\dfrac{1 - |\Gamma_L|}{1 + |\Gamma_L|}\right) + 1} \tag{13.13}$$

Minimum efficiency occurs when R_p is equal to infinity and the phase of Γ_L' is 180°:

$$\eta_{min} = \frac{1}{\left(\dfrac{1}{\eta_0} - 1\right)\left(\dfrac{1 + |\Gamma_L|}{1 - |\Gamma_L|}\right) + 1} \tag{13.14}$$

These limits may also be obtained, in reverse order, by putting R_s equal to zero instead of R_p equal to infinity. The efficiency must be within the range given by equations 13.13 and 13.14. The closer the value of η_0 is to one, the smaller is the uncertainty range. Reference [38] gives an alternative treatment in terms of scattering parameters and presents the results in normalised graphical form.

13.3 References

1 Skilton, P.J., 'A technique for determination of loss, reflection and repeatability in waveguide flanged couplings', *IEEE Trans.*, **IM-23**, 4, pp. 390–395, Dec. 1974.
2 Skilton, P.J., 'Correction to "A technique for determination of loss, reflection and repeatability in waveguide flanged couplings"', *IEEE Trans.*, **IM-24**, 3, p. 279, Sept. 1975.
3 Soderman, R.A., 'Application of precision connectors to high-frequency measurements', *IEEE Trans.*, **IM-16**, 1, pp. 62–68, Mar. 1967.

4 IEEE G-IM Subcommittee on Precision Coaxial Connectors, 'Precision coaxial connector coupling mechanisms, contact designs, and higher mode resonances', *IEEE Trans.*, **IM-17**, 3, pp. 219–222, Sept. 1968.

5 Sladek, N.J., and Jesch, R.L., 'Standardization of coaxial connectors in the IEC', *Proc. IEEE*, **74**, 1, pp. 14–18, Jan. 1986.

6 'IEEE Standard for precision coaxial connectors', *IEEE Trans.*, **IM-17**, 3, pp. 206–218, Sept. 1968.

7 Weinschel, B.O., 'Standardization of precision coaxial connectors', *Proc. IEEE*, **55**, 6, pp. 923–932, June 1967.

8 Woods, D., 'A coaxial connector system for precision rf measuring instruments and standards', *Proc. IEE*, **108B**, 38, pp. 205–213, Mar. 1961.

9 Huber, F.R., and Neubauer, H., 'The Dezifix connector – a sexless precision connector for microwave techniques', *Microwave J.*, **6**, 6, pp. 79–85, June 1963.

10 Powell, R.C., 'Precision coaxial connectors', Advances in Microwaves, Vol. 6, pp. 1–28, Academic Press 1971.

11 Neubauer, H., and Huber, F.R., 'Higher modes in coaxial rf lines', *Microwave J.*, **12**, 6, pp. 57–66, Jun. 1969.

12 Mackenzie, T.E., and Sanderson, A.E., 'Some fundamental design principles for the development of precision coaxial standards and components', *IEEE Trans.*, **MTT-14**, 1, pp. 29–39, Jan. 1966.

13 Donecker, B., 'Accuracy predictions for a new generation network analyzer', *Microwave J.*, **27**, 6, pp. 127–141, June 1984.

14 Adam, S.F., Kirkpatrick, G.R., Sladek, N.J., and Bruno, S.T., 'A new 34 GHz 3.5 mm low-cost utility coaxial connector featuring low leakage, low standing-wave ratio and long life', *IEEE Trans.*, **MTT-24**, 12, pp. 995–997, Dec. 1976.

15 Kirkpatrick, G.R., Pratt, R.E., and Chambers, D.R., 'Coaxial components and accessories for broadband operation to 26.5 GHz', *HP Journal*, **28**, 10, pp. 10–16, June 1977.

16 Estin, A.J., 'Scattering parameters of SMA coaxial connector pairs', *IEEE Trans.*, **IM-25**, 4, pp. 329–334, Dec. 1976.

17 Oldfield, W.W., 'Comparing miniature coaxial connectors, *Microwaves and RF*', **24**, 9, pp. 171–174, Sept 1985.

18 Browne, J., 'Team effort yields 50-GHz connectors', *Microwave and RF*, **25**, 4, p. 131, April 1986.

19 Bergfried, D., and Fischer, H., 'Insertion loss repeatability versus life of some coaxial connectors', *IEEE Trans.*, **IM-19**, 4, pp. 349–353, Nov. 1970.

20 Jesch, R.L., 'Repeatability of SMA coaxial connectors', *IEEE Trans.*, **IM-25**, 4, pp. 314–320, Dec. 1976.

21 Pakai, P., Terek, A., and Fedorov, A.M., 'Method and results of a reproducibility study of the losses in coaxial connectors', *Meas. Tech.*, **20**, 3, pp. 425–427, Mar. 1977.

22 Pakay, P., and Török, A., 'Analysis of insertion loss repeatability of coaxial connectors', *Radio Electron. Eng.*, **47**, pp. 315–319, July 1977.

23 Pakai, P., and Fedorov, A.M., 'Losses in coaxial junctions and reproducibility in connectors', *Meas. Tech.*, **23**, 4, pp. 338–339, April 1980.

24 Bradshaw, J.D., and Griffin, E.J., 'Report on repeatability of APC7 connectors', British Calibration Service, BCS/2.2/83/197, Sept. 1983.

25 Somlo, P.I., and Hunter, J.D., 'Microwave impedance measurement', Peter Peregrinus, 1985.

26 Somlo, P.I., 'The computation of coaxial line step capacitances', *IEEE Trans.*, **MTT-15**, 1, pp. 48–53, Jan. 1967.

27 Jurkus, A., 'Computation of step discontinuities in coaxial line', *IEEE Trans.*, **MTT-20**, 10, pp. 708–709, Oct. 1972.

28 Skilton, P.J., 'A technique for measuring the efficiency of waveguide to coaxial line adaptors', *IEEE Trans.*, **IM-27**, 3, pp. 231–234, Sept. 1978.

29 Engen, G.F., 'Coaxial power meter calibration using a waveguide standard', *J. Res. NBS*, **70C**, pp. 127–138, April–June 1966.

30 Sucher, M, and Fox, J. (eds), 'Handbook of microwave measurements', (3 vols), Polytechnic Press, 1963.

31 Cullen, A.L., 'Measurement of microwave transmission efficiency', *Wireless Engrg.*, **26**, 11, pp. 255–258, 1949.

32 Mathis, H.F., 'Experimental procedures for determining the efficiency of four-terminal networks', *J. Appl. Phys.*, **25**, 8, pp. 982–986, 1954.

33 Beatty, R.W., 'Efficiencies of microwave 2-ports from reflection coefficient measurements', *IEEE Trans.*, **MTT-20**, 5, pp. 343–344, May 1972.

34 Engen, G.F., 'Calibration technique for automated network analyzers with application to adapter evaluation', *IEEE Trans.*, **MTT-22**, 12, pp. 1255–1260, Dec. 1974.

35 Almassey, G., 'A first-order correction to sliding short behaviour with application to the problem of measuring small losses', *IEEE Trans.*, **IM-20**, 3, pp. 162–169, 1971.

36 Kerns, D.M., and Beatty, R.W., 'Basic theory of waveguide junctions and introductory microwave network analysis', Pergamon Press, 1967.

37 Beatty, R.W., 'Applications of waveguide and circuit theory to the development of accurate microwave measurement methods and standards', NBS Monograph 137, Aug. 1973.

38 Clark, R.F., 'Maximum limits of efficiency for low-loss reciprocal adapters', *IEEE Trans.*, **IM-25**, 4, pp. 312–314, Dec. 1976.

Part D

Other topics

Chapter 14

Instruments and techniques for pulsed power measurements

14.1 Introduction

This chapter presents a description of the chief features of pulsed power meters and the main ways in which they can be calibrated in a traceable way from fundamental electrical measurement standards. The chief quantity of interest is usually the peak power, but one might also wish to determine other properties such as the shape of the pulse envelope or the energy per pulse. The commonly encountered range of pulse durations (defined to be between the half power points [1]) is 0.1 to 10 μs with duty factors ranging from 0.0001 to 0.01 and pulse repetition rates from 50 Hz to 10 kHz. With a few exceptions traceability is via continuous wave (CW) power standards and the parameter of prime importance is the CW/pulse transfer error. In a few cases national standards laboratories have set up calibration services for pulsed power meters [2], although these instances are comparatively rare.

In the early days of radar development – during the late 1930s and early 1940s – pulsed power measurements tended to be carried out on an ad hoc basis by assembling collections of components whenever the need arose [3]. Subsequently some of this equipment came to be packaged in the form of pulsed power meters and sold commercially. The simplest pulsed power meter is the peak-reading analogue instrument, consisting essentially of a diode envelope detector followed by a peak detector [4] (Fig. 14.1). At the more sophisticated end of the commercially available range are microprocessor-driven digital instruments [5] (Fig. 14.2). The uncertainties in pulsed power meters are comparatively large in relation to those for power meters intended for CW only – at best 3 percent, with typical commercial meters falling in the range 5 to 25 percent depending on frequency and other parameters. One factor which is relevant to the development of instruments is the fact that sensors which are sufficiently fast to respond to pulses are far from ideal in other respects. Early microwave detectors, of the cat's whisker type, were neither stable nor reliable, and for many years the calorimeter, the metal wire barretter, and the thermistor were regarded as the only reliable power indicators. In 1955 Henning [6,7] described how a barretter could be used for pulsed power measurements if the output was differentiated (Fig. 14.3). The differentiation is necessary because the time constant of the barretter element is much longer than the duration of the pulses to be measured, and therefore the temperature rise of the element is proportional to the time integral of the power. The sensitivity of a barretter used in this way depends on its thermal mass, which is related to the time constant. This method is known as the barretter-integration-differentiation technique. Commercial instruments using this principle have been produced [8].

The improvements in semiconductors in the 1960s resulted in reliable low barrier height Schottky barrier diodes with controllable characteristics. It is this type of diode which is now common in both CW and pulsed power meters. However, corrections are still necessary to overcome other well known disadvantages of diodes, namely temperature sensitivity and departure from square law. Increased interest in the use of sampling techniques in the 1960s led to the application of such methods to pulsed power measurements. One benefit of sampling is that it can overcome bandwidth limitations. For example, a sampling oscilloscope may operate up to at least 12 GHz, whereas a conventional oscilloscope is limited to below 1 GHz unless special techniques such as the use of travelling wave structures are employed.

Calibration methods for pulsed power meters fall into three groups. The oldest is the average power method [7] (Fig. 14.4). If the pulse can be assumed to be rectangular, then the peak power can be calculated from the average power and the duty cycle. This assumption may be sufficiently valid for rough measurements, but otherwise a correction is required for non-rectangular pulse shapes. The correction is derived by viewing the detected pulse on an oscilloscope. The second calibration approach uses a fast switch to gate a known CW power. The most obvious way to implement this idea would be to use the gate to produce a pulse of known characteristics, which could then be applied to the instrument to be calibrated. In practice a more elegant method known as the notch wattmeter is used. This technique, which dates from 1946 [9,10], uses both a pulse signal and a CW signal. The CW signal is gated off, or notched, during the pulse (Figs 14.5 and 14.6). Comparison of the heights of the two signals is carried out by adding them together, detecting, and viewing the video output on an oscilloscope. The third approach is the sampling comparison method [11–13] (Figs 14.7 and 14.8). An unknown pulse and a known CW signal are fed in turn to a detector via a gate. A switching pulse of duration less than that of the unknown pulse is applied to the gate, which thus converts both the known and the unknown signals into a pulse which is much narrower than the original unknown pulse.

In dealing with pulse-modulated signals, there are some features which pose additional problems over and above those which occur for CW measurements.

For example, the signal occupies a finite bandwidth and components should ideally have a flat frequency response and be free from dispersion. Directional couplers do not actually fulfil this requirement, although usually the resulting error is significant only for short pulses. Some measurement systems require a repetitive signal, whilst others will measure a single pulse. In the former case variations in pulse repetition rate may cause errors. Leakage poses a particular problem in pulsed power measurements, for two main reasons. Firstly, measurements of high values of attenuation are often called for. Secondly, the modulators used to gate the signals may have poor isolation in the off condition. Harmonics are also more serious for pulsed measurements than for CW signals, because diode sensors often operate above their square-law region, in which case voltages, rather than powers, at the fundamental and harmonic frequencies tend to add.

The early literature on pulsed power measurements contained warnings against the use of barretters for measuring the average power of a pulsed signal. This results in errors because the time constant of a barretter, unlike that of a thermistor, is not long enough to average over the repetition period. Nowadays barretters are less likely to be considered for the average power method as their use has decreased in general. On the other hand attention must be drawn to a problem which arises if instruments incorporating chopper amplifiers are used to measure average powers of pulsed signals. Beating may occur between the chopper frequency and the pulse repetition frequency, both of which are likely to be in the audio range [4]. This situation applies to many thermoelectric and diode CW power meters.

14.2 Instruments

14.2.1 Meters using diode sensors

Diode instruments can be used directly as pulse/CW transfer instruments. The simplest consist of an envelope detector of adequate speed followed by an amplifier and peak detector (Fig. 14.1). Provision is usually made to view the pulse envelope on an external oscilloscope, since this not only allows the shape to be seen but also gives better accuracy compared with the use of the built-in peak detector. Because the sensor diode usually operates in the transition region, that is between the square law and the linear detection regions,

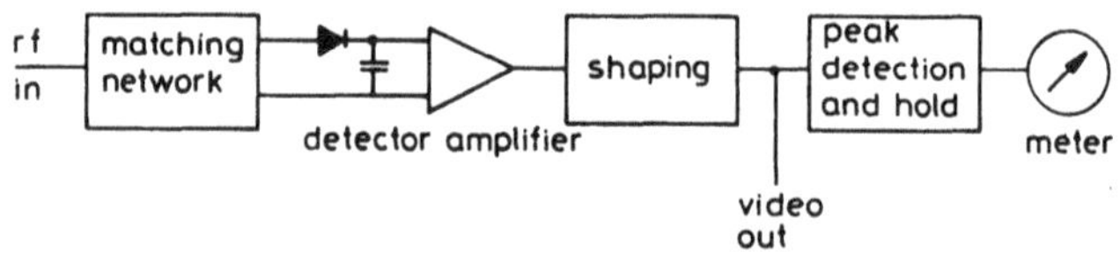

Fig. 14.1 *Analogue peak power meter employing a diode sensor. The shaping circuit compensates for the diode characteristic*

compensating or shaping circuits are included in the electronics to correct for the amplitude characteristic of the diode. A typical instrument [4] operates from 100 MHz to 18 GHz at power levels of 3 mW to 100 mW. Pulses with duration down to 1 μs can be measured using the built-in peak detector, or 100 ns using an external oscilloscope. The pulse/CW transfer error claimed by the maker is less than 0.2 dB for the peak detector mode and 0.1 dB for the oscilloscope mode of operation. The overall uncertainty is quoted as 0.7 dB up to 12 GHz and 1 dB above this frequency.

A more complicated pulse power meter incorporating a cathode ray tube is described in refs. [5,14] (Fig. 14.2). It is a digital instrument employing video sampling. Two microprocessors are used, one for the measurement process and one for the input/output operations. A programmable read only memory (PROM) holds the correction data both for square law deviations and for the frequency response of the diode sensor. The use of an amplifier with a video bandwidth of up to 50 MHz enables pulses of duration down to 15 ns to be measured. Three cursors on the cathode ray display enable the value of the pulse envelope at any selected point on the waveform to be obtained in digital form. The claimed uncertainty for this instrument is 0.3 dB at frequencies from 75 MHz to 18 GHz.

14.2.2 Barretter type meters

In the barretter-integration-differentiation technique [7,15] (Fig. 14.3) a constant dc bias current is applied to the barretter element. Application of an rf pulse causes a step increase in temperature proportional to the energy in the pulse. The resulting change in barretter resistance causes a change in voltage, which is amplified and then differentiated so as to restore the original pulse shape. The peak amplitude of the restored pulse is then measured by a peak voltmeter. The calibration depends on knowing the pulse sensitivity of the barretter element (in $\Omega\ mW^{-1}\ s^{-1}$), the effective efficiency of the barretter-mount assembly, the bias current, and the characteristics of the amplifier, differentiator and peak voltmeter. The pulse sensitivity figure of the barretter element is usually supplied by the manufacturer and it is assumed that this applies to all elements of a given type, due allowance being made for the effect of tolerances on the uncertainty. This is possible because barretters, unlike thermistors, are relatively reproducible. The electronic part of the instrument is calibrated immediately prior to the measurement by means of a built-in dc pulse generator whose output is passed through a network of the same time-constant as the barretter element. The effective efficiency of the barretter and mount is assumed to be the same as for CW power.

Barretter instruments are suitable for pulse durations of 0.25 to 10 μs. The maximum duration is set by the requirement that the duration should be small compared with the time constant of the barretter element, which is typically 300 μs. The shortest pulse duration is limited by the capacitance of the mount, which must be less than approximately 100 pF, and by the amplifier rise time,

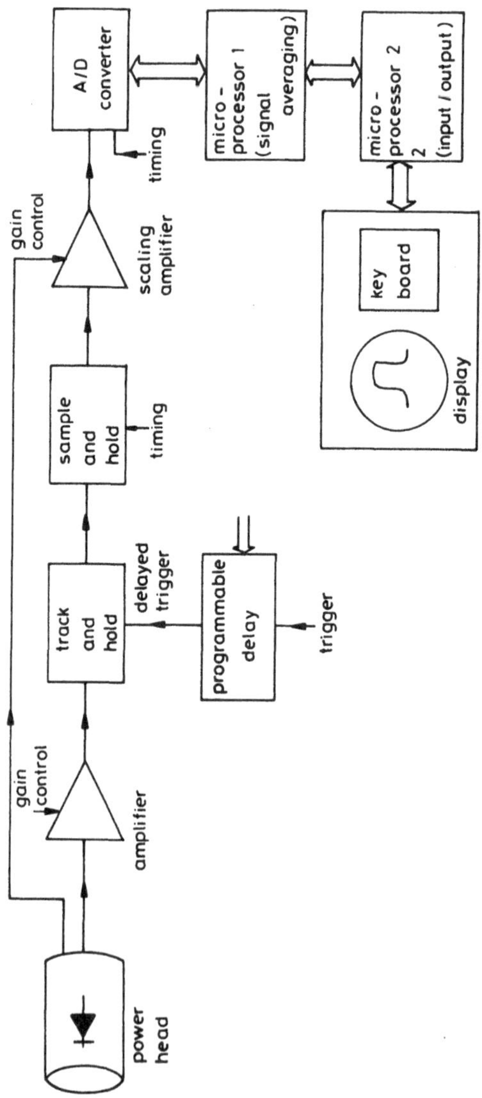

Fig. 14.2 *Digital pulsed power meter incorporating a cathode ray display (greatly simplified). Movable cursors enable the value at any point on the envelope waveform to be read off in digital form*

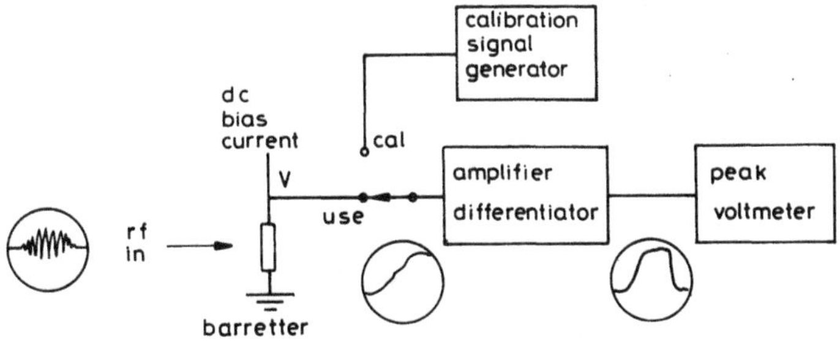

Fig. 14.3 *Barretter-type pulsed power meter. The barretter voltage V is differentiated in order to restore the original pulse envelope*

which should be no more than 0.03 μs. The peak power levels which can be measured depend on the pulse duration: for a 10 μs pulse the range is 10–300 mW. With shorter pulses the levels are increased in inverse proportion to the duration of the pulse, since the important quantity is the energy per pulse. There is also an overriding restriction that the average power must not be greater than 10 mW, although this limit is relevant only if the pulses are so close together that the element does not have time to cool between one pulse and the next. The special barretters which are manufactured for pulse measurements have pulse sensitivities in the range 0.02–0.04 $\Omega\ mW^{-1}\ s^{-1}$ and manufacturing tolerances of a few percent. Manufacturers' claims for uncertainty are in the range 5–10 percent for instruments operating in the range 500 MHz to 18 GHz.

14.2.3 Other pulsed power measuring instruments

Apart from the rectifying diode and the barretter, several other power meter types offer possibilities for pulsed power measurement. For example, the thermoelectric diode (Chapter 7) is sufficiently fast for this purpose, as is the pyroelectric detector [16,17] (Chapter 10). If the power level is sufficiently high force-operated instruments can be used ballistically [18,19]. Other possibilities are mentioned in refs. [20–30].

14.3 Calibration methods

14.3.1 Average power method

The average power method as originally described by Henning [7] (Fig. 14.4), was applied to the measurement of transmitter peak power. The power to be measured is coupled through a directional coupler to a hybrid junction, where it divides between a calibrated thermistor, which measures average power, and a diode detector and oscilloscope which enable the pulse envelope shape to be displayed. In this way the advantages of the fast response of the oscilloscope

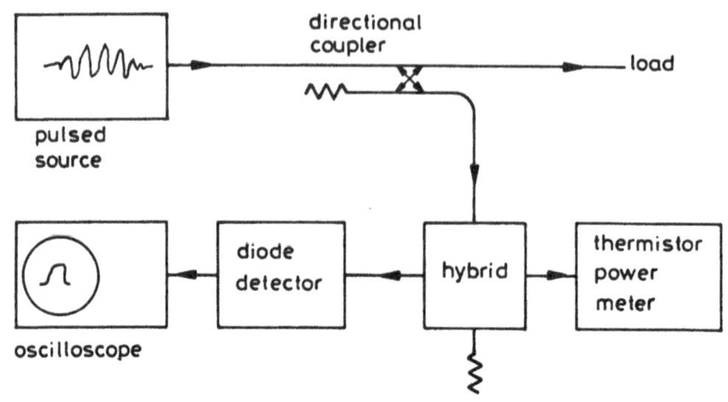

Fig. 14.4 *Average power method of peak power measurement. The thermistor measures average power, which can be related to the peak power in the load by the envelope shape, the pulse repetition rate, and the characteristic of the coupler and hybrid*

display and the relative stability of the thermistor are combined. The time constant of the CW power meter must be long compared with the pulse repetition period. The necessary operations are:

1. Determination of the correction for the detector law
2. Average power measurement using the thermistor
3. Repetition rate measurement using the oscilloscope
4. Pulse width measurement using the oscilloscope
5. Determination of pulse shape correction using the oscilloscope, that is determination of the relative height of a rectangular pulse of the same area as the actual pulse
6. Calculation of peak power.

The average power method assumes that the display, after correction, is a faithful reproduction of the pulse envelope, which is virtually equivalent to assuming zero pulse/CW transfer error.

14.3.2 The notch wattmeter

An implementation of the notch wattmeter technique, as used by the National Bureau of Standards, is described in ref. [31]. Fig. 14.5 shows the arrangement for low power measurements; that for high powers is similar except for a re-arrangement of couplers. The system was implemented in coaxial line up to 4.4 GHz and waveguide for the band 8.2 to 12.4 GHz, with a peak power range of 10 μW to 10 kW. The uncertainty was 3 percent for the coaxial version and 4-6 percent for the waveguide version. The operation is as follows. The output from the rf pulse generator is applied to the instrument under test through the main arm of a directional coupler A. A known fraction of the power is coupled to the side arm of coupler A and this passes on to the side arm of coupler B. A

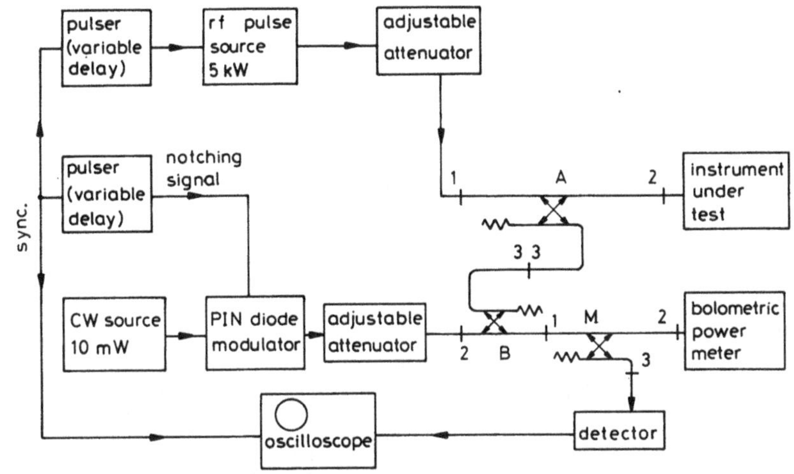

Fig. 14.5 *The notch wattmeter technique. A gated CW signal is adjusted to be equal in amplitude to the pulse to be measured*

sample of this signal is then coupled to the main arm of coupler B, continues to the monitor coupler M, and finally is absorbed by the bolometer mount.

A low-power notched CW signal (zero to 10 mW) is delivered to the input of the main arm of coupler B and is thus added to the existing pulsed signal. The notch is generated by a fast PIN diode switch modulator which is gated to turn off the CW for a short interval each time an rf pulse is generated. The duration of the notch is adjusted to be slightly greater than the pulse duration. A sample of the composite signal, which includes the pulse and the notched CW, is coupled to the side arm of coupler M, detected by a diode and then viewed on an oscilloscope. By means of adjustable time delays the pulse can be positioned within the notch. The pulse and notched signal heights are adjusted to be equal (Fig. 14.6). The value of the power is read from the bolometer, and from this power reading together with the coupling factors and attenuations of the couplers the peak power at the instrument under test can be deduced.

A summary of the uncertainty contributions for the notch wattmeter is as follows. The dead time, that is the period when neither of the two sources is gated on, is usually no greater than 1 μs per pulse. For a pulse repetition rate of 1000 pulses per second the result is a 0.1 percent maximum error, which could if necessary be corrected for. The precision of the pulse/CW intercomparison is 0.16 percent and the uncertainty in CW power measurement approximately 1.2 percent. Leakage of the CW power through the modulator results in an error of less than 0.01 percent. There is no leakage from the pulse power generator, as this does not oscillate during the off period. Leakage of CW into the device being tested, which depends on the directivities of the couplers, was shown to be negligible. The coupling factors of the couplers were amongst the

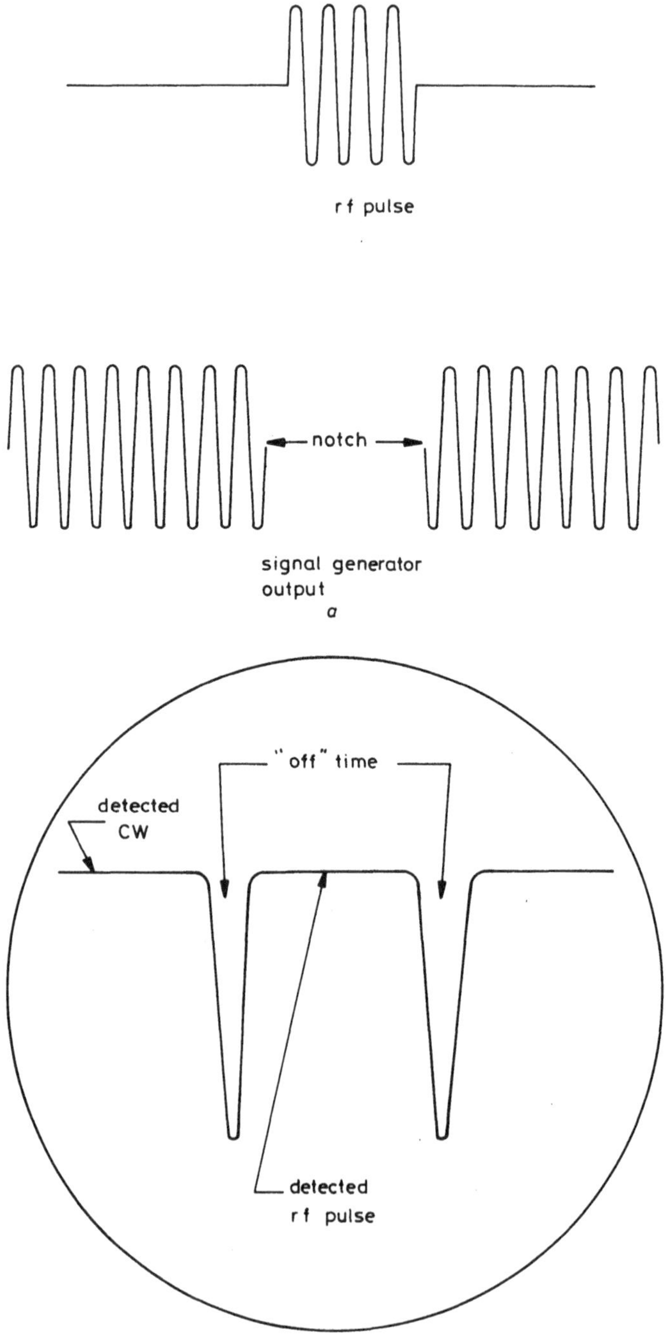

b

Fig. 14.6 *Rf signals (a) and video output (b) for the notch wattmeter.*

larger sources of uncertainty and these were determined to between 0.5 and 3 percent depending on value (10 dB–40 dB). Errors due to mismatch were up to 0.4 percent for the coaxial version and 1 percent for the waveguide version. The rise time of the system was measured as 8 ns using a tunnel diode pulse generator. A test was also performed to determine the effect on the pulse of the frequency response and dispersion of the couplers. It was concluded from this test that for pulse durations of 100 ns or longer the error was negligible, but at 14 ns the error would be 0.75 percent. It is of course necessary to ensure that the pulse and CW signals are at the same frequency.

14.3.3 Sampling-comparison Technique

As with the notch wattmeter, the sampling-comparison method [11,12] is based on comparison between pulse power and CW power. This comparison is carried out by means of identical samples in time extracted from both signals, using a fast semiconductor diode switch. Fig. 14.7 shows the version developed at the National Bureau of Standards [12]. The manually operated coaxial switch at point X allows either the CW or the pulsed signal to be selected. The

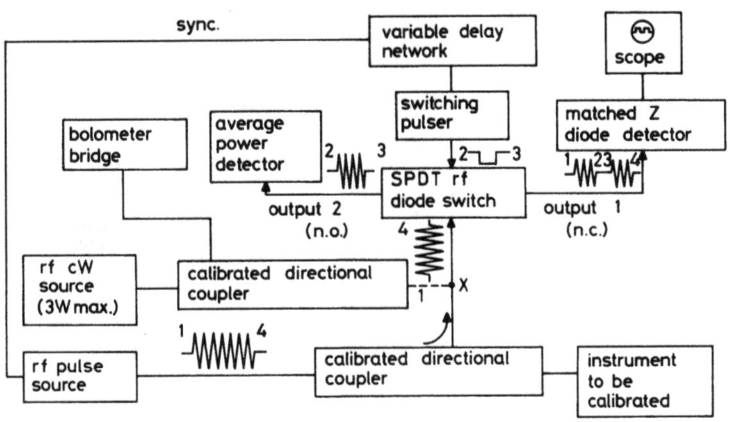

Fig. 14.7 *Sampling comparison method. Samples are taken from the pulse to be measured and from an adjustable CW signal*

pulsed signal may come directly from the source or via a directional coupler. In making the measurement the pulsed power is applied to the diode switch input, from where it goes to output 1, which is normally closed. A switching or gating pulser switches the input to output 2 for a period of time less than the rf pulse duration. The gating pulser is synchronised, through a variable delay network, with the pulser modulating the rf source. Thus the switch is gated each time an rf pulse appears at the switch input. The sample of the rf pulse at output 2 is detected by means of a thermocouple or diode detector. The pulsed power is then removed and CW power applied to the diode switch input via a calibrated directional coupler. The CW power is adjusted so that the detector output is the same as previously recorded. The CW power incident upon the switch is then measured and equated to the peak pulse power. The diode which terminates output 1 on the switch may be used with an oscilloscope to view the pulse envelope shape and to aid in positioning the sampling pulse at the peak of the main rf pulse. The range of the basic system extends from 0.02 W to 2 W, and the limit of uncertainty is 2 percent. An additional uncertainty of 1 percent is added for the couplers which serve to extend the range. The single pole double throw semiconductor diode switch was of a special design which enabled lower VSWR to be achieved than was available with commercial switches. Once set up it is possible to make rapid precise measurements with the sampling-comparison system provided that the pulse repetition rate is constant. However, the minimum length of the pulses which can be measured is not as low as for the notch wattmeter.

A modification of the comparison technique which does not require a constant pulse repetition rate is described in ref. [32]. The method employs sequential comparison between the pulsed rf power and a variable CW reference level. With reference to Fig. 14.8, a known fraction of the rf pulse power is coupled to the secondary arm of a 4-port calibrated coupler, from

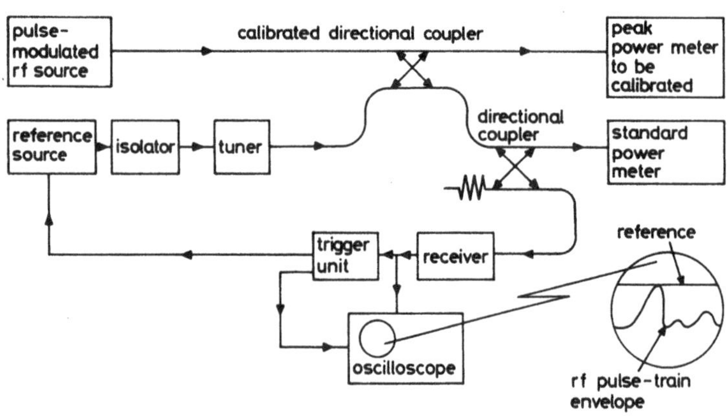

Fig. 14.8 *Sequential sampling comparison method of Bradstreet and Griffin. This method does not require a constant pulse repetition rate*

where the sample is fed to a detector through a second coupler. The video output of the detector goes to an oscilloscope and a trigger unit which gates a CW reference source on for a predetermined time in the interval between pulses. The output of the gated CW source is connected to a standard power meter through the side arm of the first coupler and the main arm of the second coupler. A portion of the CW power is coupled to the receiver by the second coupler, and its envelope is displayed on the oscilloscope along with the pulse envelope. The trigger unit causes the timebase to be initiated twice in each pulse-modulation period. Thus the CW and pulse envelopes are superimposed on the CRT, allowing precise adjustment of the reference level to that of the pulse. The peak pulse power is given by the power level as measured by the standard meter multiplied by the coupling coefficient of the calibrated coupler and the factor $(n + 1)$, where n is the off-on ratio of the reference source. An uncertainty limit of 3 percent was claimed by the developers of the system.

Weinert and Weinschel [33] described a sampling comparison system in which the sampling was carried out in the video channel, making leakage less of a problem, although it assumes zero pulse/CW transfer error for the diode, which must be checked by examining the response speed.

14.3.4 Other techniques

Two CW/pulse mixing techniques, somewhat akin to the notch wattmeter, have been described in the literature. They are the homodyne [34] and heterodyne methods [35]. The CW and pulse signals are applied simultaneously to a mixer and the resultant output viewed on an oscilloscope. The principle of both versions is based on the fact that it is easy to recognise, from the form of the display, the condition when the peak pulse power is four times the CW power.

14.4 References

1 Stuckert, P.E., 'The IEEE/IEC pulse standards – basic tools for waveform analysis', *IEEE Trans.*, **IM-31**, 3, pp. 192–198, Sept. 1982.

2 'NBS peak-pulse power calibration initiated', *IEEE Trans.*, **MTT-14**, 1, pp. 47–48, Jan. 1966.

3 Evendorf, S., 'Microwave peak power measurement', *J. ISA*, **12**, pp. 71–75, May 1965.

4 Minck, J.L., 'Pulsed-microwave-signal peak-power measurement improves', *IEE Electron. and Power*, **30**, 9, pp. 715–718, Sept. 1984.

5 'Peak meter enlists CRT to analyze power', *Microwaves and RF*, **23**, 3, pp. 174–175, Mar. 1984.

6 Henning, R., 'Microwave peak power measurement techniques', Conference on High-Frequency Measurements, Washington, D.C., Great Neck, N.Y., Jan. 1955.

7 Henning, R.E., 'Microwave peak power measurement techniques', *Sperry Engineering Review*, **8**, 3, pp. 10–15, May/June 1955.

8 Narda Microwave Corporation, Catalogue, 1985.

9 Bowman, D.F., 'The notch wattmeter', *Proc. N.E.C.*, **2**, pp. 361–371, Chicago, Oct. 1946.

10 'Use of the notch wattmeter with HP signal generators', *HP Journal*, **7**, 4, Dec. 1955.

11 Denny, C.A., Mavis, C.L., and Still, C.J., 'Microwave pulse-power measurements', *IRE Trans.*, **I-11**, 3/4, pp. 276–280, Dec. 1962.

12 Hudson, P.A., Ecklund, W.L., and Ondrejka, A.R., 'Measurement of RF peak pulse power by a sampling comparison method', *IRE Trans.*, **I-11**, pp. 280–284, Dec. 1962.

13 Hudson, P.A., 'Measurement of RF peak pulse power', *Proc. IEEE*, **55**, 6, pp. 851–855, June 1967.

14 Browne, J., 'RF power meters: Measurement workhorses', *Microwaves and RF*, **23**, 1, pp. 65–70, 92, 143, Jan. 1984.

15 Pelykh, N.A., 'Bolometric meter of pulsed UHF power', *Measurement Techniques*, pp. 226–229, Feb. 1966.

16 Klyuchnik, I.I., 'Calibration of a pyroelectric peak power meter at microwaves', *Radiotekhnika*, **69**, pp. 94–97, 1985 (in Russian).

17 Klyuchnik, I.I., 'Measurement of pulse microwave power using a pyroelectric receiver', *Radiotekhnika*, **73**, pp. 72–75, 1985 (in Russian).

18 Kukush, V.D., and Zakharov, I.P., 'Measurement of energy of single SHF pulses of random shape by ponderomotive wattmeter', *Radiotekhnika*, **59**, pp. 95–100, 1981 (in Russian).

19 Zakharov, I.P., 'Dynamic error in measuring energy of single microwave pulses using ponderomotive transducers', *Radiotekhnika*, **70**, pp. 116–120, 1984 (in Russian).

20 Rando, J.F., 'Power measurements by means of thermally excited mechanical vibrations', *IEEE Trans.*, **IM-15**, 1–2, pp. 59–61, March/June 1966.

21 Otkidach, V.A., Rozhkov, I.A., and Red'kin, B.E., 'Digital-reading instrument for measuring single and rarely repeating pulses of microwave power', *Izmer. Tekh.*, pp. 1732–1735 (in Russian) (translation in *Meas. Tech.*, **11**, pp. 66–68, Nov. 1970).

22 Bokrinskaya, A.A., and Vuntesmeri, V.S., 'The use of ferromagnetic films for the measurement of peak SHF power', *Radiotekhnika*, 23, pp. 41–44, 1972 (in Russian).

23 Balakov, V.F., Bogolyubov, V.N., Puchkov, V.S., and Radchenko, V.F., 'An SHF peak wattmeter', *Radiotekhnika*, **23**, pp. 44–48, 1972 (in Russian).

24 Shakirzyanov, F.N., and Apenina, G.V., 'A selective ferrite sensor-convertor for the measurement of high and super-high pulsed SHF power', *Radiotekhnika*, **23**, pp. 55–57, 1972 (in Russian).

25 Pauzha, A.S., and Pozhela, Yu.K., 'A sensor for measuring through pulsed SHF power', *Radiotekhnika*, **23**, pp. 64–70, 1972 (in Russian).

26 Kovbasa, A.P. and Shelamov, G.N., 'Errors in measuring UHF pulse power with ferrite heads', *Meas. Tech.*, **16**, 7, pp. 1040–1042, July 1973.

27 Alferov, V.N., 'Digital measurement of pulsed high-frequency power', *Prib. Tekh. Eksp.*, **18**, 2, Pt. 2, pp. 128–129, March/April 1975 (translation in *Instrum. and Exp. Tech.*, **18**, 2, Pt. 2, pp. 483–484, March/April 1975).

28 Kats, L.I., Sinitsyn, E.V., and Somov, A.Yu., 'Millimeter-wave magnetoresistive pulse-power measuring instrument', *Prib. Tekh. Eksp.*, **21**, 2, Pt. 2, pp. 171–172, March/April 1978, (translation in *Instrum. and Exp. Tech.*, **21**, 2, Pt. 2, pp. 448–449, March/April 1978).

29 Loza, O.T., and Tsopp, L.E., 'Detection of single high-power pulses of microwave radiation', *Sb. Kratk. Soobshch. Fiz. AN SSSR Fiz. Inst.* P.N. Lebedeva, No. 1 pp. 8–11, 1982 (translation in *Sov. Phys.-Lebedev Inst. Rep.*, No. 1, pp. 5–7, 1982).

30 Mul'nikov, A.V., 'Precision wattmeter for pulsed SHF power', *Izmer. Tekh.*, **25**, 8, pp. 64–6, Aug. 1982 (translation in *Meas. Tech.*, 25, 8, pp. 708–712, Aug. 1982).

31 Simpson, P.A., and Hudson, P.A., 'Implementation of the notch technique as an RF peak pulse power standard', Rep No. NBS-TN-682, Nat. Bur. Stand., Washington, July 1976.

32 Bradstreet, D.W.G., and Griffin, E.J., 'Measurement of pulse-modulated RF power by sequential comparison with a reference', *IEE Electronics Lett.*, **2**, pp. 316–317, Aug. 1966.

33 Weinert, F.K., and Weinschel, B.O., 'Microwave peak power measurement independent of detector characteristics by comparison of video samples', *IEEE Trans.*, **IM-21**, 4, pp. 474–479, Nov. 1972.

34 Griesheimer, R.N., 'Microwave power measurements', Chapter 3 in 'Technique of Microwave Measurements', ed. C.G. Montgomery, McGraw-Hill, 1947, p. 220.

35 Schwartz, L.S., 'Pulse power measurement by a heterodyne method', *Commun.*, **29**, 2, pp. 26–27, Feb. 1949.

Chapter 15

Voltage and current measurements

The difficulties associated with voltage and current measurements increase with frequency and most voltmeters and ammeters are restricted to the region below 1 GHz [1–4]. Above 1 GHz voltages and currents in transmission lines can be deduced from a knowledge of the power flowing in the line, the characteristic impedance, and the reflection coefficient of the load. This approach enables the concepts of voltage and current to be applied at frequencies well above those for which these quantities could be measured directly [5], provided that only one mode is present in the line. Because of this restriction, it is only in regions which are free from unwanted modes, including evanescent modes generated by discontinuities, that one can talk about voltages and currents in high frequency transmission lines.

Many of the instruments resemble those used for power measurement. The main classes are bolometers, thermoelement instruments, diodes, and force-operated instruments. Thermoelement instruments are common in calibration laboratories whilst diodes are used in most commercial voltmeters. A number of older instruments, including the photoammeter and the differential air thermometer milliammeter, were once common but are no longer used [1,6–11]. The six-port reflectometer also has a role to play in the calibration of voltmeters and ammeters. This is discussed in Chapter 12.

Measurement standards for rf voltage and current can be realised in one of two main ways. The first is to derive rf voltage measurements from dc voltage standards and rf current measurements from dc current standards, using rf/dc transfer instruments. Such instruments are satisfactory up to about 30 MHz, but at higher frequencies stray reactances and the skin effect make evaluation difficult. The second approach is to dissipate a known amount of rf power in a known impedance and to calculate the voltage and current. This technique, which is referred to as the power/impedance method, may be used at any frequency and it is therefore useful as a check on the accuracy of rf/dc transfer instruments. It can also be used as a means of extending the frequency range of these rf/dc transfer instruments by determining correction factors for the higher frequencies.

15.1 Definition of voltage and current

Voltage and current must be defined more precisely at rf and microwave frequencies than for dc. The dc voltage between two points is the work done in moving a unit charge from one point to the other. For alternating signals this definition is not unique because the work done depends on the path taken and, moreover, the choice of path becomes more important as the frequency increases. For coaxial and other TEM lines this obstacle is fairly easily removed by restricting the permissible paths to those which lie in the transverse plane of the line, as shown in Fig. 15.1. For non-TEM transmission lines the situation is less simple and in order to arrive at a unique definition of voltage one must specify one particular path [12].

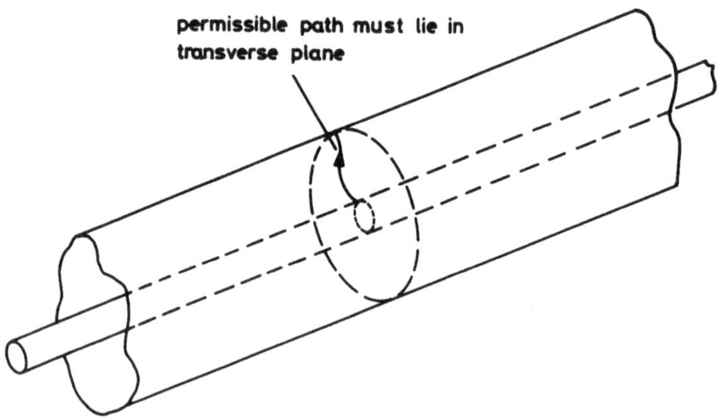

Fig. 15.1 *Definition of voltage in a TEM transmission line. The path must lie in the transverse plane*

15.2 Bolometric instruments

15.2.1 Thermistors and barretters

The principle of the thermistor type of rf/dc voltage transfer standard [13,14] is shown in Fig. 15.2. It employs two ideally identical thermistor elements which are connected as part of a self-balancing bridge. The operating principle is broadly similar to that of a coaxial thermistor power meter (see Chapter 6), although there are some significant differences. The rf voltage V_{rf} at the output of the instrument may be expressed in terms of the power P_{rf} dissipated in each thermistor element and the rf conductance G_{rf} of each element. On the assumption that the two thermistors have effective efficiencies of unity (that is, the absorbed power P_{rf} is equal to the dc substituted power P_{dc}), then:

$$V_{rf} = \sqrt{\frac{P_{rf}}{G_{rf}}} = \sqrt{\frac{P_{dc}}{G_{rf}}} = \frac{1}{4}\sqrt{\frac{G_{dc}}{G_{rf}}(V_0^2 - V^2)} \qquad (15.1)$$

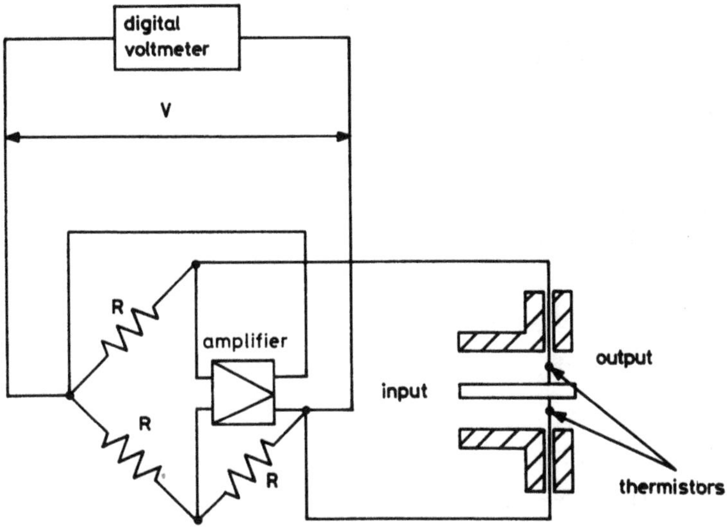

Fig. 15.2 *Principle of the thermistor voltage standard*

where G_{dc} is the dc conductance of a single thermistor, V is the bridge voltage in the presence of the rf signal and V_0 is the bridge voltage in the absence of the rf signal. The absolute values of G_{rf} and G_{dc} are not required, since it is only their ratio which is of importance. This ratio is close to unity at the lower end of the frequency range, but at higher frequencies the inductance of the thermistors and their lead wires causes the rf conductance to differ from the dc value. This is a major source of error. For non-identical thermistors, a correction to the formula of equation 15.1 is required owing to the fact that the thermistors are in series for dc and in parallel for the rf signal. Ref. [13] presents a treatment which is applicable to this case, although it is assumed that the ratio of the two thermistor resistances is independent of the applied rf signal level. A further error arises from the fact that the effective efficiency of a thermistor is not in fact equal to unity as assumed above. However, there are no mount losses to be taken into consideration, in contrast to a thermistor power meter, and the effective efficiency is therefore higher. The equivalent source impedance of a voltage standard must be as close to zero as possible and in order to achieve this the thermistor elements must be placed as near as possible to the output plane of the connector. It is also essential that the thermistor leads should be kept short, both to minimise the inductance and to avoid power dissipation in them. To this end the outer diameter of the line is reduced at the point where the thermistors are attached. Fig. 15.3 shows one design which has a substitution error of 0.3 percent or less at levels between 0.3 and 1 volt and frequencies up to 1.5 GHz [14]. The accuracy may be checked by the power/impedance method [15,16]. This is usually done at the highest operating frequency, it being assumed that the error increases with frequency.

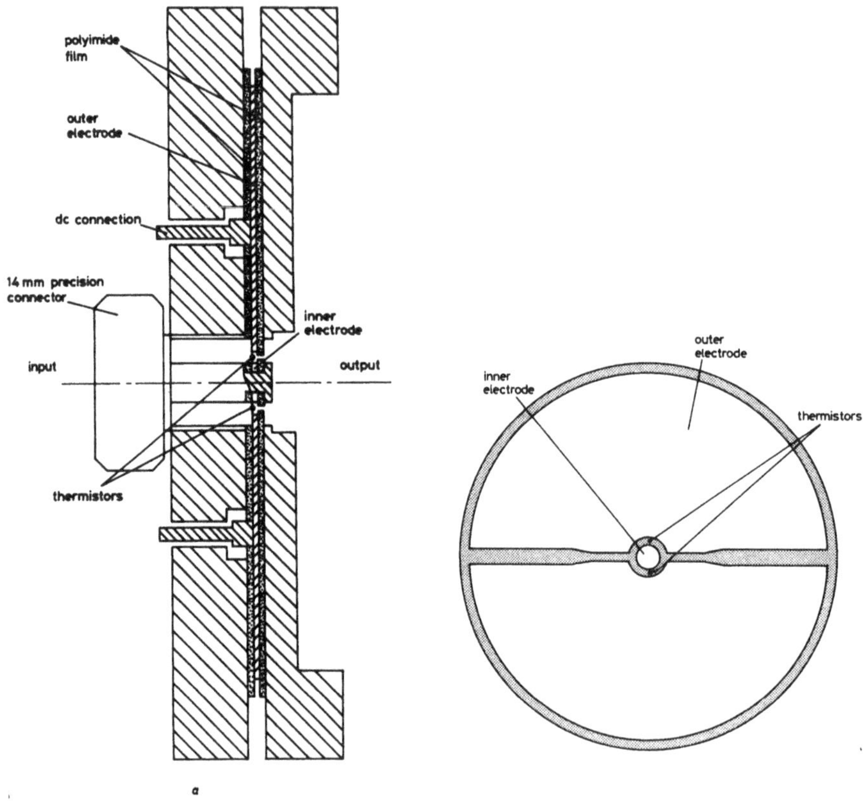

Fig. 15.3 *Practical details of a thermistor voltage standard*

Wire barretters [17] have been used in place of thermistors but, owing to their greater physical size, the inductance presents a more severe problem.

Thermistors and barretters are not well suited to measurement of current because of the difficulty of connecting dc leads. Widdis [18] has considered the indirectly heated thermistor as a current-measuring device at low frequencies.

15.2.2 Thin film bolometers

In principle a transverse thin film bolometer should make an excellent coaxial rf/dc voltage transfer standard, since the conductance can be shown theoretically to be constant up to frequencies for which the film thickness is comparable with the skin depth. This conclusion is reached by regarding the film as a very short length of transmission line with a complex characteristic impedance. The flatness and uniformity of the film ensures that no higher modes are generated and therefore there is no inductance. Early work on transverse thin films in coaxial line was carried out by Norton [19], but the instrument which he constructed was not intended as a standard. Selby [20–25] further developed the idea by using a split-film bolometer element consisting of two half discs of platinum or nickel on a 0.05 mm plastic substrate (Fig. 15.4). The two half discs were connected in parallel for rf and in series for dc. Provided that good contact is maintained around the whole of the circumference of the film, the rf and dc current distributions will be similar. This instrument was known as a BOLOVAC (bolometric voltage and current). The film resistance was between 1 and 10 Ω for the two half discs in parallel, corresponding to a surface resistivity of 8 to 80 Ω per square. However, because of practical difficulties such as contact potentials the performance of the instrument was not as good as the theory suggested; the agreement which was obtained with other methods was apparently no better than a few percent above 1 GHz [26], whereas the theoretical error was less than 0.1 percent up to 20 GHz.

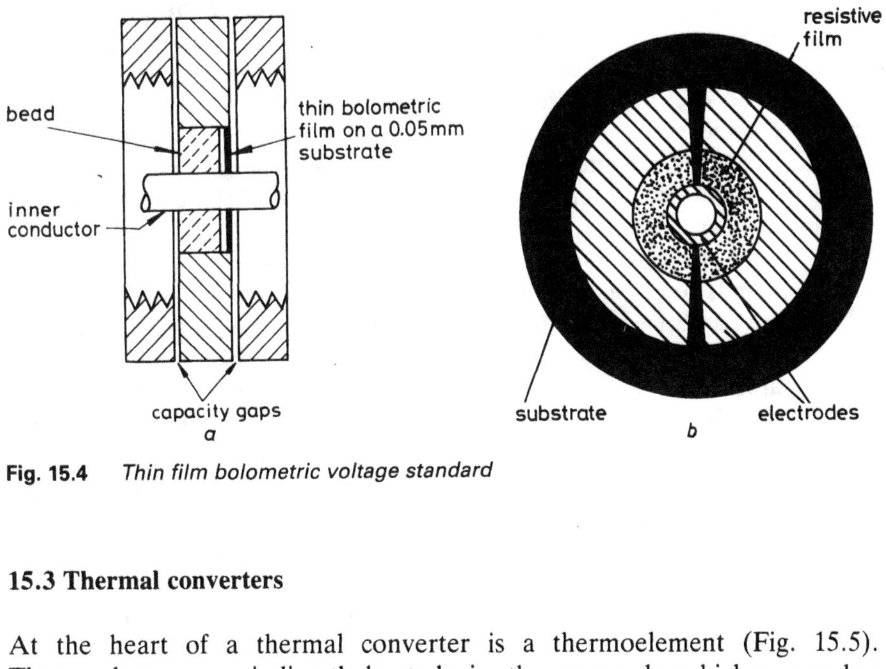

Fig. 15.4 *Thin film bolometric voltage standard*

15.3 Thermal converters

At the heart of a thermal converter is a thermoelement (Fig. 15.5). Thermoelements are indirectly heated wire thermocouples which are used as the basis of rf/dc current transfer devices. The heater is a straight length of wire made of nickel-chromium or some other alloy with a low temperature coefficient of resistance and isolated electrically from the thermocouple junction by means of an insulating bead. Heaters of different diameter may be used to produce units with a variety of current ratings from 1 A down to 2 mA, the most popular values being 5 and 10 mA. The thermocouple itself is usually of the copper-constantan variety. Both heater and thermocouple are mounted

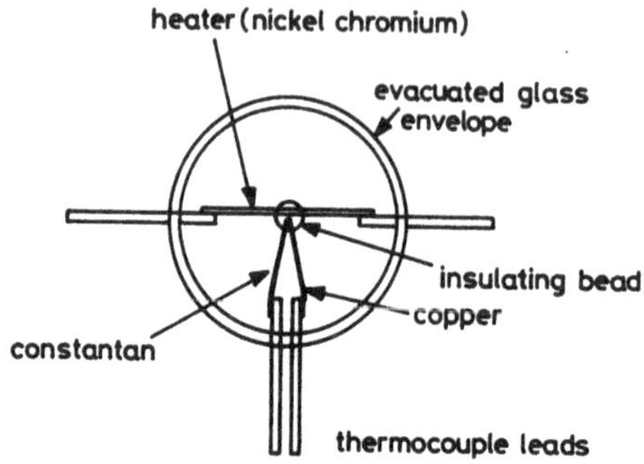

Fig. 15.5 *Construction of a typical thermoelement intended for use at radio frequencies*

in an evacuated glass envelope, giving a temperature rise of about 170 deg C at rated current, which corresponds to an output voltage of 7 mV. One undesirable feature of thermoelements is the inability of the heater to withstand even moderate overloads. There is also a tendency for dielectric breakdown to occur in the insulating bead, so that the thermocouple wires become connected to the heater. For currents above 1 A thermocouple ammeters having the heater exposed to the atmosphere have been designed, but such instruments present special difficulties and the rf/dc transfer error is correspondingly greater [4].

The useful dynamic range of a thermoelement when used for precise measurement is no more than 3 to 1 in current. Consequently in order to cover the range of currents and voltages commonly encountered in practice a set of instruments of each type is required.

Thermal converters are also used for ac/dc transfer measurements at audio frequencies [27–30]. The highest accuracy (approximately 1 part per million) is obtained from the multi-junction version [31–33], but this type is not generally suitable for frequencies above 100 kHz because of its relatively high reactance. An exception is the multi-junction rf ammeter described by Scott [34], in which the temperature rise of the heater is detected by focussing the infra-red radiation on to a thermopile.

Voltage-measuring instruments employing thermoelements are of two main types, known as thermal voltage converters [35] and micropotentiometers [36]. One form of thermal voltage converter is shown in Fig. 15.6. It is a feedthrough instrument and consists essentially of a thermoelement, a series resistor, and a T-junction. The resistor is chosen to suit the thermoelement rating and the voltage to be measured, which is usually in the range 0.2 to 200 V. One can calculate the approximate frequency response of the resistor by treating it as the inner conductor of a lossy coaxial line, the outer of which is formed by the cylindrical housing. The effects of capacitance and inductance of such a line can be arranged to cancel by a suitable choice of outer conductor

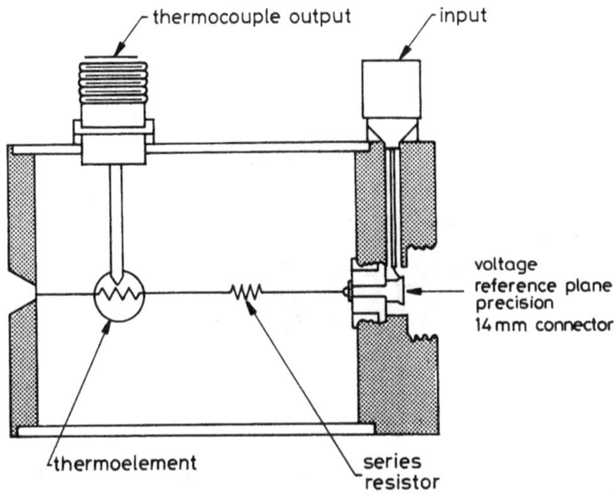

Fig. 15.6 *Thermal voltage converter incorporating a thin T-junction*

diameter (compare with the tractorial load described in Appendix 3), although the range of resistances giving reasonable values for this diameter is very limited. The major source of error is usually the frequency dependence of the thermoelement impedance. This may be investigated by comparing the readings of instruments having identical thermoelements but different resistance values. The T-junction must be very thin in relation to the wavelength [37]. A disadvantage is that any unwanted evanescent modes generated in the T-junction (caused through the disturbance to the circular symmetry of the line) will extend into the device under test owing to the closeness of the T-junction to the measurement plane. This effect can invalidate the assumption of single mode propagation and can cause errors.

The micropotentiometer [36,38] (Fig. 15.7) enables voltages down to microvolt levels to be established. It is a feed-through instrument consisting of a thermoelement in series with a transverse thin film disc resistor, the latter having theoretically a frequency-independent real impedance for small film thicknesses. The voltage is developed across the disc resistor and the rf/dc transfer error is less dependent on the thermoelement impedance than for the thermal voltage converter. A disadvantage of the micropotentiometer is that it does not behave as a true voltage source, since the equivalent source impedance is not zero.

Fig. 15.8 shows the frequency responses of a thermal voltage converter and a micropotentiometer, determined by the power/impedance method [39]. It will be seen that the micropotentiometer is the flatter of the two and that the corrections required for the thermal converter become very large above 300 MHz.

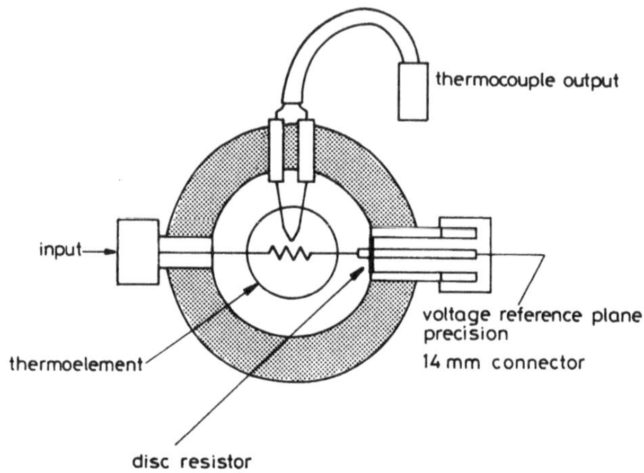

Fig. 15.7 *Micropotentiometer. The voltage is produced by the current flowing through a disc resistor*

15.3.1 Substitution errors in thermal converters

The substitution error in a thermal converter is due to a number of effects [30]. The inductance and capacitance of the thermoelement force some of the current to by-pass the heater at high frequencies. Skin effect results in an increasing resistance with frequency, causing an enhancement of the heating effect of the current. This effect is most serious for the large heater diameters required for ammeters with current ratings above 1 A, although it can be reduced by the use of tubular heaters [7] or plated insulators [8,34]. Induced currents in the thermocouple circuit may cause unwanted heating not present for dc. This can be minimised by arranging for the thermocouple wires to lie in a plane perpendicular to the heater axis. There is also the possibility of dielectric heating in the insulating bead. Some thermoelements have lead-in wires of magnetic alloys chosen to match the temperature coefficient of the glass bulb. The skin effect in these leads can cause additional ac heating [9]. The errors mentioned above, with the exception of skin effect, increase approximately as the square of the frequency and their combined effect may be positive or negative. Peltier and Thomson heating on the other hand produce errors which are independent of frequency [40,41]. The most easily detectable of these errors is the dc reversal error, that is the change in thermocouple output voltage when the polarity of the dc current is reversed. This change can be as much as 0.2 percent in terms of current, but it occurs only if there is some asymmetry in the thermoelement. For example, the thermocouple junction may not be exactly at the mid-point of the heater. The reversal error may be eliminated by taking the mean of two measurements carried out with opposite polarities. However, the absence of a reversal error does not imply the complete absence of an rf/dc transfer error. Hermach [27] showed that the

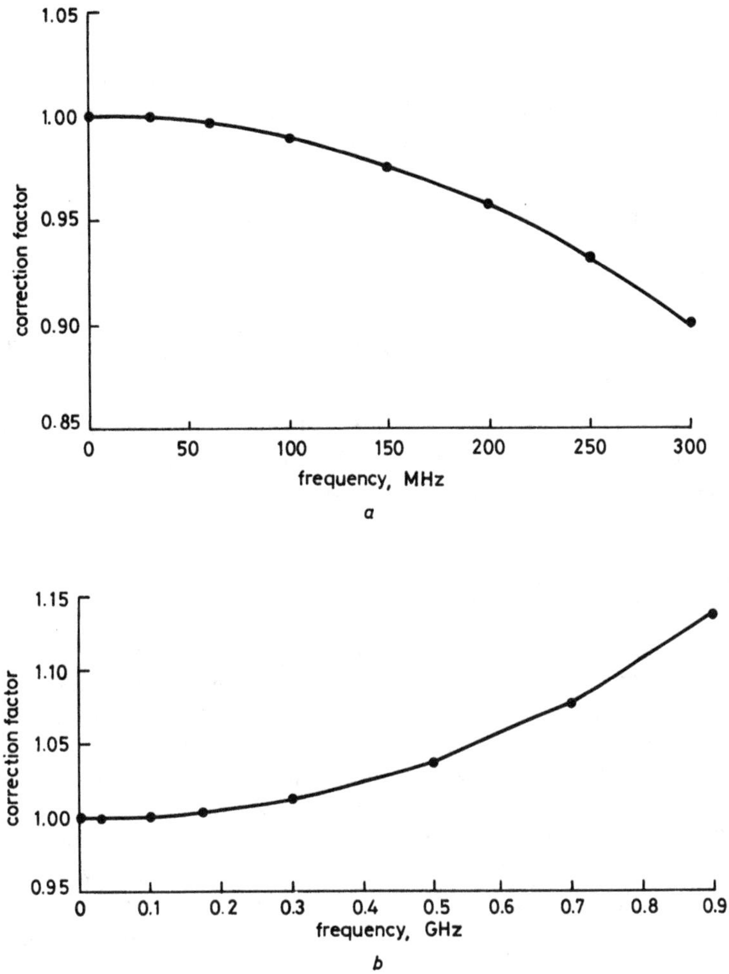

Fig. 15.8 *Correction factor for a 2-volt thermal converter (a) and a 1/4 volt micropotentiometer (b)*

presence of Thomson heating in the heater during the dc part of the measurement will cause a change in the mid-point temperature which is not present during the rf measurement. This occurs whether or not there is a reversal error. Widdis [42] extended Hermach's analysis by taking into account the heat conducted away from the heater via the thermocouple, which is significant for heater current ratings of less than 100 mA. He calculated the Thomson error for nickel-chromium against copper clad nickel iron support lead wires as being less than 1 part in 10^5, a figure which may be considered negligible in relation to other errors at frequencies above 1 MHz. The corresponding figure for a constantan heater is much higher at around 50 parts in 10^5. The Peltier effect also plays a role in this process, but the magnitude of the corresponding error is only about one tenth of that due to Thomson heating. The way in which the above frequency-independent errors arise has been summarised by Inglis [43]. The four main mechanisms are:

(*a*) Thomson heat in the heater, leading to second order Thomson heat
(*b*) Peltier heat at the heater/heater-support junctions, leading to Thomson heat in the heater
(*c*) Peltier heat at the heater/heater-support junctions, leading to second order Peltier heat
(*d*) Peltier heat at the heater/heater-support junctions, leading to Thomson heat in the support leads.

The results of calculations of errors should be treated with some caution because of the possible effect of the manufacturing process on the thermoelectric properties of the materials.

15.4 Diode instruments

Most of the principles described in Chapter 8 for diode power meters apply also to diode voltmeters. In the voltmeter case, however, the diode is housed in a feed-through mount which is inserted into the transmission line at the point where the voltage is required to be measured. Fig. 15.9 shows an example. Many commercial instruments are of the diode type [44–49]. Loading of the transmission line by the diode is minimised by operating the diode as a peak detector rather than in its square law region. A second advantage of peak detection is that the detected voltage is less dependent on the diode characteristics. Further reductions in both loading and diode-dependence can be obtained by the use of the slideback technique, in

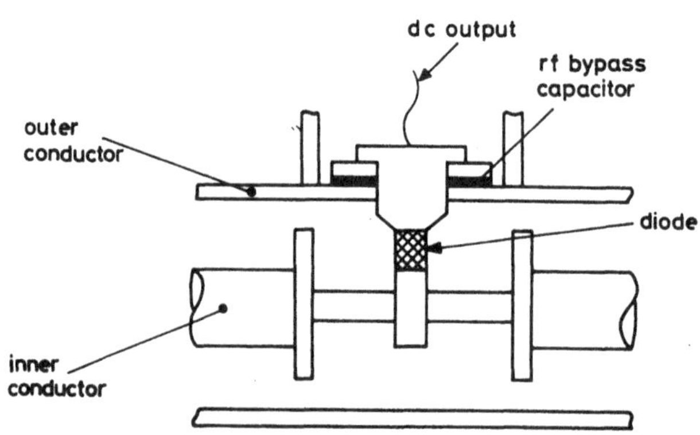

Fig. 15.9 *Details of a voltage mount incorporating an encapsulated diode*

which the output voltage of a detector is backed off by a dc voltage which is adjusted so as to reduce the current drawn from the detector to zero.

Diodes have been employed in voltage transfer standards intended for the calibration of rf voltmeters. Fig. 15.10 shows the principle of a diode instrument which resembles the thermistor voltage standards discussed in Section 15.2. In one practical implementation a pair of rf diodes was built into

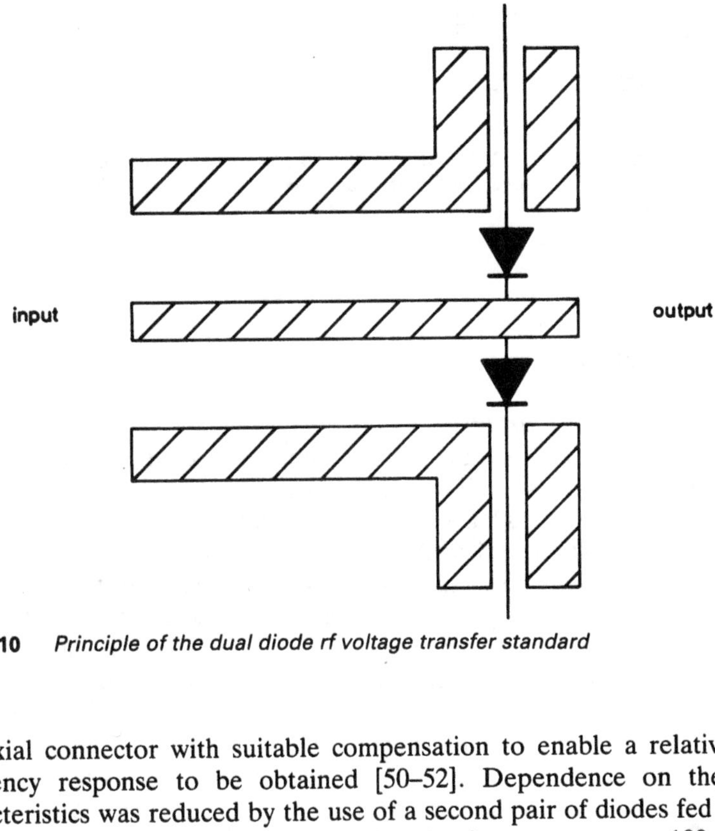

Fig. 15.10 *Principle of the dual diode rf voltage transfer standard*

a coaxial connector with suitable compensation to enable a relatively flat frequency response to be obtained [50–52]. Dependence on the diode characteristics was reduced by the use of a second pair of diodes fed with an audio signal. A commercial version for the frequency range 100 kHz to 10 MHz was produced. An alternative to simple detection is the sampling voltmeter [53–56], of which an example is shown in Fig. 15.11. This instrument is a vector voltmeter using coherent sampling, which preserves phase information. It is a dual channel instrument which measures the amplitudes of two signals to 2–17 percent and the phase difference to 0.1°. The sampler is of the four-diode kind, which is suitable for frequencies up to about 1 GHz. The faster samplers which are used in higher frequency sampling oscilloscopes and network analysers are of the two-diode kind, or alternatively of the type known as the diode-and-delay sampler [57].

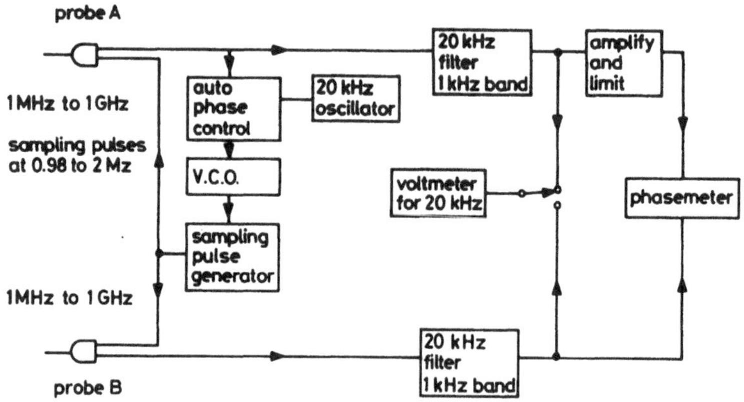

Fig. 15.11 *Vector voltmeter. This instrument uses sampling to measure the amplitudes of the two voltages applied to the probes and the phase difference between them*

15.5 Force-operated instruments

15.5.1 Electrostatic instruments

Force-operated voltage-measuring instruments have a long history of use as ac/dc transfer standards in the audio frequency range, although nowadays thermal voltage converters have largely taken over this role. A typical device is the electrostatic voltmeter described in reference [58]. Some attempts were made in the 1930s to construct high-frequency electrostatic instruments which would operate satisfactorily up to frequencies in excess of 30 MHz. Selby refers to some of this work in his review paper [2]. For high accuracy the rf and dc field distributions must closely resemble one another and this requires the structure to be small in relation to the wavelength. Chodakov [59] used an open line consisting of two 10 μm diameter platinum wires spaced 1 mm apart. One of the wires was stretched tightly, while the other was kept under relatively low tension in order to obtain a sufficiently large deflection. An uncertainty of 1.5 percent was claimed between 30 MHz and 100 MHz. Peterson [60] used a single suspended platinum wire of 13 μm diameter. The frequency error of this instrument was said to be 3 percent at 300 MHz and 33 percent at 1 GHz.

In the 1940s Collard [61] constructed a gold leaf electroscope for use in the microwave region. This instrument served for the purpose of calibrating diode voltage probes. The gold leaf was located in a slit in the outer conductor of a short-circuited coaxial line at a point where the magnetic field was zero. In this way the force due to the magnetic field, which would oppose that due to the electric field, was eliminated. The deflection was viewed with the aid of a microscope. Two versions were built, one operating in the 10 cm band and the other in the 3 cm band (Fig. 15.12), with a minimum working voltage level of about 20 V.

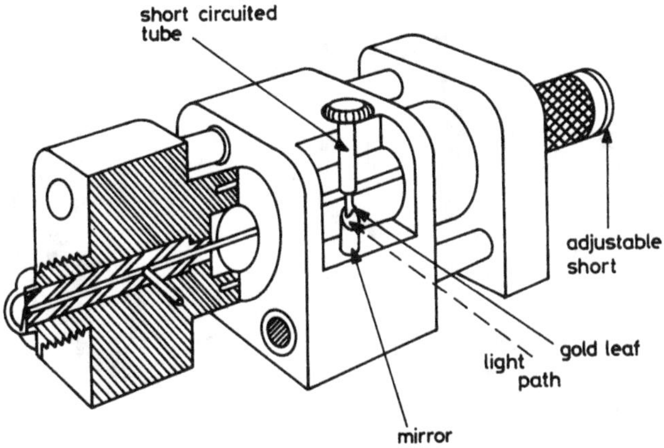

Fig. 15.12 *Exploded view of a microwave version of the gold leaf electroscope, operating at 3 cm wavelength. The gold leaf is mounted in a half wavelength short circuited tube which allows the leaf to be mechanically long whilst electrically short. The diode to be calibrated is mounted opposite the gold leaf*

A different approach was followed by Norton [19], who designed a force-operated feed-through instrument employing a thin ribbon as the moving element. Gaps were incorporated in the outer conductor in order to suppress rf currents in the ribbon and so avoid magnetic forces. High sensitivity was achieved by modulating the rf signal and by using a capacitor microphone to detect the deflection of the ribbon. When calibrated against thermal instruments the sensitivity varied by ±10 percent between 1 GHz and 10 GHz. Neither Norton nor Collard gave estimates of the theoretical magnitude of the rf/dc transfer error.

In subsequent work on force-operated instruments the rf/dc principle was forsaken and efforts centred on constructing absolute devices which would enable the measurement quantities to be derived directly from mass, length and time. This work was carried out mainly in connection with power measurement and is described in Chapter 9.

15.5.2 Electrodynamic instruments

The short-circuited ring electrodynamic ammeter [4] is shown in Fig. 15.13. It is an absolute instrument operating in the frequency range 1 MHz to 1 GHz with an uncertainty of roughly 1 percent. The principle was described as early as 1887, but the later form was developed during the 1950s and 1960s in the United States [62], the Soviet Union [63], and Japan [64]. The instrument

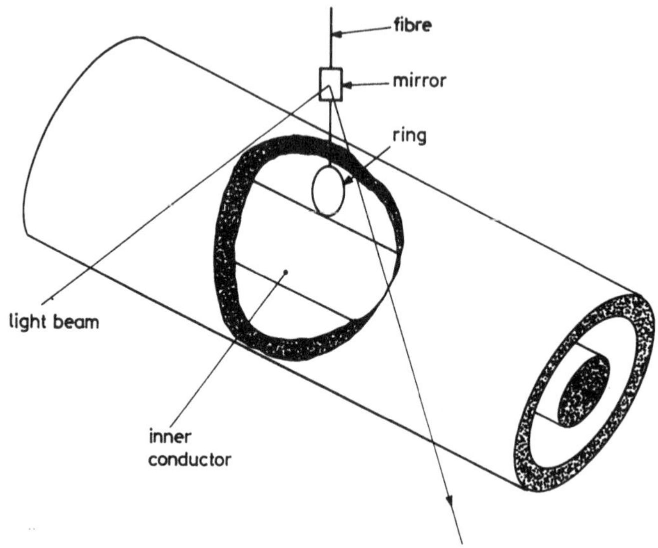

Fig. 15.13 *Short circuited ring ammeter. The magnetic field associated with the current causes an angular deflection of the ring*

consists of a conducting ring suspended between the inner and outer conductors of a large diameter coaxial line by a quartz fibre. The torque on the ring, which is proportional to the square of the current, arises from the interaction of the magnetic field in the line and the induced current in the ring. Currents from 1 A to at least 100 A can be measured in this way. The main intended application was for the calibration of high current thermocouple ammeters.

There are two ways of obtaining the torque/current relation for the short-circuited ring ammeter, one theoretical and the other experimental. The theoretical method, as described by Frederick [62], requires simplifying assumptions to be made. The inner conductor of the coaxial line is represented by a long straight current filament and the short-circuited ring by a filamentary current loop. The presence of the outer conductor is ignored and the calculation is carried out as if the ring were in free space. From energy considerations it can be shown that the torque is given by:

$$t = \frac{M\dfrac{\partial M}{\partial \theta}}{L\left(1 + \dfrac{1}{Q^2}\right)} I^2 \tag{15.2}$$

in which M is the mutual inductance, θ the angle between the ring and the plane which contains the long straight filament and the ring suspension, L is the self inductance of the ring, and Q its Q-factor ($= \omega L/R$, where R is the resistance). The self inductance L is made up of three contributions: (1) the internal inductance due to the magnetic fields contained within the surface of the ring, (2) the low frequency external inductance due to the fields outside the ring, and (3) a frequency-dependent contribution which arises when the length of the circumference of the ring is a significant fraction of a wavelength. The Q-factor includes the effect of radiation resistance. Although both L and Q are frequency-dependent, the variation in the sensitivity of the instrument with frequency is not large. For Frederick's instrument the calculated current corresponding to a given value of torque changed by only 1 percent between 1 MHz and 1 GHz.

The experimental method is based on Cullen's general theory of force operated instruments, which is explained in Chapter 9. The instrument is converted into a coaxial cavity by fitting each end with a sliding short circuit. According to the resonator action theorem, the work done in slowly deforming a lossless resonating cavity is zero if the resonant frequency is unchanged by the deformation. In applying the theorem [62] the ring is given a small measured angular deflection and the sliding short circuits are displaced by a measured distance in such a way that there is no change in the resonant frequency. By equating the total work to zero an expression is obtained for the torque on the ring in terms of the force on the short circuits. The force on the short circuits can be expressed in terms of the current in the line from a knowledge of the field distribution for a coaxial transmission line or by applying the formula for radiation pressure in a TEM transmission line. In this way the torque/current relation of the instrument is obtained. The accuracy is limited mainly by the uncertainty in the measurement of the small displacements. A comparison of the theoretical and experimental methods resulted in a discrepancy of 1.74 percent in current at 300 MHz. It was suggested that the chief cause of this difference was the shielding effect of the outer conductor of the coaxial line, which was ignored in the theoretical method.

15.6 Voltage and current from power and impedance

The simplest way of calibrating a feed-through voltmeter is to connect to its output a standard power meter whose impedance has previously been accurately determined (Fig. 15.14). The voltage V is then given by the expression:

$$V = \sqrt{\frac{P}{G}} \tag{15.3}$$

where P is the power absorbed by the power meter and G the real part of its input admittance [65–67]. Since errors in the admittance and power parts of the measurement procedure contribute equally to the error in the voltage,

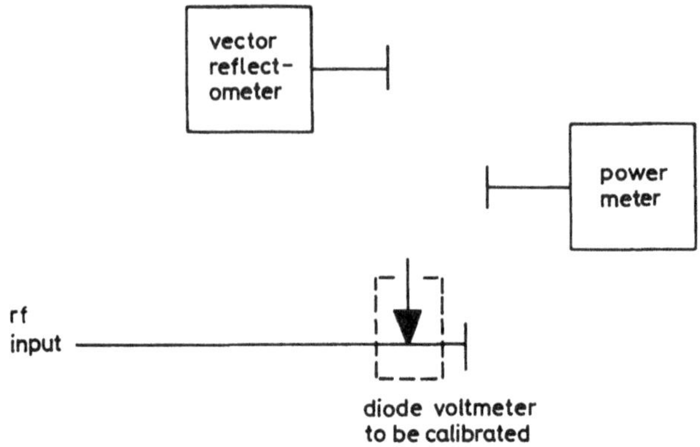

Fig. 15.14 *The power/impedance method for calibrating a feed-through voltmeter. The voltage at the input to the power meter is calculated from the absorbed power and the measured input impedance of the power meter*

equal attention must be paid to these two aspects. The measurement of the impedance can be carried out using a slotted line, an rf bridge, or a vector reflectometer which has previously been calibrated against impedance standards.

An alternative procedure which avoids the need for the prior measurement of the power meter impedance uses a quarter wavelength line [68,69], as shown in Fig. 15.15. The power meter is first connected directly to the

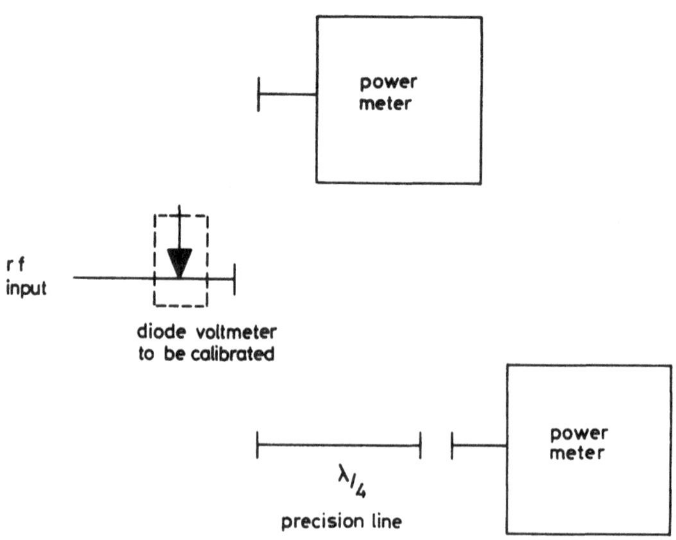

Fig. 15.15 *Alternative version of the power/impedance method using a precision quarter wave line. By averaging the results obtained with and without the quarter wave line the mismatch of the power meter is made to cancel approximately*

instrument to be calibrated. Then the quarter-wave line is interposed and the measurement repeated. The mean of the two results is calculated. This technique causes the mismatch of the power meter with respect to the quarter wave line to cancel to a first approximation, so that the characteristic impedance of the quarter wave line can be used in place of the actual input impedance of the power meter. In practice corrections may be necessary for the power loss in the precision lines [70]. A more accurate version uses two precision lines of different lengths, ideally a sixth wavelength and a third wavelength.

The precision lines which are used as impedance standards above 300 MHz can be made with mechanical tolerances of around one micrometre in the radial dimensions, representing an uncertainty of about 0.05 percent in the characteristic impedance Z_0 for 14 mm line [71,72]. For a material of infinite conductivity the value of Z_0 is given by:

$$Z_0 = \frac{1}{2\pi}\sqrt{\frac{\mu_0\mu_r}{\varepsilon_0\varepsilon_r}}\ln\left(\frac{b}{a}\right) \tag{15.4}$$

where a and b are the radii of the inner and outer conductors, ε_0 and μ_0 are the permittivity and permeability of free space, and ε_r and μ_r are the relative permittivity and permeability of the medium filling the line. For air-filled lines at a pressure of 760 mm and relative humidity of 50 percent, ε_r is 1.000649 and μ_r is taken as unity [66]. The dielectric loss of air may be ignored provided that the relative humidity is less than 80 percent. Equation 15.4 needs to be modified for practical materials in order to take account of the finite conductivity [69,73–75]. This has two effects, both of which are related to the skin depth. Firstly, the penetration of the magnetic field into the conductor material causes an increase in the inductance at low frequencies, resulting in a value of Z_0 which is slightly higher than predicted by equation 15.4. Secondly, the distributed series resistance of the line, which increases roughly as the square root of frequency, causes the value of Z_0 to become complex [76].

Below 300 MHz standard air lines become too long to be convenient. They also become very lossy, because although the loss per unit length decreases as the frequency decreases, this effect is more than off-set by the need for longer line lengths. Consequently, at the lower frequencies capacitors are used as impedance standards. These usually consist of a length of open circuited coaxial line with over-sized inner.

A typical uncertainty in impedance measurement is 0.25 percent at 1 GHz, corresponding to 0.12 percent in voltage or current calculated from equation 15.3. The uncertainty in the power measurement may produce an equal contribution, giving a total uncertainty in voltage or current of 0.25 percent if the uncertainties are added arithmetically. At lower frequencies the uncertainties are smaller. For example, at 1 MHz an uncertainty as low as 0.02 percent has been quoted [77].

15.7 References

1 Strutt, M.J.O., and Knol, K.S., 'Measurement of currents and voltages down to a wavelength of 20 centimeters', *Proc. IRE*, **27**, 12, pp. 783–789, Dec. 1939.
2 'High-frequency voltage measurements', NBS Circular 481, 1949.
3 Selby, M.C., 'The system of electromagnetic quantities at 30 kHz to 1 GHz', *Metrologia*, **2**, 1, pp. 37–45, Jan. 1966.
4 Scott, W.W., and Frederick, N.V., 'The measurement of current at radio frequencies', *Proc. IEEE*, **55**, 6, pp. 886–891, June 1967.
5 Woods, D., 'Concepts of voltage waves, current waves and power waves in S parameter definitions and measurements', *Proc. IEE*, **119**, 12, pp.1764–1765, Dec. 1972.
6 Schmitz, W., 'Differential air thermometer for ac current measurement', *Z. Hochfrequenz-tech.*, **27**, pp. 18–19, 1926.
7 Miller, J.H., 'Thermocouple ammeters for ultra-high frequencies', *Proc. IRE*, **24**, 12, pp. 1567–1572, Dec. 1936.
8 Wallace, J.D., and Moore, A.H., 'Frequency errors in radio-frequency ammeters', *Proc. IRE*, **25**, 3, pp. 327–339, March 1937.
9 Gainsborough, G.F., 'Experiments with thermocouple milliammeters at very high radio frequencies', *J. IEE*, **91**, Pt. 3, 15, pp. 156–161, Sept. 1944.
10 Behrent, L.F., 'Precision calibration of RF vacuum-tube voltmeters', NBS Tech. Note No. 121, Dec. 1961.
11 Hoffman, E., and Weber, K.H.R., 'On measuring high current in the frequency range of 0.1 to 10 MHz', *Tech. Mitt. RFZ*, **6**, pp. 129–132, Sept. 1962.
12 Kerns, D.M., 'Definitions of v, i, Z, Y, a, b, Γ, and S', *Proc. IEEE*, **55**, 6, pp. 892–900, June 1967.
13 Selby, M.C., and Behrent, L.F., 'A bolometer bridge for standardizing radio frequency voltmeters', *J. Research NBS*, **44**, 15, RP2055, 1950.
14 Uchiyama, F., Yamamura, K., and Yokoshima, I., 'Precision rf voltage standard using a thermistor bridge covering the HF-UHF range', *IEEE Trans.*, **IM-27**, 4, pp. 385–389, Dec. 1978.
15 Janik, D., 'Rf voltage accurate measuring methods – results of an experimental comparison', *IEE Conf. Pub.* **152**, pp. 27–28, 1977.
16 Janik, D., 'On the voltage dependence of the ac-dc transfer difference in h.f. voltage transfer standards', PTB Publication PTB-E-24, Section 8, June 1983.
17 Rozhdestvenskaya, T.B., Galakhova, O.P., Aknaev, R.F., Fyodorov, A.M., and Krestovsky, V.V., 'Standard apparatus for the volt realization in the frequency range from 20 Hz to 1500 MHz', *IEEE Trans.*, **IM-25**, 4, pp. 538–541, Dec. 1976.
18 Widdis, F.C., 'The indirectly-heated thermistor as a precise ac-dc transfer device', *Proc. IEE*, **103B**, 12, pp. 693, Nov. 1956.
19 Norton, L.E., 'Broadband power measuring methods at microwave frequencies', *Proc. IRE*, **37**, 7, pp. 759–766, 1949.
20 Selby, M.C., 'Bolometric voltage and current (BOLOVAC) standard for high and microwave frequencies', *J. Res. Nat. Bur. Stand.*, **72C**, 1, pp. 61–79, Jan./Mar. 1968.
21 Selby, M.C., 'Bolovac application for HF and microwave power measurement and standardization', J. Res. Nat. Bur. Stand. **74C**, 3 and 4, pp. 123–133, July/Dec. 1970.
22 Selby, M.C., 'The bolovac and its applications', *IEEE Trans.*, **IM-19**, 4, pp. 324–331, Nov. 1970.
23 Selby, M.C., 'BOLOVAC systems for measuring electrical quantities from 0.5 MHz through microwaves', NBS Monograph 123, Jan. 1972.
24 Anon., 'Bolovac reassessed', NBS Tech. Bull., p. 227, Sept. 1972.
25 Selby, M.C., 'Microwave two-port coaxial power and voltage mounts', *IEEE Trans.*, **IM-22**, 2, pp. 166–173, June 1973.

26 Selby, M.C., 'Voltage measurement a high and microwave frequencies in coaxial systems', *Proc. IEEE*, **55**, 6, pp. 877–882, June 1967.
27 Hermach, F.L., 'Thermal converters as AC-DC transfer standards for current and voltage measurements at audio frequencies', *J. Res. NBS*, **48**, p. 121, 1952.
28 Hermach, F.L., and Williams, E.S., 'Thermal converters for audio-frequency voltage measurements of high accuracy', *IEEE Trans.*, **IM-15**, 4, pp. 260–268, Dec. 1966.
29 Dix C.H., and Bailey, A.E., 'Electrical standards of measurement, part 1, DC and low-frequency standards', *Proc. IEE*, **122**, 10R, pp. 1018–1036, Oct. 1975.
30 Hermach, F.I., 'AC-DC comparators for audio-frequency current and voltage measurements of high accuracy', *IEEE Trans.*, **IM-25**, 4, pp. 489–494, Dec. 1976.
31 Wilkins, F.J., Deakon, T.A. and Becker, R.S., 'Multijunction thermal convertor – an accurate DC/AC transfer instrument', *Proc. IEE,* **112**, 4, pp. 794–805, April 1965.
32 Wilkins, F.J., 'Theoretical analysis of the ac/dc transfer difference of the NPL multijunction thermal converter over the frequency range DC to 100 kHz', *IEEE Trans.*, **IM-21**, 4, pp. 334–340, Nov. 1972.
33 Martin, P., and Knight, R.B.D., 'Components and systems for ac/dc transfer at the ppm level', *IEEE Trans.*, **IM-32**, 1, pp. 63–72, Mar. 1983.
34 Scott, W.W., 'New coaxial rf-dc ammeter', *IEEE Trans.*, **IM-19**, 4, pp. 318–323, Nov. 1970.
35 Hermach, F.L., and Williams, E.S., 'Thermal voltage converters for accurate voltage measurements to 30 megacycles per second', *Trans. AIEE, Part I, Commun. and Elect.*, **79**, pp. 200–206, July 1960.
36 Selby, M.C., 'Accurate radio-frequency microvoltages', Trans. AIEE, **72**, Commun. and Elect., Paper 52-33, pp. 158–163, May 1953.
37 Selby, M.C., 'Coaxial T for radio-frequency voltmeter calibrations', NBS Tech. Note 263, June 1965.
38 Selby, M.C., 'Radio frequency micropotentiometers', NBS Tech New Bull., 35, p. 33, Mar. 1951.
39 Clark, R.F., and Jurkus, A.P., 'Calibration of thermal transfer standards of RF voltage', *IEEE Trans.*, **IM-16**, 3, pp. 232–237, Sept. 1967.
40 Inglis, B.D., 'Errors in AC-DC transfer arising from a DC reversal difference', *Metrologia*, **17**, 4, pp. 111–117, 1981.
41 Inglis, B.D., and Franchimon, C.C., 'Current-independent AC-DC transfer errors in single-junction thermal converters', *IEEE Trans.*, **IM-34**, 2, pp. 294–301, June 1985.
42 Widdis, F.C., 'Theory of Peltier- and Thomson-effect errors in thermal ac-dc transfer devices', *Proc. IEE*, **109C**, 15, pp. 328–334, Mar. 1962.
43 Inglis, B.D., 'AC-DC transfer standards – present status and future directions', *IEEE Trans.*, **IM-34**, 2, pp. 285–290, June 1985.
44 Oliver, B.M., 'Which ac voltmeter?', Application Note 60, Hewlett-Packard, 1967.
45 Vifian, H., David, F.K., and Frederick, W.L., 'A voltmeter for the microwave engineer', *Hewlett-Packard J.*, **24**, 3, pp. 2–8, Nov. 1972.
46 Blankenburg, K.H., 'The new URV-RF-DC millivoltmeter for versatile application', *News Rohde and Schwarz*, **13**, 60, pp. 9–12, 1973.
47 Blankenburg, K.L., 'Waveform weighting for RF-voltage measurements using RF-DC millivoltmeter URV', *News Rohde and Schwarz*, **16**, 75, pp. 22–24, 1976.
48 Schermer, J., 'Simple yet versatile RF millivoltmeter URV 3', *News Rohde and Schwarz*, **18**, 83, pp. 10–12, 1978.
49 Betz, T., Kohler, D., and Reichel, T., 'RF Millivoltmeter URV 5 – voltage and power measurement into the gigahertz range', *News Rohde and Schwarz*, **24**, 106, pp. 16–18, 1984.
50 Driver, L.D., and Arthur, M.G., 'A wideband RF voltmeter-comparator', *IEEE Trans.*, **IM-17**, 2, pp. 146–150, June 1968.
51 Driver, L.D., Ries, F.X., and Rebuldela, G., 'NBS voltage comparator', *Nat. Bur. Stand. Tech. Rep. NBSIR* 78-871, Dec. 1978.
52 Driver, L.D., and Ries, F.X., 'A wideband rf voltage comparator', Conference on Precision Electromagnetic Measurements, Braunschweig, June 1980, pp. 487–488.

53 Colwell, J.M., 'The rf vector voltmeter – an important new instrument for amplitude and phase measurements from 1 MHz to 1000 MHz', *Hewlett-Packard J.*, **17**, 9, May 1966.
54 Weinert, F.K., 'The RF vector voltmeter for amplitude and phase measurements from 1 MHz to 1000 MHz', *Hewlett-Packard J.*, **17**, 9, pp. 2–12, May 1966.
55 Wenninger, F.W., 'A sensitive new 1 GHz sampling voltmeter with unusual capabilities', *Hewlett-Packard J.*, **17**, 11, July 1966.
56 Grove, W.M., 'Sampling for oscilloscopes and other RF systems: DC through X-band', *IEEE Trans.*, **MTT-14**, 12, pp. 629–635, Dec. 1966.
57 Akers, N.P., and Vilar, E., 'RF sampling gates: a brief review', *Proc. IEE*, **133**, A, 1, pp. 45–49, Jan. 1986.
58 Spilsbury, R.S.J., and Felton, A., 'The electrostatic voltmeter as a dc-ac transfer instrument', *J. IEE*, 89, Part II, pp. 129–132, April 1942 (Discussion pp. 133–136).
59 Chodakov, A.L., 'An electrometer for measuring voltages in the meter wave range', *J. Tech. Phys.*, **11**, 8, p. 767, 1941 (in Russian).
60 Peterson, A.P.G., 'The measurement of voltages at ultra-high frequencies', Thesis submitted to Massachusetts Institute of Technology, May 1941.
61 Collard, J., 'The measurement of voltage at centimetre wavelengths', *J. IEE*, **93**, Pt IIIA, 9, pp. 1393–1398, 1946.
62 Frederick, N.V., 'A new high-frequency current standard', *IEEE Trans.*, **IM-17**, 4, pp. 285–290, Dec. 1968.
63 Lopan, V.L., 'Electrodynamic ammeter for measuring high-frequency currents', *Meas. Tech.*, **1**, pp. 85–90, 1958.
64 Aida, Y., 'Research on the primary standard of current using short-circuited ring for high frequencies', *Res. Electrotech. Lab.*, **639**, 1963 (in Japanese).
65 Beatty, R.W., 'Impedance measurements and standards for uniconductor waveguide', *Proc. IEEE*, **55**, 6, pp. 933–941, June 1967.
66 Jesch, R.L., and Jickling, R.M., 'Impedance measurements in coaxial waveguide systems', *Proc. IEEE*, **55**, 6, pp. 912–923, June 1967.
67 Huntley, L.E., and Jones, R.N., 'Lumped parameter impedance measurements', *Proc. IEEE*, **55**, 6, pp. 900–911, June 1967.
68 Spinney, R.E., 'The measurement of small reflection coefficients of components with coaxial connectors', *Proc. IEE*, **110**, 2, p. 396, Feb. 1963.
69 Harris, I.A., and Spinney, R.E., 'The realisation of high frequency impedance standards using air-spaced coaxial lines', *IEEE Trans.*, **IM-13**, 4, pp. 265–272, Dec. 1964.
70 Nelson, R.E., and Coryell, M.R., 'Electrical parameters of precision coaxial air dielectric transmission lines', NBS Mono. 96, June 1966.
71 MacKenzie, T.E., 'Some techniques and their limitations as related to the measurement of small reflections in precision coaxial transmission lines', *IEEE Trans.*, **IM-15**, 4, pp. 365–374, Dec. 1966.
72 Mackenzie, T.E., and Sanderson, A.E., 'Some fundamental design principles for the development of precision coaxial standards and components', *IEEE Trans.*, **MTT-14**, 1, pp. 29–39, Jan. 1966.
73 Weinschel, B.O., 'Air-filled coaxial lines as absolute impedance standards', *Microwave J.*, **7**, pp. 47–50, April 1964.
74 Zorzy, J., 'Skin-effect corrections in immittance and scattering coefficient standards employing precision air-dielectric coaxial lines', *IEEE Trans.*, **IM-15**, 4, pp. 358–364, Dec. 1966.
75 Somlo, P.I., and Hunter, J.D., 'Microwave impedance measurement', Peter Peregrinus, 1985.
76 Woods, D., 'Relevance of complex normalisation in precision reflectometry', *Electronics Letters,* **19**, 15, pp. 596–598, 21 July 1983.
77 Fantom, A.E., 'International comparison of voltage at 1 MHZ', NPL Report DES 48, July 1978.

Chapter 16

International intercomparison of standards

One of the ways in which world-wide agreement of measurement standards is ensured is by intercomparison of the national standards of different countries [1–3]. Such comparisons are beneficial in a number of ways. They provide increased confidence in the capabilities of the participating laboratories, always assuming that the spread of the results is consistent with the claimed uncertainties. The results may throw light on the relative merits of the techniques and types of standard used by the participants, as well as on the stability and robustness of the travelling standards employed. This information may act as a stimulus to further work, leading to improvements in instrument design with potential applications beyond the immediate scope of the comparison. In some circumstances the results of comparisons may lead to the determination of national offsets – that is the recognition of a systematic difference between the standards of various countries. Naturally, not all of these benefits will apply to all comparisons. In a relatively new measurement area comparisons of any kind will be seen mainly as a test of the principles behind different techniques used, whereas in a well established area they will be seen mainly as giving confidence that the techniques generally regarded as being the best ones are being correctly implemented.

The organisation of international comparisons is one of the functions of the International Bureau of Weights and Measures (BIPM), together with the International Committee on Weights and Measures (CIPM) to which BIPM is responsible. BIPM possesses laboratory facilities at Sèvres, near Paris, where it maintains electrical standards for dc voltage and resistance [4], but it has no rf and microwave standards. The CIPM is advised on electrical matters by the Consultative Committee on Electricity (CCE), which in turn is assisted by a number of working groups. Comparisons of rf and microwave quantities are the concern of the RF Working Group, set up in 1965 when CIPM took over responsibility for this function from the International Scientific Radio Union (URSI). Information on the progress of comparisons is given in the Proceedings of CIPM [5]. The results are usually, but not always, published in the appropriate specialist journals. There are of course many other comparisons, both international ones and those between establishments in the same country, which take place outside the BIPM-CIPM framework, and the results of the more significant of these are likewise published in the literature. In this chapter a summary of the more important comparisons of power and voltage is given and a selection of the results is presented.

16.1 Power comparisons

Microwave quantities were first compared internationally in 1957 when URSI initiated a series of bilateral power comparisons at 9.375 GHz involving the United States, Japan and Britain [6,7]. The travelling standards were waveguide barretter mounts, for which each country was required to determine the effective efficiency at a power level of 10 mW. The comparison was therefore a test of the abilities of the laboratories to relate rf power to dc standards, rather than a true power comparison. Subsequent comparisons have followed a similar pattern, the advantages being twofold. Firstly, since effective efficiency is a property of the mount alone rather than of the whole power meter, it is likely to be a relatively stable parameter. Secondly, it is more convenient to transport a barretter mount from one laboratory to another than a complete power meter, which includes a precision self-balancing bridge and associated electronics. In theory, comparison of effective efficiency is all that is required to demonstrate agreement in absolute rf power measurements, since dc standards are already compared internationally. Indeed, according to Selby [1] it was once argued that such rf comparisons were in any case unnecessary, since the rf measurements would always agree if they were correctly derived from dc standards. Experience has shown otherwise, for the occurrence of discrepancies between laboratories, whilst rare, is by no means negligible.

The results of the 9.375 GHz URSI power comparisons, which lasted several years, were of special interest because of the variety of the techniques and types of standard used by the participants. The United States and Japan possessed microcalorimeters, and both also repeated some of the measurements using the impedance method. The British standards comprised a water calorimeter, a torque vane wattmeter of the double vane type, and a film bolometer. Generally speaking, agreement was achieved to within the combined quoted uncertainties. The two microcalorimeters agreed with each other to 0.7 percent (better than 0.3 percent in most cases), and both agreed with the film bolometer to 1 percent or better. The water calorimeter and torque vane wattmeter gave values for the effective efficiency of the travelling standards of approximately 1 percent below the United States values, but the difference was within the uncertainties. Part of this difference may have been due to the calibrated attenuator which was necessary to bridge the gap between the operating level of the barretter mounts and those of the water calorimeter and torque vane wattmeter. A more serious discrepancy of several percent between the microcalorimeters and the impedance method as carried out in Japan was later resolved [8]. The claimed uncertainties were 0.2–0.7 percent for the microcalorimeters, 1 percent for the film bolometer, 0.7–2 percent for the impedance method, and 1.5 percent for the water calorimeter and the torque vane wattmeter. This series of comparisons helped to establish the microcalorimeter as probably the most accurate waveguide power standard. A later series of comparisons between Japan, Sweden and Italy in 1963–1968 resulted in agreements within 0.6 percent at 9.375 GHz and 10 GHz [9,10].

The URSI program also included comparisons at 1 GHz and below. Power comparisons carried out from 1959 to 1965 at frequencies of 0.3, 0.4 and 1 GHz [11] produced agreements ranging from 0.1 percent to 2.25 percent, the participants being the United States, Britain, Japan and Canada. The results of these and other URSI comparisons were summarised by Selby [9].

The first round of comparisons carried out after the CIPM assumed responsibility included coaxial power at 3 GHz and a new series of 10 GHz comparisons in waveguide, including some laboratories which did not participate in those under URSI. The results of the 3 GHz comparison [12] are summarised in Table 16.1, from which it can be seen that the spread (i.e. maximum difference between the standards of any two laboratories) after averaging the results for several travelling standards was 0.66 percent. No uncertainties were given. The 10 GHz results [13] are summarised in Table 16.2. Only three of the ten travelling standards survived the five year long comparison without suffering damage or being lost, a problem which is not uncommon in spite of the care taken by participants. Some laboratories therefore measured a greater number of travelling standards than others and

Table 16.1 *CIPM international intercomparison of coaxial power at 3 GHz. Figures are deviation of effective efficiency measurements of each laboratory from international mean after averaging over several travelling standards .*

	United Statyes (NBS)	Canada (NRC)	Soviet Union (VNIIM)	Japan (ETL)	United States (NBS) Impedance method
average deviation (%)	−0.23	0.23	0.20	−0.43	0.2

Table 16.2. *CIPM international intercomparison of waveguide power at 10 GHz. Figures are deviation of effective efficiency measurements for each laboratory from international mean after averaging over several travelling standards*

	Japan (1) (ETL)	United States (NBS)	Soviet Union (IMPR)	Canada (NRC)	Italy (IEN)	Japan (2) (ETL)
average deviation (%)	0.07	0.06	−0.07	0.03	−0.02	−0.09

this fact made interpretation of the results more complicated than would otherwise have been the case. It appeared, however, that when averaged over several travelling standards the spread amongst the laboratories was 0.16 percent. Again, no uncertainties were quoted. A second comparison at 10 GHz took the form of a bilateral comparison between the Soviet Union and France [14].

Further comparisons initiated by CIPM in 1972 were those for power at 6 GHz in coaxial line [15] and at 34.5/35 GHz in waveguide. For the 6 GHz comparison three thermistor transfer standards fitted with GR900, 7 mm and N type connectors, together with adapters, were circulated. Table 16.3 shows the results for the GR900 thermistor. For this travelling standard the

Table 16.3 *International intercomparison of power in coaxial line at 6 GHz. Measured effective efficiency η_{e9} for thermistor with 14 mm connector. (NB. NOM result is calculated using adapter efficiency)*

	NBS	NML	NPL	LCIE	RIND
η_{e9}	0.937	0.941	0.936	0.939	0.938
	NRC	PTB	IEN	NOM	Average
η_{e9}	0.938	0.937	—	0.936	0.938

maximum difference between any two laboratories was 0.5 percent. For the N type standard the maximum measured difference was 1.4 percent and for the 7 mm 1.1 percent, but most of these were measured through adapters and the errors in the evaluation of the adapters probably account for the larger differences. The final report on this comparison did not appear until ten years after the start of the measurements. Fig. 16.1 shows the measured effective efficiency at 34.5 GHz for one of the three thermistor travelling standards used in the 34.5/35 GHz comparison [16]. Systematic differences between the four

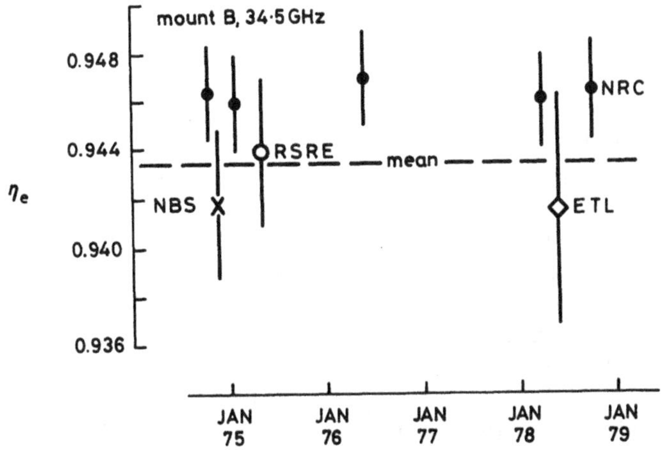

Fig. 16.1 *CIPM International Intercomparison of Waveguide Power at 34.5/35 GHz. The plot shows the effective efficiency η_e of thermistor mount B at 34.5 GHz. The vertical bars represent the estimated uncertainty of each measurement*

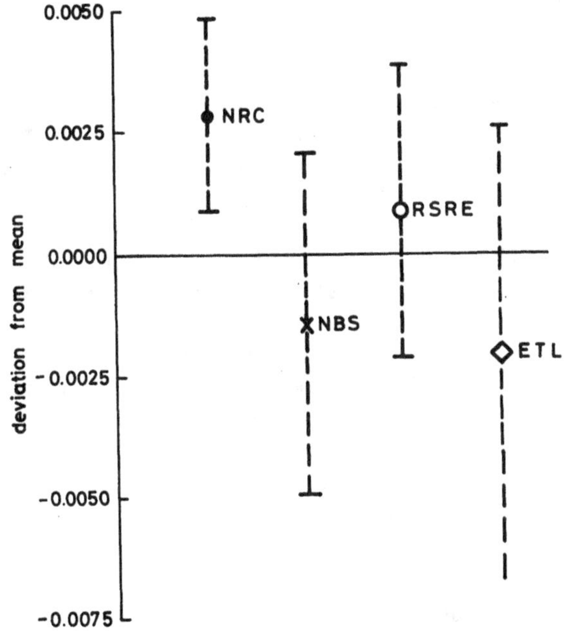

Fig. 16.2 *CIPM International Intercomparison of Waveguide Power at 34.5/35 GHz. The plot shows the average deviation of each laboratory from the international mean. Vertical bars represent the estimated uncertainty of the measurements*

participating laboratories were obtained by averaging the measured values for the three thermistors at the two measurement frequencies, and these are displayed in Fig. 16.2. The spread here is 0.49 percent, which is consistent with the quoted uncertainties of between 0.2 percent and 0.5 percent. The primary standards of all the laboratories were microcalorimeters and it was originally expected that the laboratories would calibrate the thermistors directly in their calorimeters. This expectation was only partly fulfilled. The organisation of the comparison followed the 'star' scheme, in which the travelling standards were remeasured by the pilot laboratory after each measurement by one of the other laboratories. Although such a procedure increases the total time, it produces more reliable results than a straight forward circulatory or round robin scheme, as any changes due to aging, corrosion or damage in the travelling standards can be detected more easily. Furthermore, if a change does become apparent the earlier measurement results may still be usable.

A waveguide power comparison at 15 GHz was initiated in 1975 [17]. Three pairs of travelling standards were circulated, each pair of a different design. Two of the pairs were of the tuned wire barretter type. The third pair consisted of untuned commercial thermistor mounts. However, one of the tuned barretter designs proved not to be sufficiently stable for use as a travelling standard. Fig. 16.3 shows the results for the more satisfactory barretter design. The weighted mean and its confidence interval (calculated by the method described in ref. [18]) are also shown. When all the results were taken into account it was concluded that there was good agreement to within ±0.15 percent amongst four of the participating laboratories and in only one case was a systematic deviation of 0.5 percent or greater apparent. A comparison of power at 50 MHz in 50 Ω coaxial line was completed in 1986. The uncertainties quoted by the different participants varied considerably, but amongst the laboratories which quoted the lower uncertainty levels the spread in results was 0.15 percent [19]. Other power comparisons which were in progress in 1986 were those in 50 Ω coaxial line at 12–18 GHz, in 75 Ω coaxial line at 500 MHz, and in waveguide at 94 GHz.

Some power comparisons arranged independently of CIPM are reported in refs. [20–24]. As most of the primary standards are now microcalorimeters or dry load calorimeters, comparisons including both of these two types, such as those described in refs. [12,25], represent one of the few opportunities for cross-checking standards based on independent designs, although both types are of course thermal instruments.

16.2 Voltage comparisons

Voltage comparisons have been confined to frequencies of 1 GHz and below. The travelling standards are normally thermal voltage converters or micro-potentiometers (see Chapter 15), for which the participants are required to determine the rf/dc difference. Under URSI a bilateral comparison between the United States and Sweden at various frequencies and voltage levels resulted in differences between laboratories of between 1 percent and

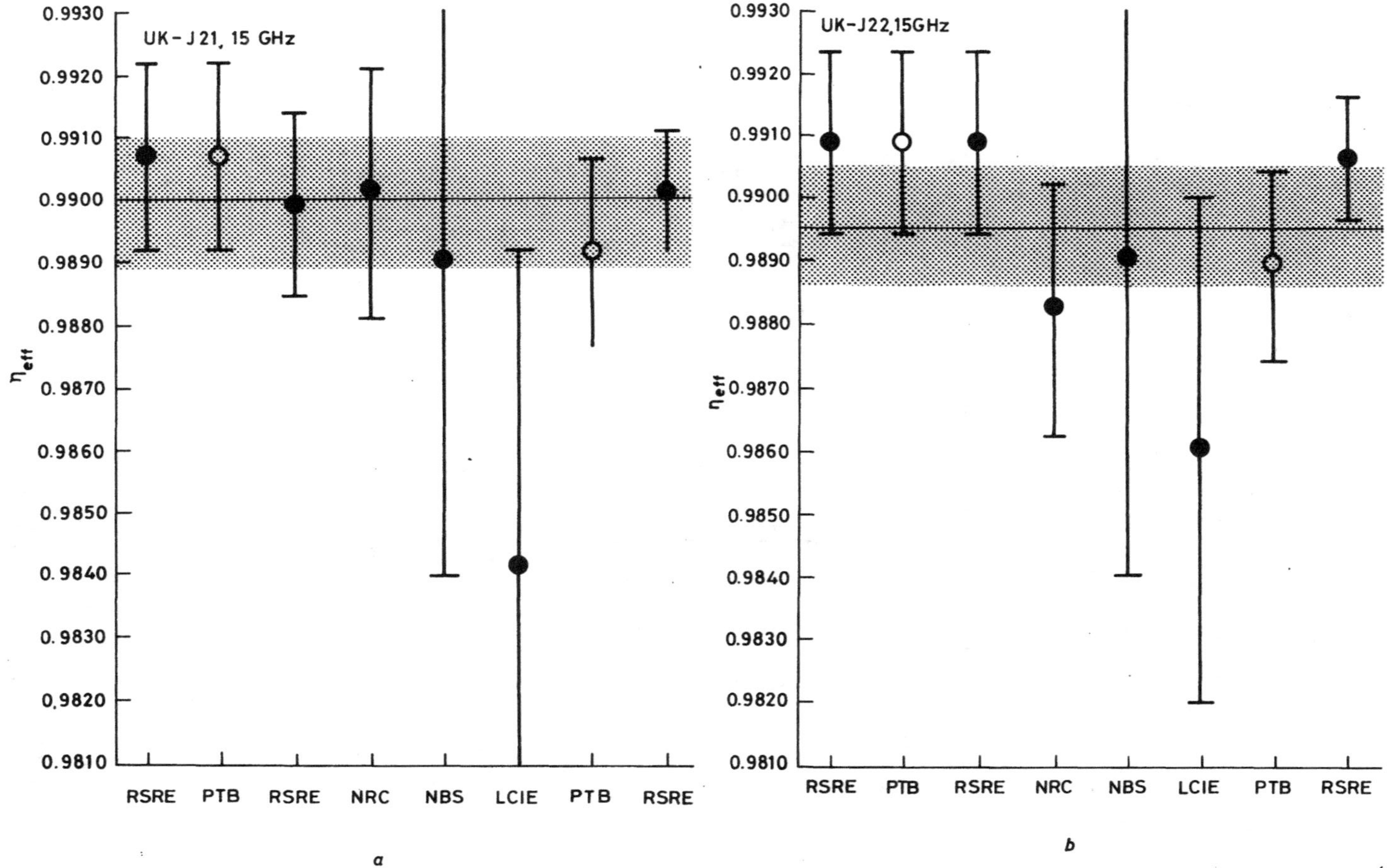

Fig. 16.3 *CIPM International Intercomparison of Power at 15 GHz. The plots show the effective efficiencies for two tuned metal wire barretter mounts. The weighted mean values and their uncertainty intervals are indicated*

15 percent [1,9]. In the period up to 1986 CIPM completed three comparisons. Results for the first of these [26], at 1 GHz, are shown in Figs. 16.4. The maximum difference between any two laboratories was 0.8 percent for the thermal voltage converter and 0.7 percent for the micropotentiometer. Quoted uncertainties were 0.35–2 percent for the thermal voltage converter and 0.5–5 percent for the micropotentiometer. The second comparison, which took place at 1 MHz and employed two thermal voltage converters as the travelling standards, yielded a spread of 0.02 percent in the measured values of rf/dc voltage difference, excluding the measurements of one laboratory which afterwards discovered a fault due to a defective component [27,28]. The

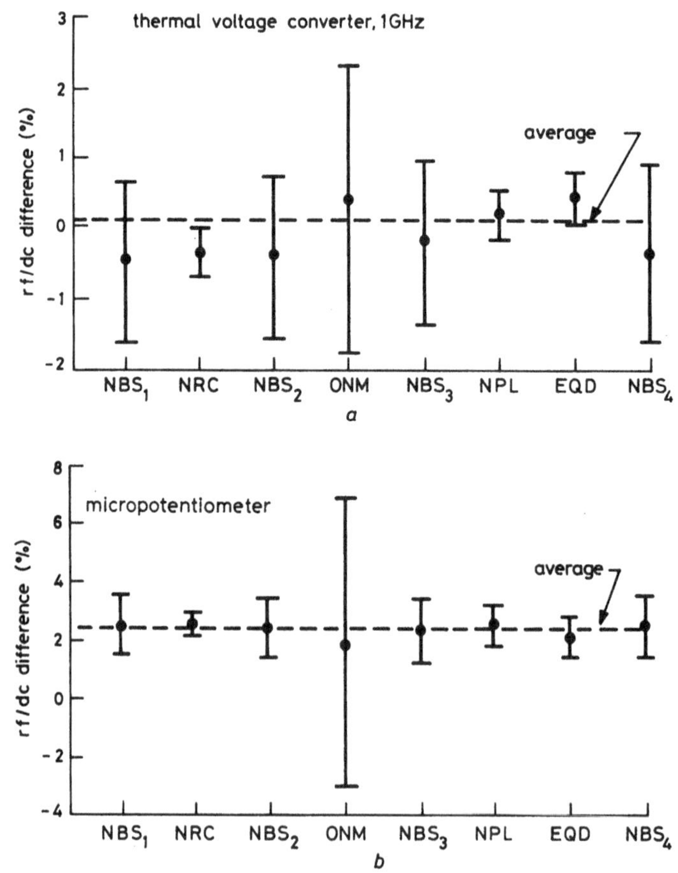

Fig. 16.4 *CIPM International Intercomparison of Voltage at 1 GHz. The plots show the measured rf/dc difference of a thermal voltage converter and a micropotentiometer. Vertical bars represent estimates of each laboratory for its own limit of uncertainty. The participants were the United States (NBS), Canada (NRC), Hungary (ONM), Britain (NPL and EQD)*

quoted uncertainties were in the range 0.01–0.05 percent. One of the conclusions drawn from this comparison was that at 1 MHz the rf/dc difference of the travelling standards, after correcting for dc reversal error, was probably too small to be detected with current uncertainty levels. The third CIPM voltage comparison [29,30] covered the four frequencies 100 MHz, 250 MHz, 500 MHz and 1 GHz. Four travelling standards were used of which two were resistance mounts and two were thermal converters. Results at 100 MHz and 1 GHz for one of the travelling standards are shown in Fig. 16.5. The plotted points represent the distances from the weighted mean value. The uncertainty interval for the weighted mean is shown by the shaded area. On analysing all of the results at all four frequencies it was concluded that the measurement values of all participants were in good agreement with the uncertainties they themselves calculated. The deviations were in all cases within ±1.5 percent and in most cases within ±0.5 percent. The organisation of this comparison followed the 'petal' scheme, the travelling standards being returned to the pilot laboratory several times during the course of the measurements but less frequently than in the star scheme. A voltage comparison at 30 MHz was still underway in 1987. Other voltage comparisons are reported in ref. [31].

16.3 Discussion

Apart from the earliest comparisons, which took place at a time when basic techniques were still being assessed, the main role of both power and voltage comparisons has been that of providing confidence in the implementation of the techniques and in ensuring international consistency. From the results which are available it appears that the claims of those national laboratories which have taken part have in the main been substantiated, although such comparisons cannot of course detect errors which are common to all laboratories. Regarding power measurements, one can say that uncertainties in effective efficiency of 0.2–0.5 percent can be achieved up to 8 GHz in 14 mm coaxial line fitted with precision connectors and up to at least 35 GHz in waveguide. These figures must be combined with the uncertainty in dc power measurement (typically 0.1 percent or better) to obtain the total uncertainty for absolute rf power measurement. The importance of coaxial power standards at the higher frequencies has increased, but some standards laboratories still derive coaxial power above 8 GHz from waveguide standards via calibrated adapters, which increases the uncertainties. As regards voltage measurement, one may say that, using 14 mm line fitted with precision connectors, uncertainties in rf/dc voltage difference of 0.01–0.05 percent can be achieved at 1 MHz and 0.3–0.4 percent from 100 MHz to 1 GHz.

The time taken to carry out the measurements in an international comparison can be four years or more. This has been a constant source of concern, although typical of this type of endeavour. In the absence of a central

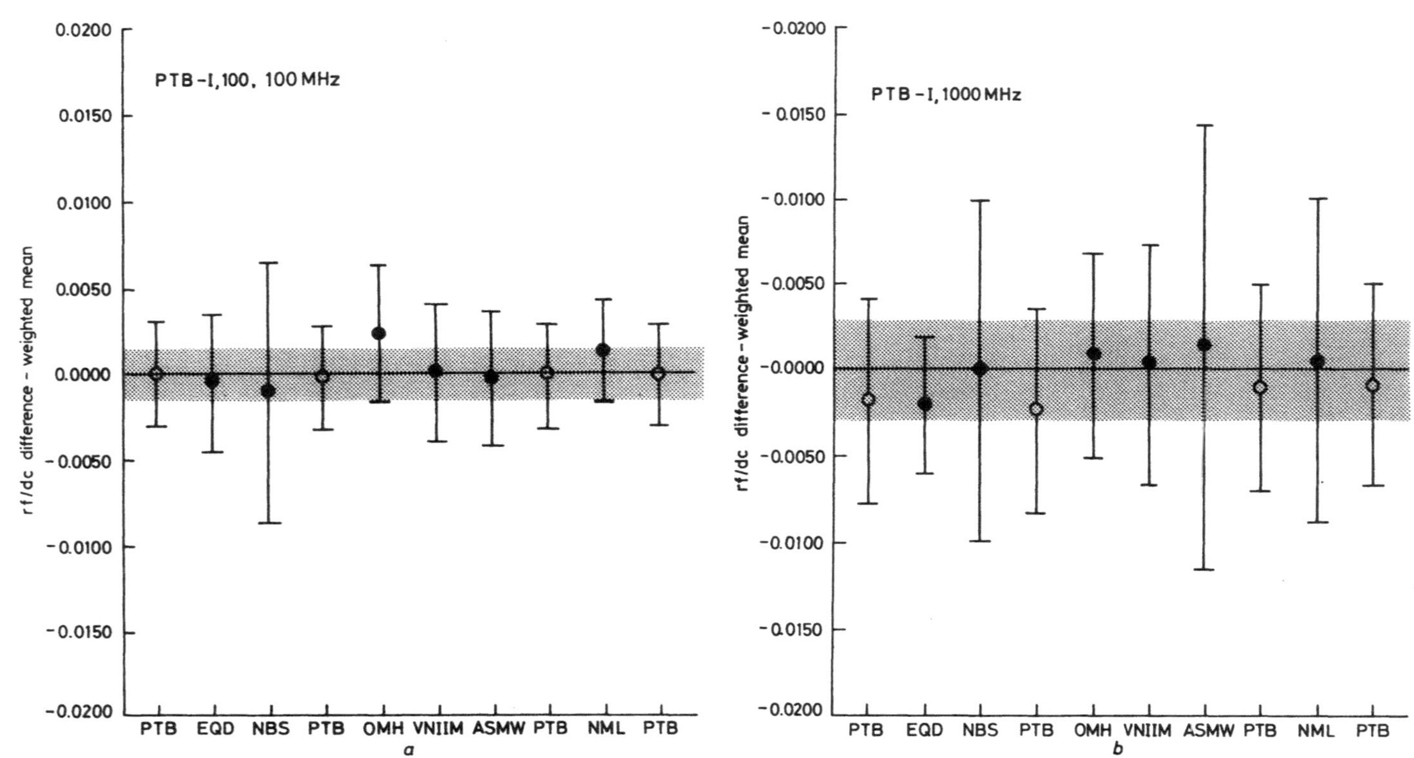

Fig. 16.5 *CIPM International Comparison of Voltage at 100 MHz, 250 MHz, 500 MHz, and 1 GHz. The plot shows the rf/dc difference for four travelling standards. Vertical bars represent the uncertainty of each measurement. The participants were the German Federal Republic (PTB), Britain (EQD), the United States (NBS), Hungary (ONM), the Soviet Union (IMM), the German Democratic Republic (ASMW), Australia (NML)*

laboratory for rf and microwaves at BIPM or elsewhere, it was envisaged that one of the benefits of international comparisons would be to enable those countries which had not developed independent standards for all the required quantities to have access to calibrations based on the most accurate standards available. This has happened to a limited extent but, in view of the long time scales, such an arrangement is clearly an unsatisfactory way of obtaining traceability. The tables and graphs reproduced in this chapter do not necessarily show laboratories which have participated on this basis.

An area where international standardisation has not yet been achieved is the method of expressing uncertainties. In many of the earlier reports on international comparisons no indication is given of the method by which the various contributions to the total uncertainty have been combined (e.g. arithmetically or in quadrature) or of the confidence level (see Appendix A2).

16.4 References

1 Selby, M.C., 'International comparison of measurements at high frequencies', *IEEE Spectrum*, **3**, 1, pp. 89–98, Jan. 1966.

2 Bailey, A.E., 'International harmonisation of microwave standards', *Proc. IEE*, **127**, Pt. H, 41, pp. 70–73, April 1980.

3 Bailey, A.E., Hellwig, H.W., Nemoto, T., and Okamura, S., 'International organisation in electromagnetic metrology and international comparison of RF and microwave standards', *Proc. IEEE*, **74**, 1, pp. 9–14, Jan. 1986.

4 Wilt, T.J., and Reymann, D, 'The maintenance and comparison of standards of electromotive force at the BIPM', *IEEE Trans.*, **IM-32**, 1, pp. 260–266, Mar. 1983.

5 'Proces Verbaux des Seances', CIPM, published annually by BIPM, Sèvres, France. (The rf working group met in 1965, 1968, 1972, 1975, 1978, 1983, and 1986.)

6 Engen, G.F., 'International intercomparison of standards for microwave power measurement', Acta IMEKO, pp. 337–345, 1964.

7 Engen, G.F., 'International intercomparison of standards for microwave power measurement', *IEEE Trans.*, **MTT-13**, 5, p. 713–715, Sept. 1965.

8 Tamaru, T., 'A note on bolometer mount efficiency measurement technique by impedance method in Japan', *IEEE Trans.*, **MTT-14**, 9, p. 437, Sept. 1966.

9 Selby, M.C., 'International comparison of HF and microwave electromagnetic quantities', *Proc. IEEE*, **55**, 6, pp. 745–747, June 1967.

10 Radio Frequency Section, Standards Division, Electrotechnical Laboratory, 'International comparison of rf electromagnetic quantities', *Bulletin Electrotechnical Lab.*, **33**, 4, pp. 462–72, 1969 (in Japanese).

11 Harris, I.A. and Fowler, C.S., 'Some comparisons of power measuring standards at 300 and 400 Mc/s', *Proc. IEE*, **111**, 9, pp. 1597–1601, Sept. 1964.

12 Engen, G.F., and Hudson, P.A., 'International comparison of power standards at 3 GHz', *IEEE Trans.*, **MTT-19**, 4, pp. 411–413, April 1971.

13 Ishige, R., 'First international intercomparison of microwave low power standards at 10 GHz', CIPM Document GT-RF/72-5, 1972 (French translation in Procès verbaux des Séances, CIPM 1972).

14 Blouet, J., and Erard L., Report on the international comparison of power standards at 10 GHz, CIPM Document GT-RF/75-6, 1975 (in French) (also published in Proces verbaux des seances, CIPM 1975).

15 'CIPM international power comparison at 6 GHz', to be published.
16 Clark, R.F., Griffin, E.J., Inoue, T., and Weidman, M.P., 'An international intercomparison of power standards in WR-28 waveguide', *Metrologia*, **17**, 1, pp. 27–31, Mar. 1981.
17 Bayer, H., 'Results of an international intercomparison of microwave power in waveguide R 140 (WG 18, WR 62) at 15GHz among some standardizing laboratories', *IEEE Trans.*, **IM-35**, 2, pp. 116–120, June 1986.
18 Bayer, H., 'International intercomparison of power standards at 15 GHz in waveguide R140', BIPM Monograph 84/3, Sevres, France, 1984.
19 Jurkus, A., 'International comparison GT-RF 78-14: power (1 mW) at 50 MHz in 50 Ω coaxial line', *IEEE Trans.*, **IM-37**, 1, pp. 160–162. Mar. 1988.
20 Acs, E., Szokol, H., and Tarma, I., 'Comparison of microwave power standards', IMEKO Symp. on Microwave Measurement, Budapest, pp. 18–20, 1966.
21 Gergei, A., Mekhannikov, A.I., Sokol, Kh., Terek, A., and Chuiko, V.G., 'Collation of standards for SHF power in the USSR and the Hungarian People's Republic (VNR)', *Izmeritel'naya Tekhnika*, **2**, pp. 82–83, Feb. 1972.
22 Bostrom, R., and Fantom, A.E., 'International comparison of coaxial power standards at 3.03 GHz', *Electronics Letters*, **8**, 11, p. 283, 1 June 1972.
23 Fedorov, A.M., Morozova, T.B., Krzhimovskii, V.I., Grigor'ev, A.V., and Sergeev, I.A., 'International intercomparison of power standards in coaxial channels at 3 GHz', *Meas. Tech.*, **19**, 7, pp. 1075–1076, July 1976.
24 Aleksandrov, E., Mladenov, I., Gergei, A., Terek, A., Devitts, G., Bart. K., Sorokovskii, S., Kornatski, L., Prune, S., Dzhukhat, V., Fedorov, A.M., Sergeev, I.A., Bubliak, P., and Petrash, I., 'Results of comparisons of the standards of the unit of power among the COMECON countries in coaxial channels at high frequencies', *Meas. Tech.*, **21**, 8, pp. 1157–1159, Aug. 1978.
25 Skilton, P., and Fantom, A.E., 'A comparison of the United Kingdom standards of microwave power in waveguide and coaxial lines', *IEEE Trans.*, **IM-27**, 3, pp. 297–298, Sept. 1978.
26 Ries, F.X., 'An international intercomparison of voltage standards at 1 GHz in coaxial transmission line', *IEEE Trans.*, **IM-25**, 3, pp. 254–255, Sept. 1976.
27 Fantom, A.E., 'International comparison of voltage at 1 MHZ', NPL Report DES 48, July 1978.
28 Fantom, A.E., 'International comparison of voltage at 1 MHz and 1 volt', CIPM Document GT-RF/75-9, 1975.
29 Bayer, H., and Janik, D., 'Results of an international intercomparison of RF voltage at some frequencies between 100 MHz and 1 GHz in coaxial guide systems', *Proc. INSYMET*, pp. 202–206, 1982 (Sixth Int. Symp. on Metrology, Bratislava).
30 Bayer, H., and Janik, D., 'International intercomparison of rf voltage at 100 MHz, 250 MHz, 500 MHz and 1000 MHz', BIPM Monograph 84/1, 1984.
31 Mladenov, I., Shashch, Ya., Yun, D., Kornatshy, L., Voinuyu, O., Fedorov, A.M. and Bubliak, P., 'Results of comparisons of the high-frequency voltage measurement standards of member countries of the Council of Mutual Economic Assistance (COMECON)', *Meas. Tech.*, **17**, 6, pp. 862–865, June 1974.

Appendices

A1 Network parameters

The properties of linear n-port electrical networks can be expressed in a number of different ways [1,2]. At low frequencies the impedance matrix is often the most appropriate form. This enables the voltages at the ports to be calculated from the currents. For a two-port network the equations are

$$\begin{aligned} v_1 &= z_{11}i_1 + z_{12}i_2 \\ v_2 &= z_{21}i_1 + z_{22}i_2 \end{aligned} \tag{A1.1}$$

where z_{11}, z_{12}, z_{21}, z_{22} are the elements of the impedance matrix, and subscripts 1 and 2 refer to ports 1 and 2. Inverting the impedance matrix gives the admittance matrix, which enables the currents to be expressed in terms of the voltages:

$$\begin{aligned} i_1 &= y_{11}v_1 + y_{12}v_2 \\ i_2 &= y_{21}v_1 + y_{22}v_2 \end{aligned} \tag{A1.2}$$

For reciprocal networks, both the impedance matrix and the admittance matrix are symmetrical about the leading diagonal. Thus for a two-port network

$$\begin{aligned} z_{12} &= z_{21} \\ y_{12} &= y_{21} \end{aligned} \tag{A1.3}$$

Other descriptions of two-port networks are ABCD (chain) parameters and h parameters.

At higher frequencies the scattering matrix is more useful. The signal at each port is expressed as the sum of an incident wave with complex voltage a and an

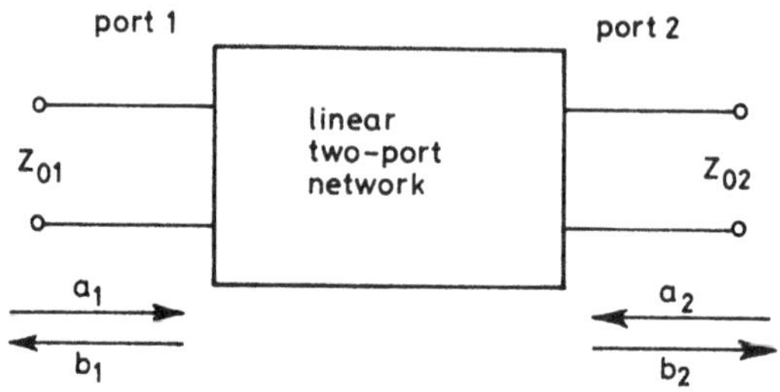

Fig. A1.1 *Definition of voltage waves for two-port network. a_1 and a_2 are the incident waves, b_1 and b_2 the emergent waves*

emergent wave with complex voltage b (Fig. A1.1). The scattering matrix enables the values of b to be found from those of a. For a two-port network

$$\begin{aligned} b_1 &= s_{11}a_1 + s_{12}a_2 \\ b_2 &= s_{21}a_1 + s_{22}a_2 \end{aligned} \tag{A1.4}$$

where the scattering parameters s_{11}, s_{12}, s_{21}, s_{22} are the elements of the scattering matrix. The values of the scattering parameters depend not only on the network itself, but also on the characteristic impedances of the transmission lines to which it is connected. These characteristic impedances will not necessarily be the same for each port. The reciprocity condition is:

$$\frac{s_{12}}{\sqrt{Z_{01}}} = \frac{s_{21}}{\sqrt{Z_{02}}} \tag{A1.5}$$

where Z_{01} and Z_{02} are the characteristic impedances of the transmission lines connected to ports 1 and 2. If Z_{01} and Z_{02} are real, which is not necessarily the case, then the incident and emergent powers P_{i1}, P_{i2}, P_{r1}, P_{r2} at the two ports are given by:

$$\begin{aligned} P_{i1} &= \frac{|a_1|^2}{Z_{01}} \\ P_{i2} &= \frac{|a_2|^2}{Z_{02}} \\ P_{r1} &= \frac{|b_1|^2}{Z_{01}} \\ P_{r2} &= \frac{|b_2|^2}{Z_{02}} \end{aligned} \tag{A1.6}$$

Sometimes a and b are normalised by dividing by the square root of the characteristic impedance. If the normalised values are represented by dashed symbols, then

$$\begin{aligned} a_1' &= \frac{a}{\sqrt{Z_{01}}} \\ a_2' &= \frac{a_2}{\sqrt{Z_{02}}} \\ b_1' &= \frac{b_1}{\sqrt{Z_{01}}} \\ b_2' &= \frac{b_2}{\sqrt{Z_{02}}} \end{aligned} \tag{A1.7}$$

The powers are then simply $|a_1'|^2$, $|a_2'|^2$, $|b'_1|^2$, $|b_2'|^2$ and the reciprocity condition becomes:

$$s_{12}' = s_{21}' \tag{A1.8}$$

The inverse of the scattering matrix is known as the gathering matrix. Circuits described by scattering parameters may be analysed by signal flow graphs. An alternative description for two-port networks at high frequencies is the wave-cascading matrix.

Scattering parameters expressed relative to one set of impedances may be recalculated for a different set of impedances using the formula:

$$S' = S - (I + S)\,\Gamma\,(I - S\Gamma)^{-1}(I - S) \tag{A1.9}$$

where S' is the scattering matrix referred to impedances ξ_{0i}, I is the identity matrix and Γ is a diagonal matrix of reflection coefficients of loads ξ_{0i} referred to the impedances Z_{0i}. This is known as matrix renormalisation and it can simplify the measurement procedure. Measurements can be carried out relative to any readily available set of impedances and the results renormalised.

A2 Uncertainty and confidence in measurements

A statement of the result of a measurement should include an estimate of the uncertainty together with the confidence level. Unfortunately there is no universally agreed method of treating uncertainties, with the consequence that comparison of the accuracies of different instruments or techniques can often be misleading. Traditionally uncertainties are divided into random and systematic ones. Random uncertainties are a consequence of errors which change when the measurement is repeated; they can be estimated from the scatter in a sufficiently large number of readings. Systematic uncertainties correspond to constant unknown errors. However, the distinction is rather artificial, since the causes of the two types of uncertainty are not necessarily different in their nature: a random uncertainty in one stage of a calibration chain may become a systematic uncertainty in the following stage. Alternative treatments which avoid the terms random and systematic have been proposed [3].

Random uncertainties which follow the normal probability distribution can be dealt with by applying Student's t factor [4] (Table A2.1). An estimate is first made of the standard deviation, using the formula:

$$\sigma_{est} = \sqrt{\frac{\Sigma_n (a_i - \bar{a})^2}{n-1}} \tag{A2.1}$$

where a_i are the results of n measurement samples and $\bar{a}$ is their mean. The uncertainty for an individual reading is then given by $t\sigma_{est}$ and for the mean of n readings by $t\sigma_{est}/\sqrt{n}$. The value of t depends on the number of samples on

Table A2.1 *Student's t distribution. Values of t for specified confidence probability as a function of the number of measurements, n*

P	0.500	0.683	0.950	0.955	0.990	0.997
n						
2	1.000	1.84	12.7	14.0	—	—
3	0.817	1.32	4.3	4.53	9.92	—
4	0.765	1.20	3.18	3.31	5.84	9.22
5	0.741	1.14	2.78	2.87	4.60	6.62
6	0.727	1.11	2.57	2.65	4.03	5.51
7	0.718	1.09	2.45	2.52	3.71	4.90
8	0.711	1.08	2.36	2.43	3.50	4.53
9	0.706	1.07	2.31	2.37	3.36	4.28
10	0.703	1.06	2.26	2.32	3.25	4.09
11	0.700	1.05	2.23	2.28	3.17	3.96
12	0.697	1.05	2.20	2.25	3.11	3.85
13	0.695	1.04	2.18	2.23	3.05	3.76
14	0.694	1.04	2.16	2.21	3.01	3.69
15	0.692	1.04	2.14	2.20	2.98	3.64
16	0.691	1.03	2.13	2.18	2.95	3.59
17	0.690	1.03	2.12	2.17	2.92	3.54
18	0.689	1.03	2.11	2.16	2.90	3.51
19	0.688	1.03	2.10	2.15	2.88	3.48
20	0.688	1.03	2.09	2.14	2.86	3.45
infinity	0.675	1.00	1.96	2.00	2.58	3.00

which σ_{est} is based and on the confidence level required. For infinite n, t is approximately 2 for 95 percent confidence and 3 for 99.7 percent confidence. For finite n, t is larger than these values, as the possibility must then be taken into account that σ_{est} may differ from the true standard deviation.

Systematic uncertainties are treated either on the basis of arithmetic summation or statistically. In the arithmetic summation approach, an estimate is made of the maximum possible error (the semi-range, corresponding to 100 percent confidence) for each source. The moduli of the semi-ranges are then added arithmetically and the 99.7 percent confidence level random uncertainty is added to this sum. In the statistical approach each source of systematic uncertainty (which for infinite n corresponds to three times the standard deviation) is assigned an equivalent standard deviation and treated as if it were a random uncertainty. Usually the exact probability distribution is not known and a conservative assumption of a rectangular distribution is made, that is an occurrence of errors equally probable over the estimated range. In this case the standard deviation is taken as $a/\sqrt{3}$, where a is the semi-range. Uncertainties from different sources are added in quadrature. If a sufficient number of independent sources of comparable magnitude are present, the combined probability distribution will be roughly normal and Student's t factor can be applied, taking the values corresponding to n = infinity. A confidence level of 95 percent is usually chosen for this method. In some cases there are theoretical grounds for assigning a different form of distribution. This applies to mismatch uncertainties. The distribution in this case is U-shaped and the standard deviation is taken as $a/\sqrt{2}$. Further details, including limitations, are given in ref. [5]. The two approaches to systematic uncertainties may yield uncertainty figures which differ by a factor of 2 or more. Confidence levels of less than 95 percent are not usual in electrical measurement, but levels as low as 50 percent are in use in other measurement areas.

A3 Theory of the tractorial load

The evolution of the tractorial load [6] is illustrated in Fig. A3.1. A simple load consisting of a cylindrical resistor in a cylindrical outer (Fig. A3.1*a*) does not provide a good match at high frequencies owing to reflections from the discontinuities at its two ends. An improvement in match can be obtained by introducing one or more steps in the outer conductor (Fig. A3.1*b* and A3.1*c*). A multi-step load can be analysed by imagining it to be divided into a number of elements connected by uniform lines. For each element the reflection due to the step can be made to cancel that due to the series resistive element. This occurs when the step size is chosen so that the decrease in characteristic impedance is equal to the elemental resistance δr. In the limit of an infinite number of steps the profile must be such that the characteristic impedance at

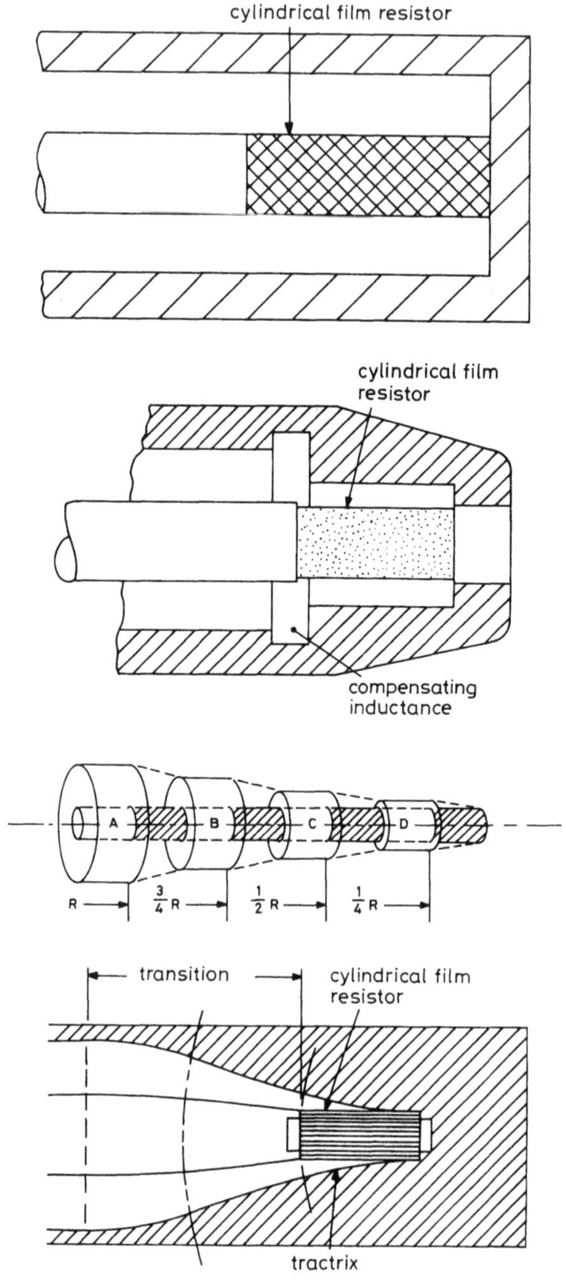

Fig. A3.1 *Stages in the evolution of the tractorial load. (a) Cylindrical outer conductor giving large mismatches at high frequencies, (b) Compensated single step outer conductor, (c) Multistepped outer conductor, (d) Tractorially tapered outer conductor*

any point along the load is equal to the resistance between that point and the short circuit.

To a first approximation the outer conductor profile can be calculated from the formula for the characteristic impedance Z_0 of a lossless coaxial line:

$$Z_0 = z\frac{1}{2\pi}\ln\left(\frac{b}{a}\right) \tag{A3.1}$$

where a and b are the inner and outer diameters and z is the impedance of the medium. This leads to the conclusion that the outer conductor must be tapered exponentially if the resistor is uniform. However, equation A3.1 assumes that the fields are purely transverse, which is not true for lossy lines. Taking into account the axial components of the electric field leads to the conclusion that the wavefront must be spherical. For zero reflection successive wavefronts must form a set of part-spheres of equal radii and with centres displaced along the axis. The profile of the outer conductor must be orthogonal to these surfaces. The curve so defined is of the type known as a tractrix (Fig. A3.1*d*), from which the name tractorial load is derived. A tractrix may be visualised as the path which is traced out if a small object is placed on a flat horizontal surface, a length of string is attached to it, and the other end of the string pulled along a straight line not directly away from the object. The load itself may be preceded by a lead-in section of line which provides a gentle transition from a plane to a spherical wavefront. Fringing fields are not entirely eliminated in the tractorial load, since there is a small effect due to the penetration of the fields into the dielectric substrate of the resistor. However, the effect can be compensated for by a slight adjustment in design, although the compensation is correct only for a particular permittivity, so that resistors on different substrate materials will not necessarily be interchangeable.

The reason why the tractorial load makes a good calorimeter load is that, within the limits of validity of the assumptions made in the theory, the distribution of power dissipated in the resistor is independent of frequency. In fact, both rf and dc powers are absorbed uniformly along the length of the resistor. This can be seen as follows. If I is the current at the input of a resistive element and $I+\delta I$ the current at the output of the same element, then the power dissipated in this element must be equal to the decrease in the transmitted power. The following equation must therefore hold:

$$I^2\delta r = I^2 Z - (I + \delta I)^2(Z + \delta Z) \tag{A3.2}$$

where δr is the resistance of the element. Putting δr and δZ equal and opposite to one another, as is required for zero reflection, leads to the result that δI is zero. Therefore the current is constant along the resistor and the power dissipation is uniform.

A4 Equivalent source reflection coefficient in terms of coupler parameters

The equivalent source reflection coefficient of a levelled source depends only on the properties of the directional coupler (or resistive power splitter) used in the levelling system. Its value may be calculated from the scattering parameters as follows. The equations describing the coupler are:

$$b_1 = s_{11}a_1 + s_{12}a_2 + s_{13}a_3 \qquad \text{(A4.1)}$$

$$b_2 = s_{21}a_1 + s_{22}a_2 + s_{23}a_3 \qquad \text{(A4.2)}$$

$$b_3 = s_{31}a_1 + s_{32}a_2 + s_{33}a_3 \qquad \text{(A4.3)}$$

where a_1, a_2, a_3, b_1, b_2, b_3 are the normalised incident and emergent waves at the ports of the coupler, as shown in Fig. A4.1. If Γ_d and Γ are the reflection

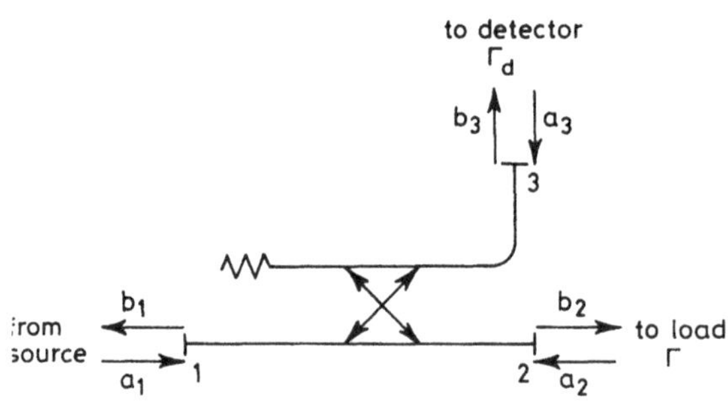

Fig. A4.1 *Incident and emergent waves for a three-port directional coupler in a revelling circuit*

coefficients of the detector connected to port 3 and the load connected to port 2, then

$$a_3 = \Gamma_d b_3 \qquad \text{(A4.4)}$$

$$a_2 = \Gamma\ b_2 \qquad \text{(A4.5)}$$

Eliminating a_1, a_2, a_3 from equations A4.2, A4.3, A4.4, A4.5 gives:

$$b_2 = \left[\left(s_{23} - \frac{s_{33}s_{21}}{s_{31}}\right)\Gamma_d + \frac{s_{21}}{s_{31}}\right]\frac{b_3}{1 - \left(s_{22} - \dfrac{s_{21}s_{32}}{s_{31}}\right)\Gamma} \qquad \text{(A4.6)}$$

in which b_3 (the wave incident on the detector) is held constant by the levelling circuit. Equation A4.6 may be simplified by collecting together the constants to give:

$$b_2 = \frac{B}{1 - \left(s_{22} - \dfrac{s_{21}s_{32}}{s_{31}}\right)\Gamma} \tag{A4.7}$$

$$P_a = |b_2|^2(1 - |\Gamma|^2) = \frac{B^2(1 - |\Gamma|^2)}{\left|1 - \left(s_{22} - \dfrac{s_{21}s_{32}}{s_{31}}\right)\Gamma\right|^2} \tag{A4.8}$$

where B is a constant. If equation A4.8 is compared with equation 11.1 in Chapter 11 it will be seen that they are of similar form but that the bracketed expression in the denominator of equation A4.8 replaces Γ_s in equation 11.1. The equivalent source reflection coefficient Γ_{se} is therefore:

$$\Gamma_{se} = s_{22} - \frac{s_{21}s_{32}}{s_{31}} \tag{A4.9}$$

A5 References to Appendices

1 Beatty, R.W., and Kerns, D.M., 'Relationships between different kinds of network parameters, not assuming reciprocity or equality of the waveguide or transmission line characteristic impedances', *Proc. IEEE,* **52**, pp. 84 and 426, 1964.

2 Somlo, P.I. and Hunter, J.D., 'Microwave impedance measurement', Peter Peregrinus, 1985.

3 Muller, J.W., 'The assignment of uncertainties to the results of experimental measurements', Second International Conference on Precision Measurement and Fundamental Constants, Gaithersburg, 1981.

4 Harris, I.A., and Warner, F.L., 'New expressions and some precise values for Student's t factor', *Proc. IEE*, **125**, 9, pp. 902–904, Sept. 1978.

5 Hinton, L.J.T., 'Uncertainty and confidence in measurements', in 'Microwave Measurement', A.E. Bailey (ed.), Peter Peregrinus Ltd, 1985.

6 Harris, I.A., 'The theory and design of coaxial resistor mounts for the frequency band 0-4000 Mc/s', *Proc. IEE*, **103C**, 3, pp. 1–10, Mar. 1956.

Index

www.ingramcontent.com/pod-product-compliance
Lightning Source LLC
LaVergne TN
LVHW011750200926
848484LV00003B/665